D1008768

MATHEMATICAL METHODS FOR SCIENTISTS AND ENGINEERS

MATHEMATICAL METHODS FOR SCIENTISTS AND ENGINEERS

Linear and Nonlinear Systems

PETER B. KAHN
Department of Physics
State University of New York at Stony Brook
Stony Brook, New York

A Wiley-Interscience Publication

JOHN WILEY & SONS

New York • Chichester • Brisbane • Toronto • Singapore

Library of Congress Cataloging-in-Publication Data:

Kahn, Peter B.
  Mathematical methods for scientists and engineers: linear and nonlinear systems / Peter B. Kahn.

    p.   cm.
  "A Wiley-Interscience publication."
  Bibliography: p.
  Includes index.
  1. Science—Mathematics. 2. Technology—Mathematics.
3. Numerical calculations. I. Title.

Q158.5.K34 1989
501.5′1—dc20 89-32383
ISBN 0-471-62305-9 CIP

Printed in the United States of America

10 9 8 7 6 5 4 3 2 1

To my friend and colleague
Arnold M. Feingold

PREFACE

As you begin to read a new textbook, it is reasonable to ask in what way this text is different from others already available. And is the difference desirable? One has a wealth of excellently presented material to draw upon in order to learn about some particular topic, so a new book is obliged to acknowledge and use this information.

Almost half of the text is devoted to the study of nonlinear systems. This acknowledges the critical new role that nonlinear phenomena are playing in modern science and technology. The path we have taken is to develop in Part I our grasp of linear systems with particular emphasis on asymptotic methods. Then, in Part II, we rely on the developed techniques when we focus our attention on weakly nonlinear oscillatory systems and nonlinear difference equations. The two parts merge so as to bring about a balance between the need to master techniques that are important for the study of traditionally included material with the necessity to be prepared for future developments. In order to keep this text to a manageable size, some time-honored topics have been omitted. This recognizes that over the past decades much of the material traditionally taught in mathematical methods courses has been assimilated into science courses, allowing the applications to be closely tied to mathematical techniques. Thus, for example, students can learn the special functions of math physics in electricity and magnetism courses, where the solution of Maxwell's equations in the standard geometries requires the use of trigonometric functions, Bessel functions, and Legendre polynomials. By contrast, material such as complex variables and matrix theory have evolved into courses in their own right. With these thoughts in mind, I decided on a selection of topics that balances techniques that are important to master, but that are not usually discussed at length in subject matter courses with methods that are so essential that I cannot omit them. I emphasize perturbation, or approximation, techniques that are relatively easy to use and that are applicable to a broad spectrum of problems that occur in science. I assume that the readers have a firm background in treating problems that admit a solution in a closed form and have become the foundation of undergraduate courses in the physical sciences and engineering. For example, I expect that readers know about the exponential function, the simple harmonic oscillator, the conservation of energy, and principles of mechanics, etc., as well as integral and differential calculus, elementary differential equations, and elementary matrix theory. I omit any discussion of complex variables and I do not develop any methods that rely on their use. This is done to make the discussion accessible to a larger audience and with the recognition that interested students usually take an independent course in complex variables.

I emphasize that this is not a math text, and thus I omit rigorous proofs and, in addition, permit a certain degree of carelessness in restrictions and conditions. For example, only matrices with distinct eigenvalues are considered since the case of degenerate eigenvalues is unusual and many times causes needless complexities in the proofs. I mean by this that there are advantages in assuming that our matrices have distinct eigenvalues since they can then be brought to diagonal form. If a particular matrix has degenerate eigenvalues, we can think of introducing a perturbation that gives a small separation to the eigenvalues. In most cases, this does not substantially change the problem.

The text is designed to help the reader develop tools of analysis that can be used to treat problems that do not admit a simple solution in a closed form, but that have their origin in solvable problems with which the reader is already familiar. I have adopted a repetitive style in which the same thing is phrased in a variety of ways, in part to help the reader catch at a later time something that was missed the first time around. It also recognizes that, in a textbook, there is a definite exchange of control between the author and the reader since the latter chooses the pace to proceed when reading the text. My choice of the material to emphasize and reemphasize serves as guideposts for the reader.

I envision the text as containing approximately enough material for a one-year course in mathematical methods for advanced undergraduates or beginning graduate students in physics, chemistry, engineering, biology, and other sciences. The examples are drawn from applications in these fields and cast in a language or form that reveals the underlying or fundamental aspects of the problem under study. There is also some discussion of incorrect techniques and inadequate methods so as to alert the reader to wrong avenues of attack or possible traps. A few basic examples are used throughout so as to enable the reader to have robust and familiar paradigms as a guide. I have used the computer software packages *Lotus 1-2-3* (Lotus Development Corp.) and *Math CAD* (MathSoft, Inc.) to generate some of the numerical and graphical solutions. In my opinion, it is essential for all of us to develop expertise at using such programs to generate solutions and to improve our calculational insight.

After presenting some preliminary material, I discuss some properties of matrices and introduce the gamma function, which forms the foundation of our calculational techniques by enabling us to evaluate the most commonly occurring integrals. This is followed by a discussion of related functions and the concept of asymptotic expansions and approximations. This leads to the Euler–MacLaurin sum expansion, a powerful technique to evaluate a large class of sums and one that has the capability to often cast results in a form usable either in conjunction with a computer or a desk calculator. I then introduce the Laplace method of evaluation of integrals that depend upon a large parameter. This technique is discussed in some detail so as to enable the interested reader to study, independently, the related method of stationary phase and the method of steepest descent, both of which require an understanding of complex variables. This section of the text concludes with a discussion of the asymptotic behavior of second-order linear differential equations and an introduction to the related perturbation theory.

I then direct the reader's attention, in the second part of the book, to a lengthy and detailed discussion of weakly nonlinear oscillating systems that have their origin in the simple harmonic oscillator. It has become common practice to devote almost all of our curricular attention to linear systems and to neglect nonlinear problems. This is done in part because a systematic discussion of linear systems is possible through the powerful techniques of superposition, classification of solutions, and final determination of coefficients.

On the other hand, nonlinear problems are interesting and also occur in diverse areas of science. They can be studied by a variety of powerful and understandable techniques that are within the capabilities of advanced undergraduate or beginning graduate students. My goal is to introduce the reader to nonlinear problems and the associated models that occur in diverse fields of science and engineering.

My approach is to first analyze the harmonic oscillator (with and without damping) by a technique that prepares us for some of the new phenomena that appear in nonlinear problems. I then go on to discuss the logistic equation, which is a simple nonlinear model that occurs in situations where one is trying to understand growth with limited resources. This prepares us for the tackling of models of weakly nonlinear oscillatory systems. I emphasize the weak character of the problems we are treating because they have a natural small parameter that we use to order terms in our perturbation expansion. I begin by discussing the classical methods of analysis due to Lindstedt and Poincaré. Then I develop the methods of averaging and multiple time scales, the two most universally applicable and commonly used methods of analysis.

I include in our discussion of nonlinear systems problems that require some preparation before they are in the proper form to be analyzed by the techniques we have developed. I conclude the study of nonlinear systems with a brief introduction to topics associated with the notion of chaos. This is accomplished through an analysis of one-dimensional quadratic maps and a derivation of the Feigenbaum numbers.

It is my experience that we are so thoroughly trained or brainwashed by linear problems that we are reluctant to let go of the intuition we have gained in this area when studying nonlinear problems, even though we know that this intuition can be both misleading and incorrect. We find, generally, that a perturbation expansion of nonlinear problems exhibits a relationship between the lowest-order approximation and the correction terms that is unfamiliar from our study of linear problems. Thus, we need some help in getting started in the right direction. For example, the simple harmonic oscillator, while being the fundamental starting point for the analysis of the Duffing oscillator, does not prepare us for the treatment of secular terms that are the new aspect of oscillatory motion, which can in this case be traced to the nonlinearity. It is essential to struggle to change our perspective from linear to nonlinear thinking if only because so much of modern science is concerned with nonlinear problems. I hope that the effort that has been put into this aspect of the text provides an adequate starting point for this venture.

Bibliographies are given at the end of each of the chapters of Part I and following Chapters 15, 16, and the Appendix.

I want to thank the students at Stony Brook, particularly Cherian Varughese, Usha Ravi, and Fernando Camilo, for sharing with me in the development of this text. It is my first extensive writing project and it took a long time, almost 20 years of classroom experience to bring it to fruition. This was in great part because my ideas regarding what was essential and what was not have changed significantly during this period. The audience has also changed and many excellent texts have appeared. My ideas firmed about ten years ago and I have used this period to try them out in the classroom and bring them into what I hope is a coherent and interesting form.

I wish to thank my colleagues Cliff Swartz, Arnie Feingold, and David Fox for their interest in this project. Cliff, an experienced author, gave me lots of much needed advice regarding how to get started and how to transform lecture notes into a text. Arnie read the first half of the manuscript with some care. He did not agree with everything either as regards emphasis or style, but he listened to my ideas and points of view and tried to help me avoid errors in logic and pedagogy as well as in substance. David read the second half and helped me to express my ideas with more clarity than I am used to. I have also greatly benefited from the comments on the second part of the manuscript by my research collaborator Juan Lin of Washington College, with whom I have talked endlessly regarding many aspects of nonlinear systems. Yair Zarmi of the Ben Gurion University of the Negev and the Jacob Blaustein Institute for Desert Research at Sede Boqer, Israel, invited me to the Institute to participate in its research program as well as to give lectures on nonlinear dynamics. This interaction led to Yair's incisive reading of the material on nonlinear oscillatory systems. I would also like to thank Madan Lal Mehta of C. E. N. Saclay, who over the past years has shared his ideas with me and both read the manuscript and gave me his valued comments. I have also enjoyed my frequent conversations with Lee Segel of the Weizmann Institute. He explained a number of things to me and encouraged me to clarify many of my ideas. Finally, I greatly appreciate the help I received from Jeffrey Kahn and Meng Zhou on all matters relating to computing; from Lois Koh who did the illustrations; from Bea Shube, recently retired from John Wiley and Sons, who encouraged me to get the book written; and from Marion Mastauskas and Heike Gustafson who worked with me to transform the typed manuscript into a form acceptable to the publisher.

PETER B. KAHN

Stony Brook, New York

CONTENTS

bounded. This is accomplished by means of the Euler–MacLaurin sum expansion.

5 Evaluation of Integrals: The Laplace Method 168

Often, we encounter integrals that depend on a parameter in such a way that the dominant or major contribution to the integral comes from a small part of the range. For example, the parameter may tend to infinity and the integrand may be sharply peaked. It is then often possible to estimate the integral by a method due to Laplace.

6 Differential Equations 207

We concentrate our attention on second-order linear differential equations since they form the basic structure for the posing and analysis of a broad spectrum of problems in science and engineering. They are usually studied in some detail in courses in differential equations, and with this in mind, we limit our discussion to a narrow aspect of the subject, i.e., asymptotic properties.

MATHEMATICAL METHODS FOR SCIENTISTS AND ENGINEERS

PART I

LINEAR SYSTEMS

CHAPTER 0

MISCELLANEOUS RESOURCES

In this chapter, we gather a few of the formulas, techniques, and miscellaneous abbreviations used in the text. In addition, we give references to sources of relevant integrals, tables, and related material.

0.1 ABBREVIATIONS

iff = if and only if
l.h.s. = left-hand side
r.h.s. = right-hand side
l.o.t. = lower-order terms
h.o.t. = higher-order terms
e.s.t. = exponentially small terms
n.l.t. = nonlinear terms

0.2 NUMERICAL RESULTS

$$\exp 3 \approx 20 \qquad \text{and} \qquad \ln 20 \approx 3 \tag{0.1}$$

$$\pi^2 \approx 10 \tag{0.2}$$

0.3 TRIGONOMETRIC IDENTITIES AND DIFFERENTIAL EQUATIONS

$$\sin(a+b) = \sin a \cos b + \sin b \cos a \tag{0.3}$$

$$\cos(a+b) = \cos a \cos b - \sin a \sin b \tag{0.4}$$

$$2(\cos a \cos b) = \cos(a+b) + \cos(a-b) \tag{0.5}$$

$$2(\sin a \sin b) = \cos(a-b) - \cos(a+b) \tag{0.6}$$

$$2(\sin a \cos b) = \sin(a+b) + \sin(a-b) \tag{0.7}$$

$$\cos^2 x = \frac{1+\cos 2x}{2} \tag{0.8}$$

$$\cos^3 x = \frac{3\cos x + \cos 3x}{4} \tag{0.9}$$

$$\cos^4 x = \frac{3 + 4\cos 2x + \cos 4x}{8} \tag{0.10}$$

$$\sin^2 x = \frac{1-\cos 2x}{2} \tag{0.11}$$

$$\sin^3 x = \frac{3\sin x - \sin 3x}{4} \tag{0.12}$$

$$\sin^4 x = \frac{3 - 4\cos 2x + \cos 4x}{8} \tag{0.13}$$

We know from the symmetry properties of the trigonometric functions that

$$\int_0^{2\pi} \sin nx\, dx = \int_0^{2\pi} \cos nx\, dx = 0; \qquad n = \pm 1, \pm 2, \pm 3, \ldots \tag{0.14}$$

$$\int_0^{2\pi} \sin nx \cos mx\, dx = 0 \tag{0.15}$$

[This result is clear if one uses Equations (0.7) and (0.14).]

$$\frac{1}{2\pi}\int_0^{2\pi} \sin^2 x \cos^2 x\, dx = \frac{1}{8} \tag{0.16}$$

$$\frac{1}{2\pi}\int_0^{2\pi} \sin^2 x\, dx = \frac{1}{2\pi}\int_0^{2\pi} \cos^2 x\, dx = \frac{1}{2} \tag{0.17}$$

$$\frac{1}{2\pi}\int_0^{2\pi} \sin^4 x\, dx = \frac{1}{2\pi}\int_0^{2\pi} \cos^4 x\, dx = \frac{3}{8} \tag{0.18}$$

There is a simple method for finding the solution to a differential equation of the form

$$\ddot{x} + x = K \cos \omega t; \qquad x(0) = A, \quad \dot{x}(0) = 0, \quad \omega \neq 1 \tag{0.19}$$

Assume a trial particular solution of the form

$$x(t) = B \cos \omega t$$

After substitution, we then have

$$B(-\omega^2 + 1) = K$$

or

$$x(t) = \frac{K}{1 - \omega^2} \cos \omega t + C \cos t \tag{0.20a}$$

where we have added a solution of the homogeneous equation to our particular solution. The constant C is then chosen so that $x(t)$ satisfies the initial conditions. This leads to the solution

$$x(t) = A \cos t + \frac{K}{1 - \omega^2} (\cos \omega t - \cos t) \tag{0.20b}$$

Example 3.1 Consider the differential equation

$$\ddot{x} + x = \cos^4 t; \qquad x(0) = A, \quad \dot{x}(0) = 0 \tag{0.21a}$$

Using Equation (0.10), we can write Equation (0.21a) as

$$\ddot{x} + x = \tfrac{3}{8} + \tfrac{1}{2} \cos 2t + \tfrac{1}{8} \cos 4t \tag{0.21b}$$

We obtain a solution by the superposition of solutions. Thus, we write

$$x(t) = x_1(t) + x_2(t) + x_3(t)$$

and form the three equations:

$$\ddot{x}_1(t) + x_1(t) = \tfrac{3}{8}$$
$$\ddot{x}_2(t) + x_2(t) = \tfrac{1}{2} \cos 2t$$
$$\ddot{x}_3(t) + x_3(t) = \tfrac{1}{8} \cos 4t$$

These are solved separately. We have

$$x(t) = \left(A - \tfrac{1}{5}\right)\cos t + \tfrac{3}{8} - \tfrac{1}{6}\cos 2t - \tfrac{1}{120}\cos 4t \tag{0.22}$$

It is simple to check that this is indeed a solution. ■

We can also use the trial solution method when the frequency of the forcing function coincides with the natural frequency of oscillation. The resulting motion is said to be aperiodic. We encounter equations with solutions of this form in Section 8.3 of Chapter 8. In addition, aperiodic terms often occur purely as the result of a particular approximation scheme. In such cases we make every effort to develop expansions that suppress terms of this form. This important point forms one of the central themes of Chapters 10–13.

EXERCISE 3.1 Consider Equation (0.19) with $\omega = 1$. Obtain the particular solution, $x(t) = (Kt/2)\sin t$, by setting $\omega = 1 - \delta$ and letting $\delta \to 0$ in Equation (0.20b). Observe that the solution is no longer periodic. ■

EXERCISE 3.2 Consider

$$\ddot{x} + x = K\sin\omega t; \qquad x(0) = \alpha, \quad \dot{x}(0) = \beta, \quad \omega \to 1.$$

Obtain $x(t)$ using the trial solution method with $\omega = 1 - \delta$ and letting $\delta \to 0$. Observe the role played by the initial conditions. (One can also obtain the solution by choosing a trial solution of the form $x(t) = At\cos t + Bt\sin t + C\cos t + D\sin t$ and solving for A, B, C, D.) ■

0.4 TAYLOR-SERIES EXPANSIONS

$$\ln(1 + x) = x - \frac{x^2}{2} + \frac{x^3}{3} - \frac{x^4}{4} + \text{h.o.t.}; \qquad |x| < 1 \tag{0.23}$$

By substituting and arranging terms, we also have

$$\ln(1 + x + x^2) = x + \frac{x^2}{2} - \frac{2x^3}{3} + \frac{x^4}{4} + \text{h.o.t.}; \qquad |x + x^2| < 1 \tag{0.24}$$

We can write

$$(1 + a)^x \quad \text{as} \quad \exp[x\ln(1 + a)] \tag{0.25}$$

Thus, using the expansion for $\ln(1 + x)$, if $a \ll 1$, it is a good approximation to write

$$(1 + a)^x \approx \exp\left[x\left(a - \frac{a^2}{2}\right)\right] \tag{0.26}$$

Example 4.1 Let $a = 1/2$, then

$$\left(\frac{3}{2}\right)^x \approx \exp\left[x\left(\frac{1}{2} - \frac{1}{8}\right)\right] = \exp\frac{3x}{8}$$

If $x = 1/3$, we have

$$\left(\frac{3}{2}\right)^{1/3} = 1.14 \qquad \text{and} \qquad \exp\frac{1}{8} = 1.13$$

The approximation is very good, indicating that the total contribution from the remaining terms is negligible. ■

If the product ax is sufficiently small, we can write

$$(1 + a)^x \approx \exp ax \approx 1 + ax$$

These approximations also enter into problems in which we have $(1 + a/n)^n$, where the exponent n tends to infinity. We then can write

$$\left(1 + \frac{a}{n}\right)^n = \exp\left[n \ln\left(1 + \frac{a}{n}\right)\right] \approx \exp\left(a - \frac{a^2}{2n}\right) \tag{0.27}$$

or $(1 + a/n)^n$ tends to $\exp a$ as n tends to infinity.

0.5 INEQUALITIES

We wish to establish the inequality

$$\frac{2}{\pi} \le \frac{\sin x}{x} \le 1; \qquad 0 \le x \le \frac{\pi}{2} \tag{0.28}$$

This result can be easily verified by plotting $y = \sin x/x$ or by observing the following:

(i) The Taylor-series expansion of

$$\frac{\sin x}{x} = 1 - \frac{x^2}{3!} + \frac{x^4}{5!} + \text{h.o.t.}$$

is a convergent alternating series. Thus, we have

$$\frac{\sin x}{x} \le 1 \qquad \text{for} \qquad 0 \le x \le \frac{\pi}{2}$$

(ii) We also know that $\sin x$ has a slope equal to 1 near the origin and has the value 1 at $x = \pi/2$. Thus, it intersects the straight line $y = 2x/\pi$ at $x = 0$ and at $x = \pi/2$.

Since it is convex to this line on the rest of the interval, we have established the inequality.

If we are given that

$$f(x) > g(x) > 0 \tag{0.29a}$$

we have

$$\exp[f(x)] > \exp[g(x)] \tag{0.29b}$$

since

$$\frac{\exp[f(x)]}{\exp[g(x)]} = \exp[f(x) - g(x)] > 1 \tag{0.29c}$$

If $f(x) > g(x) > 0$, for $x > 0$, then

$$\int_0^x f(x)\,dx > \int_0^x g(x)\,dx \tag{0.30}$$

since the area under each of the curves is positive for all x.

0.6 GRAPHS AND ANALYTICAL EXPRESSIONS

Example 6.1 Find the value of x such that

$$x \ln x = 1$$

To do this, we plot $y = 1/x$ and $y = \ln x$, and find the intersection point. See Figure 0.1.

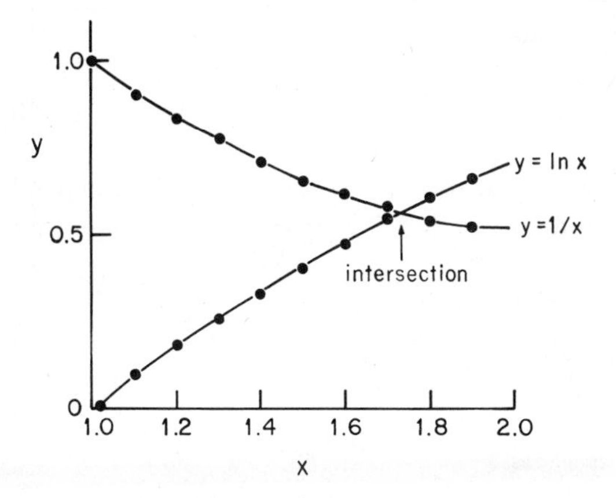

Figure 0.1 We are trying to find the value of x that satisfies the equation $x \ln x = 1$. We can guess that x is less than 2 and thus choose the scale so as to locate the intersection of the curve $y = 1/x$ and $y = \ln x$.

We can either be satisfied with the value that we read off the graph or use it as the basis for a perturbation expansion. For example, we see that the intersection point occurs when $x \approx 1.8$. Thus, we can consider

$$(1.8 + \varepsilon) \ln\left[1.8\left(1 + \frac{\varepsilon}{1.8}\right)\right] = (1.8 + \varepsilon)\left[\ln 1.8 + \ln\left(1 + \frac{\varepsilon}{1.8}\right)\right]$$

We then set this expression equal to 1 and solve for ε.

We know that $\ln 1.8 = 0.5877\ldots$; and using the first term of the Taylor series of the logarithm, we have

$$(1.8 + \varepsilon)\left(0.588 + \frac{\varepsilon}{1.8}\right) + \text{h.o.t.} = 1$$

or $\varepsilon \approx -0.04$. Thus, we would guess that $x \ln x = 1$ when $x = 1.76$. (A computation shows that $(1.76) \ln 1.76 = 0.99$.) ■

EXERCISE 6.1 Consider

$$\ln x = bx; \qquad b > 0$$

Show, graphically, that there are 0, 1, or 2 solutions depending on the value of b. Obtain, graphically, the two solutions for $b = 0.15$. Use estimates of the intersection of the two curves to obtain a simple analytical approximation, correct to two significant figures, for the points of intersection.

Answer: $x = 1.2$ and $x = 20$. ■

EXERCISE 6.2 Consider

$$x \exp(-bx) = 1 \quad \text{or} \quad x = \exp bx; \qquad b > 0$$

Plot a graph showing that there are 0, 1, or 2 solutions depending on the value of b. Find the solutions for $b = 0.23$ correct to two significant figures.

Answer: $x = 1.4$ and $x = 10$. ■

EXERCISE 6.3 Consider

$$\frac{10 \sin x}{x} = 1$$

We want to find the largest value of x for which the equation is true. It is clear that there is a solution for $x < 10$ since $\sin x \leq 1$. From a graph of $y = \sin x$ and $y = x/10$, it is seen that the largest solution for x must lie between $x = 5\pi/2$ and $x = 3\pi$. Develop an appropriate graph to obtain an estimate of the last point of intersection. Then use the Taylor-series expansion of $\sin x$ and $\cos x$ in the neighborhood of this point as a basis for an analytical approximation. Obtain a value of the intersection, correct to two significant figures.

Answer: $x = 8.4$. ■

EXERCISE 6.4 Consider

$$(\sin x)(\exp - bx) = \frac{1}{x}; \qquad x, b > 0$$

There is a condition on parameter b that determines the number of roots or solutions of the equation. Obtain an estimate, correct to two significant figures, for the least and greatest values of x for which the equation is true when $b = 0.046$. It is a good idea to find the intersection points of the curves

$$y = \frac{\exp bx}{x} \qquad \text{and} \qquad y = \sin x$$

Note that we can locate the last point of intersection by observing that $\sin x$ is bounded by 1 and $(\exp bx)/x$ is monotonically increasing for large x. Thus, we know that the last intersection comes before $(\exp bx)/x = 1$, which occurs near $x = 100$.

Answer: $x = 1.15$ and $x = 96$. ■

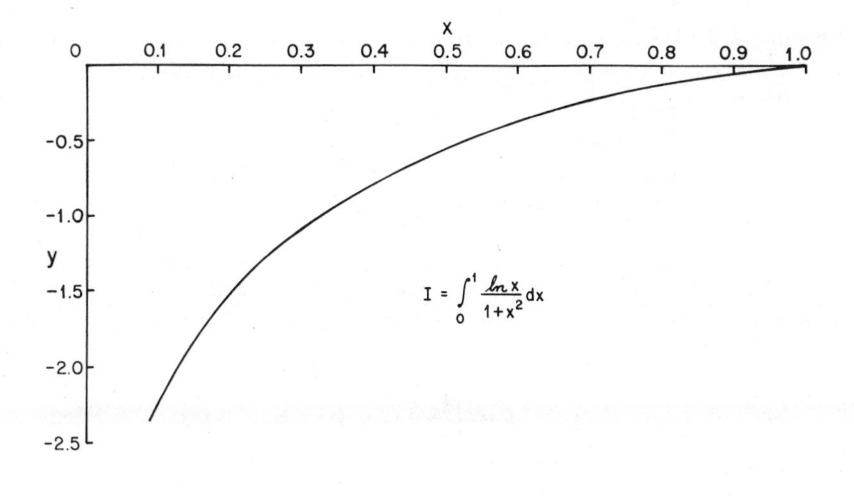
x
0
0.1
0.2
0.3
0.4
0.5
0.6
0.7
0.8
0.9
1.0
-0.5
-1.0
y
-1.5
-2.0
-2.5
$I = \int_0^1 \frac{\ln x}{1+x^2} dx$

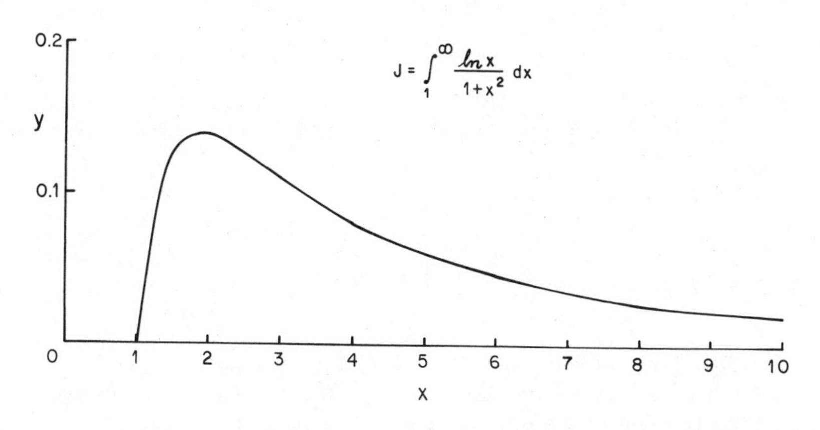
0.2
y
0.1
$J = \int_1^\infty \frac{\ln x}{1+x^2} dx$
0
1
2
3
4
5
6
7
8
9
10
x

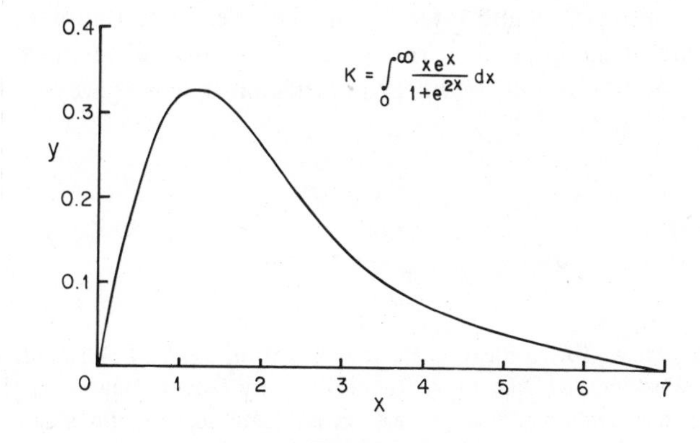
0.4
0.3
y
0.2
0.1
$K = \int_0^\infty \frac{x e^x}{1+e^{2x}} dx$
0
1
2
3
4
5
6
7
x

Figure 0.2

Example 6.2 It takes more experience to learn how to choose the scale of axes on a graph in order to evaluate an integral numerically. For example, consider the integral

$$A = \int_0^\infty \frac{\ln x}{1 + x^2}\, dx$$

We can see that $A = 0$ by the substitution $u = 1/x$ and noting that we get $A = -A$, and hence $A = 0$. Also, we can divide the integral into two parts:

$$I = \int_0^1 \frac{\ln x}{1 + x^2}\, dx$$

$$J = \int_1^\infty \frac{\ln u}{1 + u^2}\, du$$

Since the integrand of $I \leq 0$ and the integrand of $J \geq 0$, we can conclude that $I = -J$. Furthermore, we can make the substitution in integral J of $\ln u = y$ and obtain

$$K = \int_0^\infty \frac{y(\exp y)}{1 + \exp 2y}\, dy$$

We plot the integrands of integrals I, J, and K in Figure 0.2.

Notice how we have chosen different scales for the axes. Also observe that a substantial fraction of the integral J is contained in the interval $(10, \infty)$. We will return to these points and establish in Exercise 5.3 of Chapter 3 that the numerical value of the integral K is $0.916\ldots$, known as Catalan's constant.

Plot your own graphs and convince yourself that the areas are equal. ■

BIBLIOGRAPHY

Tables

1. Abramowitz, M., and I. A. Stegun, *Handbook of Mathematical Functions*, Bureau of Standards, Washington, D.C. 1964. Republished by Dover, New York, 1965. This has become the standard reference work on mathematical functions and associated tables. It is comprehensive and has been through several editions.
2. Gradshteyn, I. S., and I. W. Ryzhik, *Tables of Integrals, Series and Products*, 4th Ed., Academic Press, New York, 1965. This is a comprehensive reference that is most worthwhile.

Numerical Methods

3. Press, W. H., B. P. Flannery, S. A. Teukolsky, and W. T. Vetterling, *Numerical Recipes*, Cambridge University Press, New York, 1987. This is a veritable warehouse of numerical methods and ideas. It includes extensive computer-related techniques and a wealth of practical examples. It should be consulted when considering a computational problem and is especially useful when used in conjunction with a compendium of tables. It has both elementary and advanced techniques.

CHAPTER 1

MATRIX THEORY

We use matrices in a broad spectrum of activities, and it is possible to devote a substantial fraction of a text on mathematical methods to their study alone. However, there are a number of excellent books that discuss the subject, so that it is easy to find one at an appropriate level. With this in mind, we provide a few titles that are representative of the available literature in the Bibliography at the end of the chapter. It is our approach to just touch on a few of the topics, in part, to assure a minimum common background, and, in part, to develop the tools that we will be using. However, not all of the material is elementary. For example, the discussion of perturbation theory and stability will most likely be unfamiliar and challenging. We restrict our attention to 2×2 matrices, with just a few remarks about matrices of higher order. In this way, we can come to the heart of our development of the subject without the burden of proving many theorems in their general form. On the other hand, most of the results, properly generalized, are true for matrices of any order. Finally, the proofs of the relevant theorems are available in the titles provided.

We organize our discussion into two parts. We assume the reader has some familiarity with matrices. (Readers unfamiliar with the absolute basics of matrix algebra are advised to begin with the introductory parts of a text such as that by Davis or Schwartz.) We begin with the introduction of some notation and preliminary remarks. This is followed by a reasonably detailed discussion of eigenvalues and eigenvectors, including the Gram–Schmidt orthogonalization procedure; the Cayley–Hamilton theorem and functions of matrices; and first-order perturbation theory. We conclude with a discussion of the role of matrices in differential and difference equations.

1.1 NOTATION AND PRELIMINARY REMARKS

We introduce two matrices E and F that are used to illustrate many of the properties that are to be detailed.

$$E = \begin{bmatrix} 3 & 2 \\ 2 & 3 \end{bmatrix}; \qquad F = \begin{bmatrix} 2 & 1 \\ 3 & 4 \end{bmatrix}; \qquad EF = \begin{bmatrix} 12 & 11 \\ 13 & 14 \end{bmatrix}; \qquad FE = \begin{bmatrix} 8 & 7 \\ 17 & 18 \end{bmatrix}$$

1.1.1 Matrices and Vectors

We write matrix A as

$$A = \begin{bmatrix} a_{11} & a_{12} \\ a_{21} & a_{22} \end{bmatrix}$$

where the elements of matrix A are denoted by

$$a_{ij}; \qquad i, j = 1, 2$$

Example 1.1 The elements of matrix F are

$$f_{11} = 2, \qquad f_{12} = 1, \qquad f_{21} = 3, \qquad f_{22} = 4$$ ■

The product AB of two matrices is not necessarily equal to the product taken in the reverse order, i.e., BA.

Example 1.2 The product EF is not the same as the product FE. We say that these matrices do not commute. ■

Vectors with components x_1 and x_2 are written as

$$|x\rangle = \begin{bmatrix} x_1 \\ x_2 \end{bmatrix}$$

The adjoint vector, $\langle x|$, is written as

$$\langle x| = [x_1^* \quad x_2^*]$$

where * denotes complex conjugate.

The scalar product of two vectors $\langle y|$, $|x\rangle$, is written as

$$\langle y, x\rangle = y_1^* x_1 + y_2^* x_2$$

Two vectors $\langle y|$ and $|x\rangle$ are orthogonal if their scalar product is equal to zero.

(It is necessary to take the complex conjugate of the row elements to guarantee that the length or norm of a vector is nonnegative; $\langle x, x\rangle \geq 0$.)

1.1.2 Matrix Operations

A matrix is symmetric if $a_{ij} = a_{ji}$, for all i, j.
The matrix E is symmetric.
A matrix is antisymmetric if $a_{ij} = -a_{ji}$. Then, $a_{ii} = 0$.
The transpose of matrix A, written A^T, is formed by interchanging the rows and columns: $a_{ij} \to a_{ji}$

Example 1.3

$$F^T = \begin{bmatrix} 2 & 3 \\ 1 & 4 \end{bmatrix}$$ ■

The adjoint of matrix A, written $A^\dagger$, is formed by taking the complex conjugate of the transposed matrix. (The order of transposition and complex conjugation is immaterial.)

Example 1.4

$$S = \begin{bmatrix} 1+i & 2i \\ 3+i & 2-i \end{bmatrix}$$

$$S^\dagger = \begin{bmatrix} 1-i & 3-i \\ -2i & 2+i \end{bmatrix}$$ ■

Matrix A is Hermitian if it is equal to its adjoint. Thus,

$$A = A^\dagger \qquad \text{iff} \qquad a_{ij} = a_{ji}^*$$

(We write if and only if as "iff.")

Example 1.5 Matrix H is Hermitian.

$$H = \begin{bmatrix} a & b+ic \\ b-ic & d \end{bmatrix}; \qquad a, b, c, \text{ and } d \text{ are real}$$ ■

A matrix acting on a vector yields a vector. Thus, $A|x\rangle = |Ax\rangle$.

Example 1.6

$$E|x\rangle = \begin{bmatrix} 3 & 2 \\ 2 & 3 \end{bmatrix}\begin{bmatrix} x_1 \\ x_2 \end{bmatrix} = \begin{bmatrix} 3x_1 + 2x_2 \\ 2x_1 + 3x_2 \end{bmatrix}$$ ■

Given vector $|Ay\rangle$, the adjoint vector $\langle Ay|$ is formed by the action of the adjoint matrix $A^\dagger$ on the adjoint vector.

Example 1.7 Given the matrix

$$S = \begin{bmatrix} 1+i & 2i \\ 3+i & 2-i \end{bmatrix}$$

and the vector

$$\begin{bmatrix} y_1 \\ y_2 \end{bmatrix}$$

the adjoint vector $\langle Sy|$ is formed as follows:

$$\begin{aligned} &[y_1^* \quad y_2^*] \begin{bmatrix} 1-i & 3-i \\ -2i & 2+i \end{bmatrix} \\ &\quad = [(1-i)y_1^* - 2iy_2^* \quad (3-i)y_1^* + (2+i)y_2^*] \end{aligned}$$ ■

EXERCISE 1.1 Verify that we obtain the same result as we had in Example 1.7 if we first evaluate

$$|Sy\rangle = \begin{bmatrix} 1+i & 2i \\ 3+i & 2-i \end{bmatrix} \begin{bmatrix} y_1 \\ y_2 \end{bmatrix}$$

and then form the adjoint vector $\langle Sy|$. ■

The scalar product formed by vector $|Ax\rangle$ and adjoint vector $\langle y|$ is written $\langle y, Ax\rangle$. Similarly, the scalar product formed by adjoint vector $\langle Ay|$ and vector $|x\rangle$ is written $\langle Ay, x\rangle$.

Example 1.8 Given

$$S = \begin{bmatrix} 1+i & 2i \\ 3+i & 2-i \end{bmatrix}$$

and

$$|x\rangle = \begin{bmatrix} 1 \\ 3i \end{bmatrix}; \qquad |y\rangle = \begin{bmatrix} 2i \\ i \end{bmatrix}$$

we want to evaluate the scalar product $\langle y, Sx\rangle$. We first find $|Sx\rangle$:

$$|Sx\rangle = \begin{bmatrix} 1+i & 2i \\ 3+i & 2-i \end{bmatrix} \begin{bmatrix} 1 \\ 3i \end{bmatrix} = \begin{bmatrix} -5+i \\ 6+7i \end{bmatrix}$$

Then we compute

$$\langle y, Sx\rangle = [-2i \quad -i] \begin{bmatrix} -5+i \\ 6+7i \end{bmatrix} = 9 + 4i$$

Now let us evaluate the scalar product $\langle Sy, x\rangle$. We have

$$\langle Sy| = [\,-2i \quad -i\,]\begin{bmatrix} 1-i & 3-i \\ -2i & 2+i \end{bmatrix} = [\,-4-2i \quad -1-8i\,]$$

Then

$$\langle Sy, x\rangle = [\,-4-2i \quad -1-8i\,]\begin{bmatrix} 1 \\ 3i \end{bmatrix} = 20-5i$$

Note that in our example, scalar products $\langle y, Sx\rangle$ and $\langle Sy, x\rangle$ are not equal. A Hermitian matrix A has the property that the scalar product $\langle Ay, x\rangle$ is equal to $\langle y, Ax\rangle$. ■

1.1.3 The Trace and the Determinant

The trace of A is given by the sum of its diagonal elements or

$$\operatorname{Tr} A = \sum_i a_{ii}$$

where the trace is denoted by Tr.

Example 1.9 The trace of matrix E is $(3+3) = 6$. ■

The trace of a product of matrices possesses the cyclical property

$$\operatorname{Tr}(ABC) = \operatorname{Tr}(CAB) = \operatorname{Tr}(BCA)$$

This is established if we write the trace as the sum over the indices i, j, and k and note that they can be interchanged to read

$$\operatorname{Tr}(ABC) = \Sigma a_{ij}b_{jk}c_{ki} = \Sigma c_{ki}a_{ij}b_{jk} = \Sigma b_{jk}c_{ki}a_{ij}$$
$$= \operatorname{Tr}(CAB) = \operatorname{Tr}(BCA)$$

Example 1.10 The trace of the product EF of two matrices is equal to $(12 + 14) = 26$; the trace of the product $FE = (8 + 18) = 26$. ■

The determinant of matrix A is written as

$$\det A = \begin{vmatrix} a_{11} & a_{12} \\ a_{21} & a_{22} \end{vmatrix} \quad \text{or} \quad \det \begin{bmatrix} a_{11} & a_{12} \\ a_{21} & a_{22} \end{bmatrix}$$

and for 2×2 matrices is equal to $a_{11}a_{22} - a_{12}a_{21}$. It is also signified by the symbol Δ.

Example 1.11 The determinant of matrix F is $(2 \cdot 4 - 3 \cdot 1) = 5$. The determinant of matrix E is $(3 \cdot 3 - 2 \cdot 2) = 5$. ■

The determinant of the product of matrices A, B, and C is equal to the product of the determinants:

$$\det(ABC) = (\det A)(\det B)(\det C)$$

Example 1.12

$$\det(EF) = (12 \cdot 14 - 11 \cdot 13) = (\det E)(\det F) = 5 \cdot 5 = 25$$
$$\det(FE) = (8 \cdot 18 - 7 \cdot 17) = 25$$
■

1.1.4 The Inverse

The inverse A^{-1} of a matrix, if it exists, is unique and is such that

$$AA^{-1} = A^{-1}A = I; \qquad \det A^{-1} = \frac{1}{\det A}$$

where I is the identity matrix.

$$I = \begin{bmatrix} 1 & 0 \\ 0 & 1 \end{bmatrix}$$

It is also clear that the inverse of the inverse of a matrix is the matrix itself.

Example 1.13 The inverse of matrix E is

$$E^{-1} = \frac{1}{\Delta}\begin{bmatrix} e_{22} & -e_{12} \\ -e_{21} & e_{11} \end{bmatrix} = \frac{1}{5}\begin{bmatrix} 3 & -2 \\ -2 & 3 \end{bmatrix}$$
■

Example 1.14 The inverse of matrix F is

$$F^{-1} = \frac{1}{\Delta}\begin{bmatrix} f_{22} & -f_{12} \\ -f_{21} & f_{11} \end{bmatrix} = \frac{1}{5}\begin{bmatrix} 4 & -1 \\ -3 & 2 \end{bmatrix}$$
■

The inverse of the product of two matrices M and N is the product of the inverses in the reverse order. Thus, we have

$$(MN)^{-1} = N^{-1}M^{-1}$$

since

$$(MN)^{-1}(MN) = N^{-1}M^{-1}MN = N^{-1}IN = I$$

Example 1.15 The inverse of the product of two matrices E and F is

$$(EF)^{-1} = F^{-1}E^{-1} = \frac{1}{5}\begin{bmatrix} 4 & -1 \\ -3 & 2 \end{bmatrix} \frac{1}{5}\begin{bmatrix} 3 & -2 \\ -2 & 3 \end{bmatrix}$$
$$= \frac{1}{25}\begin{bmatrix} 14 & -11 \\ -13 & 12 \end{bmatrix}$$ ■

1.1.5 Unitary Matrices

Matrix U is unitary if its adjoint is equal to its inverse. Thus, if $U^\dagger = U^{-1}$, we have $UU^\dagger = U^\dagger U = I$.

Example 1.16 Given

$$U = \frac{1}{\sqrt{6}}\begin{bmatrix} 1 - i & \sqrt{2}(-1 + i) \\ 2 & \sqrt{2} \end{bmatrix}$$

we have

$$U^\dagger = \frac{1}{\sqrt{6}}\begin{bmatrix} 1 + i & 2 \\ \sqrt{2}(-1 - i) & \sqrt{2} \end{bmatrix}$$

and $U^\dagger U = UU^\dagger = I$. ■

1.1.6 Ill-Conditioned Matrices

A matrix is "ill-conditioned" if small changes in some of its elements produce large changes in its determinant. It is not a clearly defined notion and one has to exercise care when working with such matrices.

Example 1.17 Consider the matrix

$$A = \begin{bmatrix} 1.00 & 1.00 \\ 1.00 & 0.99 \end{bmatrix} \tag{1.1}$$

where the elements have been determined by measurements and are thus known only with the precision given. (This example is derived from the book entitled *Numerical Methods for Scientists and Engineers* by Hamming, page 128.)

(i) Show that $\det A = -0.01$. It will become clear that the small size of the determinant of A relative to the size of its matrix elements leads to problems associated with the inverse, A^{-1}, and thus leads us to call it an "ill-conditioned" matrix.

(ii) Show that

$$A^{-1} = \begin{bmatrix} -99 & 100 \\ 100 & -100 \end{bmatrix}; \qquad \det A^{-1} = -100 \tag{1.2}$$

and verify explicitly that

$$A^{-1}A = AA^{-1} = I$$

(iii) In Equation (1.2), we have found the exact inverse of matrix A given by Equation (1.1). Let us say that we have a situation where we are unable to find the exact inverse of matrix A. We use some approximation scheme to find an "approximate inverse."
For example, matrix

$$L = \begin{bmatrix} -98 & 99 \\ 100 & -100 \end{bmatrix} \tag{1.3}$$

is an "approximate left inverse" of matrix A given by Equation (1.1) since

$$LA = \begin{bmatrix} 1.00 & 0.01 \\ 0.00 & 1.00 \end{bmatrix} \tag{1.4}$$

We also have an "approximate right inverse"

$$R = \begin{bmatrix} -98 & 100 \\ 99 & -100 \end{bmatrix} \tag{1.5}$$

since

$$AR = \begin{bmatrix} 1.00 & 0.00 \\ 0.01 & 1.00 \end{bmatrix} \tag{1.6}$$

Now, comes the crucial observation. We find that these "approximate" inverses are quite unsatisfactory if used on the "wrong side" since it is easily checked that

$$AL = \begin{bmatrix} 2 & -1 \\ 1 & 0 \end{bmatrix} \tag{1.7a}$$

$$RA = \begin{bmatrix} 2 & 1 \\ -1 & 0 \end{bmatrix} \tag{1.7b}$$

We see that since det A is so small, we have $AL \neq I$ and $RA \neq I$. The point is that you have to be especially careful when working with problems in which the determinant is a small quantity compared to some of the matrix elements, since its inverse is then a large multiplier. We see another aspect of this in Example 1.18. ■

Example 1.18 Consider two equations for the quantities x and y:

$$\begin{aligned} ax + by &= 1 \\ cx + dy &= 1 \end{aligned} \tag{1.8}$$

where a, b, c, and d are parameters that determine x and y. The parameters can be considered elements of a matrix, so that we have the equation

$$A\begin{bmatrix} x \\ y \end{bmatrix} = \begin{bmatrix} 1 \\ 1 \end{bmatrix}; \qquad A = \begin{bmatrix} a & b \\ c & d \end{bmatrix} \tag{1.9}$$

Let us assume that $a = 2$ and $c = 1$, precisely; and we make measurements to establish the values of b and d from which we determine x and y. We solve for x and y and find

$$x = 1 - \frac{d}{2d - b} \qquad \text{and} \qquad y = \frac{1}{2d - b}$$

Let us now assume that we make measurements as follows:

$$d = 10 \qquad \text{and} \qquad b = 21, \qquad \text{then} \qquad x = 11 \qquad \text{and} \qquad y = -1$$

We measure these quantities in a different experiment and find:

$$d = 11 \qquad \text{and} \qquad b = 21, \qquad \text{then} \qquad x = -10 \qquad \text{and} \qquad y = +1$$

So we have a change in sign for both x and y following a 10% change in the value of d. This result follows from the fact that our matrix is "ill-conditioned" or its determinant is too small when compared to some of its elements. (The determinant changes in value from -1 to $+1$, while element d goes from 10 to 11.) ■

1.1.7 Eigenvalues and Eigenvectors

Matrix A has characteristic vectors (eigenvectors) $|x\rangle$ and characteristic values (eigenvalues) λ such that

$$A|x\rangle = \lambda|x\rangle$$

or, equivalently,

$$(A - \lambda I)|x\rangle = 0$$

This means that matrix A, when acting on its eigenvector, simply multiplies the eigenvector by a constant, λ. (We exclude the case were $|x\rangle$ is a zero vector.)

A 2×2 matrix has two eigenvalues, but not always two eigenvectors. (If eigenvalues coincide, they are counted according to their multiplicity.)

Example 1.19 Matrix E has eigenvalues equal to 5 and 1. We will see in Section 1.5 that the eigenvalues are found by solving the characteristic equation

$$g(\lambda) \equiv \det|E - \lambda I| = \begin{vmatrix} 3-\lambda & 2 \\ 2 & 3-\lambda \end{vmatrix} = \lambda^2 - 6\lambda + 5 = 0$$

It has eigenvectors

$$|x_1\rangle = c_1\begin{bmatrix} 1 \\ 1 \end{bmatrix}; \qquad |x_2\rangle = c_2\begin{bmatrix} 1 \\ -1 \end{bmatrix}$$

Then it is easy to verify that

$$\begin{bmatrix} 3 & 2 \\ 2 & 3 \end{bmatrix} c_1 \begin{bmatrix} 1 \\ 1 \end{bmatrix} = 5c_1 \begin{bmatrix} 1 \\ 1 \end{bmatrix}$$

$$\begin{bmatrix} 3 & 2 \\ 2 & 3 \end{bmatrix} c_2 \begin{bmatrix} 1 \\ -1 \end{bmatrix} = 1c_2 \begin{bmatrix} 1 \\ -1 \end{bmatrix}$$

Constants c_1 and c_2 appear on both sides of their respective equations and thus are at our disposal. Often, they are chosen so that the eigenvectors are normalized to unity, i.e.,

$$\langle x_1, x_1 \rangle = 1 \Rightarrow c_1 = \frac{1}{\sqrt{2}}$$

$$\langle x_2, x_2 \rangle = 1 \Rightarrow c_2 = \frac{1}{\sqrt{2}}$$

(These ideas are discussed in detail in Section 1.2.) ■

Example 1.20 The matrices given by Equations (1.10) and (1.11) that follow have degenerate or coincident eigenvalues and each has only one eigenvector.

$$A = \begin{bmatrix} 4 & 4 \\ 0 & 4 \end{bmatrix}; \qquad A\begin{bmatrix} 1 \\ 0 \end{bmatrix} = 4\begin{bmatrix} 1 \\ 0 \end{bmatrix} \tag{1.10}$$

$$B = \begin{bmatrix} 6 & 4i \\ i & 2 \end{bmatrix}; \qquad B\begin{bmatrix} -2i \\ 1 \end{bmatrix} = 4\begin{bmatrix} -2i \\ 1 \end{bmatrix} \tag{1.11}$$

■

1.1.8 Diagonalization

We often want to diagonalize matrix A since it then enables many problems to be transformed into a simple form.

Diagonalization can be accomplished by means of a similarity transformation S such that

$$S^{-1}AS = \Lambda$$

where Λ is the diagonal matrix whose elements are the eigenvalues of A.

Example 1.21 Consider the set of coupled differential equations:

$$\frac{dx}{dt} = 3x + 2y \tag{1.12a}$$

$$\frac{dy}{dt} = 2x + 3y \tag{1.12b}$$

with $x(0)$ and $y(0)$ specified. Set up the matrix equation

$$\frac{d}{dt}\begin{bmatrix} x \\ y \end{bmatrix} = \begin{bmatrix} 3 & 2 \\ 2 & 3 \end{bmatrix}\begin{bmatrix} x \\ y \end{bmatrix} = E\begin{bmatrix} x \\ y \end{bmatrix} \tag{1.13}$$

We want to find $x(t)$ and $y(t)$. This is often done by transforming the matrix to diagonal form. The equations are thereby decoupled and the solutions are found with no difficulty. We will learn in Section 1.2 how to construct the diagonalizing matrix; here, we simply present the results. The matrix that diagonalizes matrix E is

$$M = \begin{bmatrix} 1 & 1 \\ 1 & -1 \end{bmatrix}; \qquad M^{-1} = \frac{-1}{2}\begin{bmatrix} -1 & -1 \\ -1 & 1 \end{bmatrix}$$

Thus,

$$M^{-1}EM = \frac{-1}{2}\begin{bmatrix} -1 & -1 \\ -1 & 1 \end{bmatrix}\begin{bmatrix} 3 & 2 \\ 2 & 3 \end{bmatrix}\begin{bmatrix} 1 & 1 \\ 1 & -1 \end{bmatrix}$$

$$= \begin{bmatrix} 5 & 0 \\ 0 & 1 \end{bmatrix}$$

We then multiply Equation (1.13) by M^{-1} and insert the identity matrix in the form MM^{-1}, so that Equation (1.13) becomes

$$M^{-1}\frac{d}{dt}\begin{bmatrix} x \\ y \end{bmatrix} = \frac{d}{dt}M^{-1}\begin{bmatrix} x \\ y \end{bmatrix} = M^{-1}EMM^{-1}\begin{bmatrix} x \\ y \end{bmatrix}$$

or

$$\frac{d}{dt}\begin{bmatrix} z_1 \\ z_2 \end{bmatrix} = \begin{bmatrix} 5 & 0 \\ 0 & 1 \end{bmatrix}\begin{bmatrix} z_1 \\ z_2 \end{bmatrix}$$

where

$$\begin{bmatrix} z_1 \\ z_2 \end{bmatrix} = M^{-1}\begin{bmatrix} x \\ y \end{bmatrix}$$

$$z_1 = \tfrac{1}{2}(x + y)$$

$$z_2 = \tfrac{1}{2}(x - y)$$

The equations are decoupled so that the solutions can be found by inspection:

$$z_1(t) = z_1(0) \exp 5t$$

$$z_2(t) = z_2(0) \exp t$$

We now recover $x(t)$ and $y(t)$ by using the inverse transformations:

$$\begin{bmatrix} x \\ y \end{bmatrix} = M\begin{bmatrix} z_1 \\ z_2 \end{bmatrix} = \begin{bmatrix} 1 & 1 \\ 1 & -1 \end{bmatrix}\begin{bmatrix} z_1 \\ z_2 \end{bmatrix}; \qquad \begin{bmatrix} x \\ y \end{bmatrix} = \begin{bmatrix} z_1 + z_2 \\ z_1 - z_2 \end{bmatrix}$$

$$\begin{aligned} x(t) &= z_1(0) \exp 5t + z_2(0) \exp t \\ &= \tfrac{1}{2}[x(0) + y(0)] \exp 5t + \tfrac{1}{2}[x(0) - y(0)] \exp t \\ y(t) &= z_1(0) \exp 5t - z_2(0) \exp t \\ &= \tfrac{1}{2}[x(0) + y(0)] \exp 5t - \tfrac{1}{2}[x(0) - y(0)] \exp t \end{aligned}$$

These ideas are discussed again when we introduce the Cayley–Hamilton theorem in Section 1.5. ■

Not all matrices can be brought to diagonal form by an invertible matrix. For example, consider the matrices given by Equations (1.10) and (1.11):

$$\text{Neither } A = \begin{bmatrix} 4 & 4 \\ 0 & 4 \end{bmatrix} \quad \text{nor} \quad B = \begin{bmatrix} 6 & 4i \\ i & 2 \end{bmatrix}$$

can be diagonalized. (Note that the eigenvalues of matrix B are coincident and equal to 4 even though all the matrix elements differ.) It is, however, easy to verify that

$$S = \begin{bmatrix} 1 & 1 \\ 0 & 0 \end{bmatrix} \tag{1.14}$$

and

$$T = \begin{bmatrix} 1 & 1 \\ \tfrac{1}{2}i & \tfrac{1}{2}i \end{bmatrix} \tag{1.15}$$

are such that $AS = S\Lambda_A = S4I$ and $BT = T\Lambda_B = T4I$. Neither S nor T have an inverse due to the vanishing of their respective determinants. (Notation: Λ_A denotes a diagonal matrix whose elements are the eigenvalues of matrix A.)

1.1.9 Similarity Transformations

Large classes of matrices can be brought into diagonal form by similarity transformations. In particular, any matrix of arbitrary order with distinct (different) eigenvalues can be diagonalized. Since, quite often, the elements of a matrix are derived from measurements, it is not unreasonable to conclude that the matrix elements and hence the eigenvalues can be changed by arbitrarily small amounts without affecting the underlying problem. On the other hand, sometimes symmetry considerations lead directly to degenerate (coincident) eigenvalues, and one is then led to matrices that cannot be diagonalized. In our work, we will always assume that our matrices are diagonalizable.

We can see the origin of the term similarity transformation by observing that the trace and the determinant are invariant under such transformations. This means that if we have

$$A = SBS^{-1}$$

then

$$\operatorname{Tr} A = \operatorname{Tr}(SBS^{-1}) = \operatorname{Tr}(S^{-1}SB) = \operatorname{Tr} B \tag{1.16}$$

and

$$\begin{aligned} \det A &= (\det S)(\det B)(\det S^{-1}) = (\det S^{-1})(\det S)(\det B) \\ &= (\det S^{-1}S)(\det B) = \det B \end{aligned} \tag{1.17}$$

We show, in Section 1.5, that matrices that are connected by a similarity transformation have the same eigenvalues. However, if matrices have the same eigenvalues, there need not exist a similarity transformation that connects them. For example, matrices A and B given by Equations (1.10) and (1.11) have coincident eigenvalues equal to 4. They cannot be diagonalized and, hence, neither is connected to a diagonal matrix by a similarity transformation. A matrix that has a complete set of linearly independent eigenvectors can be brought to diagonal form by a similarity transformation. Consider two such diagonalizable matrices C and D with the same eigenvalues. There exists a matrix S such that

$$S^{-1}CS = \Lambda_C$$

There also exists a matrix P such that

$$P^{-1}DP = \Lambda_D$$

We are given that

$$\Lambda_C = \Lambda_D$$

Then

$$S^{-1}CS = P^{-1}DP \qquad \text{or} \qquad C = (SP^{-1})D(PS^{-1})$$

and if we write $M = PS^{-1}$, we have $C = M^{-1}DM$.

Example 1.22 Consider

$$A = \begin{bmatrix} 0 & 1 \\ -1 & 0 \end{bmatrix}; \qquad B = \begin{bmatrix} \frac{3}{4} & \frac{5}{4} \\ -\frac{5}{4} & -\frac{3}{4} \end{bmatrix}$$

$$\operatorname{Tr} A = 0 = \operatorname{Tr} B$$

$$\det A = 1; \qquad \det B = -\tfrac{9}{16} + \tfrac{25}{16} = 1$$

Since A and B both have the eigenvalues $\pm i$, they have the same trace and the same determinant. ■

1.2 EIGENVALUES AND EIGENVECTORS

We want to understand the interrelationship between eigenvalues and eigenvectors on the one hand and symmetry and diagonalizability on the other hand.

1.2.1 Eigenvalues and Eigenvectors

In Section 1.1.7, we stated that if we have the equation

$$A|x\rangle = \lambda|x\rangle \quad \text{or} \quad A\begin{bmatrix} x_1 \\ x_2 \end{bmatrix} = \lambda\begin{bmatrix} x_1 \\ x_2 \end{bmatrix}; \qquad A = \begin{bmatrix} a_{11} & a_{12} \\ a_{21} & a_{22} \end{bmatrix}$$

we say that $|x\rangle$ is an eigenvector and λ is an eigenvalue. In general, if a matrix operates or acts on a vector, it yields a vector that is not a simple multiple of the vector. But a matrix operating on one of its eigenvectors simply gives a multiple of the vector.

Real-symmetric matrices have real elements such that $a_{ij} = a_{ji}$. They have real eigenvalues and identical left and right eigenvectors with identical elements:

$$\langle x_j|A = \lambda_j\langle x_j| \qquad \text{and} \qquad A|x_i\rangle = \lambda_i|x_i\rangle$$

Furthermore,

$$\langle x_j, x_i\rangle = 0 \qquad \text{if} \quad \lambda_i \neq \lambda_j$$

However, if a matrix is not real-symmetric, it may have complex eigenvalues and a set of left eigenvectors that is different from its set of right eigenvectors.

Although we will discuss nonsymmetric matrices, we have no need to consider their left eigenvectors.

Thus, we will use the term eigenvector, referring to the right eigenvector, for nonsymmetric matrices. (See the texts by Joshi and Stephenson for a discussion of left and right eigenvectors.)

Example 2.1 Consider

$$A = \begin{bmatrix} 4 & 1 \\ 2 & 3 \end{bmatrix}$$

Its eigenvalues are 2 and 5. Its right eigenvectors are

$$\begin{bmatrix} -1 \\ 2 \end{bmatrix} \quad \text{and} \quad \begin{bmatrix} 1 \\ 1 \end{bmatrix}$$

Then

$$A\begin{bmatrix} -1 \\ 2 \end{bmatrix} = 2\begin{bmatrix} -1 \\ 2 \end{bmatrix}; \qquad A\begin{bmatrix} 1 \\ 1 \end{bmatrix} = 5\begin{bmatrix} 1 \\ 1 \end{bmatrix}$$

Its left eigenvectors are

$$[1 \quad -1] \quad \text{and} \quad [2 \quad 1]$$

We also have

$$[1 \quad -1]A = 2[1 \quad -1] \quad \text{and} \quad [2 \quad 1]A = 5[2 \quad 1]$$

Note that the left eigenvector corresponding to the eigenvalue λ_i is orthogonal to the right eigenvector corresponding to the eigenvalue λ_j; $\lambda_i \neq \lambda_j$.

$$[1 \quad -1]\begin{bmatrix} -1 \\ 2 \end{bmatrix} \neq 0; \qquad [1 \quad -1]\begin{bmatrix} 1 \\ 1 \end{bmatrix} = 0$$

$$[2 \quad 1]\begin{bmatrix} -1 \\ 2 \end{bmatrix} = 0; \qquad [2 \quad 1]\begin{bmatrix} 1 \\ 1 \end{bmatrix} \neq 0 \qquad ■$$

Let us understand this by relating the matrix operation to the solution of a polynomial equation. We know that a polynomial of degree n, with real coefficients, has precisely n roots that are either real or occur in conjugate complex pairs. We see the correspondence by writing

$$A|x\rangle = \lambda|x\rangle \quad \text{as} \quad (A - \lambda I)|x\rangle = 0$$

or

$$\begin{bmatrix} a_{11} - \lambda & a_{12} \\ a_{21} & a_{22} - \lambda \end{bmatrix}\begin{bmatrix} x_1 \\ x_2 \end{bmatrix} = 0 \tag{1.18a}$$

or

$$
\begin{aligned}
(a_{11}-\lambda)x_1 \quad + \quad a_{12}x_2 \quad &= 0 \\
a_{21}x_1 \quad + \quad (a_{22}-\lambda)x_2 \quad &= 0
\end{aligned}
\tag{1.18b}
$$

These are two linear homogeneous equations with two unknowns, x_1 and x_2. There is a nontrivial solution iff (if and only if) the determinant

$$
\begin{vmatrix} a_{11}-\lambda & a_{12} \\ a_{21} & a_{22}-\lambda \end{vmatrix} = 0
$$

***Example* 2.2** If we have

$$
\begin{aligned}
ax + by &= 0 \\
cx + dy &= 0
\end{aligned}
$$

then either

$$x = y = 0$$

or

$$\det\begin{bmatrix} a & b \\ c & d \end{bmatrix} = 0$$

This follows since we must have

$$\frac{x}{y} = \frac{-b}{a} = \frac{-d}{c} \qquad \text{if} \quad y \neq 0$$

or

$$\frac{y}{x} = \frac{-a}{b} = \frac{-c}{d} \qquad \text{if} \quad x \neq 0$$

■

We observe that the determinant equation is just a polynomial of degree n in the unknown parameter λ. (Furthermore, in the n-dimensional case, our polynomial has n roots and thus there are precisely n values of λ; $\lambda_1, \lambda_2, \ldots, \lambda_n$ for which there exist nontrivial solutions.) If two roots coincide, we say that we have degenerate eigenvalues.

1.2.2 Notation for the Scalar Product

There is some confusion associated with the notation for the scalar product of two vectors $|x\rangle$ and $|y\rangle$. We will try to avoid that confusion by adopting the following convention. We will write a general vector as $|x\rangle$ or

$$\begin{bmatrix} x_1 \\ x_2 \end{bmatrix}$$

Then its components are written as

$$x_1 \quad \text{and} \quad x_2$$

If we have two vectors $|x_1\rangle$ and $|x_2\rangle$, we use superscripts or "commas" to indicate the components as follows:

$$|x_1\rangle = \begin{bmatrix} x_1 \\ x_2 \end{bmatrix}^1 = \begin{bmatrix} x_{1,1} \\ x_{1,2} \end{bmatrix}$$

$$|x_2\rangle = \begin{bmatrix} x_1 \\ x_2 \end{bmatrix}^2 = \begin{bmatrix} x_{2,1} \\ x_{2,2} \end{bmatrix}$$

Thus, we write $x_{1,1}$ and $x_{1,2}$ to mean elements 1 and 2, respectively, of vector 1. And, similarly, $x_{2,1}$ and $x_{2,2}$ indicate components of vector 2. For example, if we have two eigenvectors $|x_1\rangle$ and $|x_2\rangle$ with eigenvalues λ_1 and λ_2, respectively:

$$A|x_1\rangle = \lambda_1|x_1\rangle; \qquad A|x_2\rangle = \lambda_2|x_2\rangle$$

We mean

$$\begin{bmatrix} a_{11} & a_{12} \\ a_{21} & a_{22} \end{bmatrix} \begin{bmatrix} x_{1,1} \\ x_{1,2} \end{bmatrix} = \lambda_1 \begin{bmatrix} x_{1,1} \\ x_{1,2} \end{bmatrix}$$

$$\begin{bmatrix} a_{11} & a_{12} \\ a_{21} & a_{22} \end{bmatrix} \begin{bmatrix} x_{2,1} \\ x_{2,2} \end{bmatrix} = \lambda_2 \begin{bmatrix} x_{2,1} \\ x_{2,2} \end{bmatrix}$$

In addition, we say that the vectors are orthogonal if

$$\langle x_1, x_2\rangle = [x_{1,1}^* \quad x_{1,2}^*] \begin{bmatrix} x_{2,1} \\ x_{2,2} \end{bmatrix} = x_{1,1}^* x_{2,1} + x_{1,2}^* x_{2,2} = 0 \tag{1.19}$$

Also, by complex conjugation, we have

$$\langle x_2, x_1\rangle = 0$$

1.2.3 The Concept of Linear Dependence and Linear Independence

We say that vectors $|x\rangle$ and $|y\rangle$ are linearly independent if

$$c_1|x\rangle + c_2|y\rangle = 0$$

requires that c_1 and c_2 are both identically equal to zero. Otherwise, we say that $|x\rangle$ and $|y\rangle$ are linearly dependent.

Example 2.3

$$|x\rangle = \begin{bmatrix} 0 \\ 1 \end{bmatrix}; \qquad |y\rangle = \begin{bmatrix} 1 \\ 0 \end{bmatrix}$$

These vectors are linearly independent. It is also clear that any two-dimensional vector can be written as a linear combination of these two:

$$\begin{bmatrix} \alpha \\ \beta \end{bmatrix} = \alpha \begin{bmatrix} 1 \\ 0 \end{bmatrix} + \beta \begin{bmatrix} 0 \\ 1 \end{bmatrix} \qquad \blacksquare$$

Example 2.4

$$|x_1\rangle = \begin{bmatrix} 1 \\ 3 \end{bmatrix}; \qquad |x_2\rangle = \begin{bmatrix} 3 \\ -1 \end{bmatrix}; \qquad |x_3\rangle = \begin{bmatrix} 1 \\ -1 \end{bmatrix}$$

These vectors are linearly dependent since

$$|x_1\rangle = 2|x_2\rangle - 5|x_3\rangle \qquad \blacksquare$$

In an n-dimensional space, we can have at most n linearly independent vectors. We can construct such a set as follows:

$$|x_1\rangle = \begin{bmatrix} 1 \\ 0 \\ 0 \\ \cdot \\ \cdot \\ \cdot \\ 0 \\ 0 \\ 0 \end{bmatrix}; \qquad |x_2\rangle = \begin{bmatrix} 0 \\ 1 \\ 0 \\ \cdot \\ \cdot \\ \cdot \\ 0 \\ 0 \\ 0 \end{bmatrix} \ldots |x_n\rangle = \begin{bmatrix} 0 \\ 0 \\ 0 \\ \cdot \\ \cdot \\ \cdot \\ 0 \\ 0 \\ 1 \end{bmatrix}; \qquad x_{i,j} = \delta_{i,j}$$

where $\delta_{i,j}$ is the Kronecker "δ" symbol:

$$\delta_{i,j} = \begin{cases} 0; & i \neq j \\ 1; & i = j \end{cases} \tag{1.20}$$

Example 2.5 Consider matrix E:

$$E = \begin{bmatrix} 3 & 2 \\ 2 & 3 \end{bmatrix}$$

We introduce the characteristic equation

$$(E - \lambda I)|x\rangle = 0; \qquad \det\begin{bmatrix} 3-\lambda & 2 \\ 2 & 3-\lambda \end{bmatrix} = 0$$

and observe that the eigenvalues are equal to 5 and 1. We then want to find the eigenvectors corresponding to these eigenvalues and thus we introduce the equations

$$E|x_1\rangle = 5|x_1\rangle; \qquad E|x_2\rangle = 1|x_2\rangle$$

$$\begin{bmatrix} 3 & 2 \\ 2 & 3 \end{bmatrix}\begin{bmatrix} x_{1,1} \\ x_{1,2} \end{bmatrix} = 5\begin{bmatrix} x_{1,1} \\ x_{1,2} \end{bmatrix}$$

$$\begin{bmatrix} 3 & 2 \\ 2 & 3 \end{bmatrix}\begin{bmatrix} x_{2,1} \\ x_{2,2} \end{bmatrix} = 1\begin{bmatrix} x_{2,1} \\ x_{2,2} \end{bmatrix}$$

We can solve these equations and find that

$$|x_1\rangle = c_1\begin{bmatrix} 1 \\ 1 \end{bmatrix}; \qquad |x_2\rangle = c_2\begin{bmatrix} 1 \\ -1 \end{bmatrix}$$

As previously mentioned, the free constants are often chosen so that the vectors are normalized, i.e., their total length equals 1.

$$\langle x_1, x_1\rangle = 1 \Rightarrow c_1 = \frac{1}{\sqrt{2}}$$

$$\langle x_2, x_2\rangle = 1 \Rightarrow c_2 = \frac{1}{\sqrt{2}}$$

We observe that these vectors are orthogonal, which follows because the original matrix was real-symmetric.

$$\langle x_1, x_2\rangle = \langle x_2, x_1\rangle = 0$$

■

Example 2.6 Now consider another matrix F that is not real-symmetric, but has the same eigenvalues as matrix E.

$$F = \begin{bmatrix} 2 & 1 \\ 3 & 4 \end{bmatrix} \tag{1.21}$$

Check that the eigenvalues are equal to 5 and 1.

$$F|y_i\rangle = \lambda_i|y_i\rangle; \qquad \lambda_1 = 5, \quad \lambda_2 = 1$$

We can also find the eigenvectors and normalize them to unity:

$$|y_1\rangle = d_1\begin{bmatrix} 1 \\ 3 \end{bmatrix}; \qquad |y_2\rangle = d_2\begin{bmatrix} 1 \\ -1 \end{bmatrix}$$

We now see that they are not orthogonal since

$$[1 \quad 3]\begin{bmatrix} 1 \\ -1 \end{bmatrix} = -2$$

(However, they are linearly independent since they are not multiples of one another.) ■

1.3 THE GRAM – SCHMIDT PROCEDURE

Often, we want to orthogonalize a given set of linearly independent vectors. This process is called the Gram–Schmidt method. We use these ideas to understand the concepts of orthogonality and diagonalization.

Consider a two-dimensional linear vector space with two linearly independent nonorthogonal vectors $|y_1\rangle$ and $|y_2\rangle$.

We wish to construct a set of orthogonal vectors. We choose the first one to be

$$|z_1\rangle = |y_1\rangle \tag{1.22}$$

since there is "nothing for it to be nonorthogonal to." Then we write the second vector as a linear combination of this vector and the second eigenvector as follows:

$$|z_2\rangle = |y_2\rangle - \frac{\langle y_1, y_2\rangle}{\langle y_1, y_1\rangle}|y_1\rangle \tag{1.23}$$

We see that vectors $|z_1\rangle$ and $|z_2\rangle$ are orthogonal since

$$\langle z_1, z_2\rangle = \langle y_1, y_2\rangle - \frac{\langle y_1, y_2\rangle}{\langle y_1, y_1\rangle}\langle y_1, y_1\rangle = 0 \tag{1.24}$$

Example 3.1 Given matrix F, construct orthogonal vectors from the eigenvectors:

$$F = \begin{bmatrix} 2 & 1 \\ 3 & 4 \end{bmatrix}; \qquad |y_1\rangle = \begin{bmatrix} 1 \\ 3 \end{bmatrix}; \qquad |y_2\rangle = \begin{bmatrix} 1 \\ -1 \end{bmatrix}$$

$$|z_1\rangle = \begin{bmatrix} 1 \\ 3 \end{bmatrix}$$

$$|z_2\rangle = \begin{bmatrix} 1 \\ -1 \end{bmatrix} + \tfrac{2}{10}\begin{bmatrix} 1 \\ 3 \end{bmatrix} = \tfrac{1}{10}\begin{bmatrix} 12 \\ -4 \end{bmatrix}$$

$$= (\text{constant})\begin{bmatrix} 3 \\ -1 \end{bmatrix}$$

■

Now, it is easy to see that we can choose as our two linearly independent vectors either eigenvectors $|y_1\rangle$ and $|y_2\rangle$ or orthogonal vectors $|z_1\rangle$ and $|z_2\rangle$. Then since any vector $|v\rangle$ can be written as a linear combination of two linearly independent vectors, we have

$$|v\rangle = a|y_1\rangle + b|y_2\rangle \qquad \text{or} \qquad |v\rangle = c|z_1\rangle + d|z_2\rangle \tag{1.25}$$

where we can solve for unknown coefficients a and b or c and d.

We use the eigenvectors to construct the diagonalizing matrix, and we use the orthogonal vectors as "basis" vectors. If we have an $n \times n$ real-symmetric matrix, the eigenvectors can be made mutually orthogonal, even if the roots are coincident.

1.4 DIAGONALIZATION OF MATRICES

There is a theorem that says that any $n \times n$ matrix with distinct eigenvalues can be diagonalized. Let us see how this is done by means of an example.

We use the eigenvectors of a given matrix to construct the matrix that will diagonalize the given matrix.

Example 4.1

$$E = \begin{bmatrix} 3 & 2 \\ 2 & 3 \end{bmatrix} \tag{1.26a}$$

$$\lambda_1 = 5; \qquad \lambda_2 = 1 \tag{1.26b}$$

$$|x_1\rangle = \begin{bmatrix} 1 \\ 1 \end{bmatrix}; \qquad |x_2\rangle = \begin{bmatrix} 2 \\ -2 \end{bmatrix} \tag{1.26c}$$

We construct matrix M that has the eigenvectors of matrix E as its columns:

$$M = \left[\begin{bmatrix} x_{1,1} \\ x_{1,2} \end{bmatrix} \begin{bmatrix} x_{2,1} \\ x_{2,2} \end{bmatrix} \right] = \begin{bmatrix} x_{1,1} & x_{2,1} \\ x_{1,2} & x_{2,2} \end{bmatrix} \tag{1.27}$$

Thus, for matrix E given by Equation (1.26a) with eigenvectors given by Equation (1.26c), we have:

$$M = \begin{bmatrix} 1 & 2 \\ 1 & -2 \end{bmatrix} \tag{1.28}$$

Then

$$EM = M\Lambda \qquad \text{or} \qquad \Lambda = M^{-1}EM \tag{1.29}$$

where Λ is the diagonalized matrix. Observe that the normalization of the eigenvectors does not affect their role in the diagonalization process. Different normalization "schemes" are used in different circumstances. We have chosen the particular one in this example in order to illustrate the properties of matrix P, which are discussed in what follows.

(Be careful to correctly "pair" the eigenvalues λ_1 and λ_2 and the eigenvectors $|x_1\rangle$ and $|x_2\rangle$; otherwise, matrix M is incorrectly constructed. It is, therefore, important to check your work.)

It is easy to find the inverse of 2×2 matrices and thus we will just write the result:

$$M^{-1} = \tfrac{1}{4}\begin{bmatrix} 2 & 2 \\ 1 & -1 \end{bmatrix}$$

Check that matrix M does indeed diagonalize matrix E. We could have chosen to diagonalize matrix E by a transformation P such that

$$PE = \Lambda P \qquad \text{or} \qquad \Lambda = PEP^{-1} \tag{1.30}$$

It is clear that we can choose matrix P to be the inverse of matrix M. ■

EXERCISE 4.1 Consider the matrix

$$A = \begin{bmatrix} 4 & 1 \\ 2 & 3 \end{bmatrix}$$

Find matrix G such that $AG = G\Lambda$. Find matrix H such that $HA = \Lambda H$. Find the matrix formed by the product GH. ■

EXERCISE 4.2 Consider one of the Pauli spin matrices that are used in quantum mechanics to characterize the spin of an electron.

$$\sigma = \begin{bmatrix} 0 & 1 \\ 1 & 0 \end{bmatrix}$$

(i) Establish that it has eigenvalues $+1$ and -1.

(ii) Find its eigenvectors:

$$\begin{bmatrix} 1 \\ 1 \end{bmatrix}; \quad \begin{bmatrix} 1 \\ -1 \end{bmatrix}$$

(iii) Normalize them to unity:

$$\frac{1}{\sqrt{2}}\begin{bmatrix} 1 \\ 1 \end{bmatrix}; \quad \frac{1}{\sqrt{2}}\begin{bmatrix} 1 \\ -1 \end{bmatrix}$$

(iv) Observe that they are orthogonal:

$$[1 \quad 1]\begin{bmatrix} 1 \\ -1 \end{bmatrix} = 0$$

■

Example 4.2 Consider the matrix

$$S = \begin{bmatrix} \sigma & \alpha \\ 0 & \delta \end{bmatrix}; \qquad \sigma, \alpha, \delta \neq 0$$

(i) Observe that the eigenvalues are σ and δ.

(ii) We find the eigenvectors by constructing the equations

$$\begin{bmatrix} \sigma & \alpha \\ 0 & \delta \end{bmatrix} \begin{bmatrix} x_{1,1} \\ x_{1,2} \end{bmatrix} = \sigma \begin{bmatrix} x_{1,1} \\ x_{1,2} \end{bmatrix} \tag{1.31a}$$

or

$$\begin{aligned} \sigma x_{1,1} + \alpha x_{1,2} &= \sigma x_{1,1} \\ \delta x_{1,2} &= \sigma x_{1,2} \end{aligned} \tag{1.31b}$$

Since σ and δ are arbitrary, we have $x_{1,2} = 0$. We choose $x_{1,1} = 1$.

$$\begin{bmatrix} \sigma & \alpha \\ 0 & \delta \end{bmatrix} \begin{bmatrix} x_{2,1} \\ x_{2,2} \end{bmatrix} = \delta \begin{bmatrix} x_{2,1} \\ x_{2,2} \end{bmatrix} \tag{1.32a}$$

or

$$\begin{aligned} \sigma x_{2,1} + \alpha x_{2,2} &= \delta x_{2,1} \\ \delta x_{2,2} &= \delta x_{2,2} \end{aligned} \tag{1.32b}$$

We choose

$$x_{2,1} = 1$$

then

$$x_{2,2} = \frac{\delta - \sigma}{\alpha}$$

Our eigenvectors are

$$\begin{bmatrix} x_{1,1} \\ x_{1,2} \end{bmatrix} = \begin{bmatrix} 1 \\ 0 \end{bmatrix}; \qquad \begin{bmatrix} x_{2,1} \\ x_{2,2} \end{bmatrix} = \begin{bmatrix} 1 \\ \dfrac{\delta - \sigma}{\alpha} \end{bmatrix} \tag{1.33}$$

(iii) Find the invertible matrix U that brings S to diagonal form under the condition that the eigenvalues are distinct.

$$SU = U\Lambda \qquad \text{or} \qquad \Lambda = U^{-1}SU \tag{1.34a}$$

$$U = \begin{bmatrix} 1 & 1 \\ 0 & \dfrac{\delta - \sigma}{\alpha} \end{bmatrix}; \qquad U^{-1} = \begin{bmatrix} 1 & \dfrac{-\alpha}{\delta - \sigma} \\ 0 & \dfrac{\alpha}{\delta - \sigma} \end{bmatrix} \tag{1.34b}$$

(iv) Observe that S has distinct eigenvectors iff $\sigma \neq \delta$, and we need this condition if we are to be able to construct our diagonalizing matrix U. This gives us an insight into how the diagonalization takes place. ■

Consider a general matrix with distinct eigenvalues and eigenvectors:

$$K = \begin{bmatrix} k_{11} & k_{12} \\ k_{21} & k_{22} \end{bmatrix} \tag{1.35a}$$

$$K|j_1\rangle = \lambda_1|j_1\rangle; \qquad K|j_2\rangle = \lambda_2|j_2\rangle \tag{1.35b}$$

$$|j_1\rangle = \begin{bmatrix} j_{1,1} \\ j_{1,2} \end{bmatrix}; \qquad |j_2\rangle = \begin{bmatrix} j_{2,1} \\ j_{2,2} \end{bmatrix} \tag{1.35c}$$

We show, by a step-by-step construction, how the diagonalizing matrix U is formed from the eigenvectors of K and how the various terms develop that lead to the diagonalization.

$$U = \begin{bmatrix} j_{1,1} & j_{2,1} \\ j_{1,2} & j_{2,2} \end{bmatrix} \qquad \text{and} \qquad KU = U\Lambda$$

The columns of U are the eigenvectors of K. Then

$$KU = \begin{bmatrix} \lambda_1 j_{1,1} & \lambda_2 j_{2,1} \\ \lambda_1 j_{1,2} & \lambda_2 j_{2,2} \end{bmatrix}$$

and

$$U\Lambda = \begin{bmatrix} j_{1,1} & j_{2,1} \\ j_{1,2} & j_{2,2} \end{bmatrix} \begin{bmatrix} \lambda_1 & 0 \\ 0 & \lambda_2 \end{bmatrix} = \begin{bmatrix} \lambda_1 j_{1,1} & \lambda_2 j_{2,1} \\ \lambda_1 j_{1,2} & \lambda_2 j_{2,2} \end{bmatrix}$$

In general, if we have an $n \times n$ matrix with distinct eigenvalues,

$$K|x_s\rangle = \lambda_s|x_s\rangle; \qquad \lambda_i \neq \lambda_j, \quad i, j = 1, 2, \ldots, n \tag{1.36}$$

the diagonalizing matrix U has as its columns the eigenvectors:

$$U = \begin{bmatrix} \begin{bmatrix} x_{1,1} \\ \vdots \\ x_{1,n} \end{bmatrix} & \begin{bmatrix} x_{2,1} \\ \vdots \\ x_{2,n} \end{bmatrix} & \cdots & \begin{bmatrix} x_{n,1} \\ \vdots \\ x_{n,n} \end{bmatrix} \end{bmatrix} \tag{1.37}$$

and

$$KU = U\Lambda = \begin{bmatrix} \lambda_1 \begin{bmatrix} x_{1,1} \\ \vdots \\ x_{1,n} \end{bmatrix} \lambda_2 \begin{bmatrix} x_{2,1} \\ \vdots \\ x_{2,n} \end{bmatrix} \ldots \lambda_n \begin{bmatrix} x_{n,1} \\ \vdots \\ x_{n,n} \end{bmatrix} \end{bmatrix} \tag{1.38}$$

(Note that matrices belonging to certain special classes, e.g., Hermitian, real-symmetric, have a complete set of linearly independent eigenvectors even if their eigenvalues coincide.)

Finally, we establish the result that every 2×2 real-symmetric matrix with equal roots is always diagonal. Thus, you just do not have to diagonalize it. (It is also true, that every $n \times n$ real-symmetric matrix with all roots equal is diagonal and a multiple of the unit matrix.) Consider the matrix

$$A = \begin{bmatrix} a_{11} & a_{12} \\ a_{21} & a_{22} \end{bmatrix}; \qquad a_{12} = a_{21} \tag{1.39}$$

and write the equation for the eigenvalues in the form

$$(a_{11} - \lambda)(a_{22} - \lambda) - a_{12}^2 = 0$$

$$\lambda = \frac{a_{11} + a_{22}}{2} \pm \frac{\sqrt{(a_{11} - a_{22})^2 + 4a_{12}^2}}{2} \tag{1.40}$$

Notice that if the roots are coincident, the quantity under the square-root sign must vanish. Since the terms are positive, we have the two conditions:

$$a_{11} = a_{22}; \qquad a_{12} = 0$$

This result shows "how unlikely" it is for a general real-symmetric matrix to have equal roots. This result is also true for Hermitian matrices.

EXERCISE 4.3 Consider

$$H = \begin{bmatrix} a & b + ic \\ b - ic & d \end{bmatrix} \tag{1.41}$$

with a, b, c, and d real. Show that H has real eigenvalues and that if the eigenvalues λ are coincident, $a = d$ and $b = c = 0$. Thus, a 2×2 Hermitian matrix with coincident eigenvalues is a multiple of the unit matrix. ■

Observe that 2×2 matrices of other classes may have coincident eigenvalues without being diagonal.

Example 4.3 Consider

$$A = \begin{bmatrix} 2 & 1 \\ -1 & 4 \end{bmatrix}$$

Both eigenvalues = 3. ■

Example 4.4 Consider the symmetric matrix with some elements that are not real:

$$B = \begin{bmatrix} 2i & 1 \\ 1 & 0 \end{bmatrix}$$

It has coincident eigenvalues equal to i. ■

1.5 THE CAYLEY – HAMILTON (CH) THEOREM

This theorem provides a simple connection between a power of a matrix and its characteristic equation. More precisely, it says that every matrix satisfies its characteristic equation. It is a powerful tool with diverse applications, leading us to develop our treatment at some length. After establishing the theorem, we show how it can be used to find a power of a matrix without a knowledge of its eigenvectors. It also enables us to find the inverse of a matrix by computing powers of a matrix, thereby yielding a simple algorithm. We then turn our attention to difference equations. Interest in difference equations has reemerged in recent years in part because they form the basis of many computer computations. They also have a role in the new field called "chaos," which has yielded so many new and unexpected results. (See Chapter 16.)

We then show how to use the CH theorem to find functions of a matrix. The section concludes with a discussion of matrix differential equations in cases where the matrix elements are constants.

We know that associated with any matrix A is a set of eigenvalues and eigenvectors such that

$$A|x\rangle = \lambda|x\rangle$$

We also have the result that since the characteristic equation

$$g(\lambda) \equiv \det(A - \lambda I) = 0 \tag{1.42}$$

the eigenvalues λ_i are found as the roots of a polynomial equation whose highest power is the dimension of the matrix. This means that an $n \times n$ matrix has precisely n eigenvalues that are found as the roots of the characteristic equation.

$$g(\lambda_i) = \lambda_i^n + a_1\lambda_i^{n-1} \cdots + a_{n-1}\lambda_i + a_n = 0$$

where $i = 1, 2, 3, \dots, n$.

The Cayley–Hamilton (CH) theorem says that any matrix A satisfies its characteristic equation. The proof is particularly simple for diagonalizable matrices, and is as follows.

Assume that we have a matrix A whose eigenvalues are distinct. Therefore, there exists a matrix N whose columns are the eigenvectors of matrix A and has the property that

$$N^{-1}AN = \Lambda \qquad \text{or} \qquad A = N\Lambda N^{-1}$$

We indicate the diagonal matrix by Λ:

$$\Lambda = \begin{bmatrix} \lambda_1 & & & & \\ & \lambda_2 & & & \\ & & \lambda_3 & & \\ & & & \ddots & \\ & & & & \lambda_n \end{bmatrix}$$

It has the eigenvalues of A, λ_i, on its diagonal. Then,

$$g(\Lambda) = \begin{bmatrix} g(\lambda_1) & & & \\ & g(\lambda_2) & & \\ & & \ddots & \\ & & & g(\lambda_n) \end{bmatrix} = \mathbf{0} \tag{1.43}$$

since each element is identically $= 0$.

We observe that the same matrix N that diagonalizes matrix A diagonalizes A^2. This is clear if we write

$$A^2 = A \cdot A = N\Lambda N^{-1} \cdot N\Lambda N^{-1} = N\Lambda^2 N^{-1}$$

Also, matrix N diagonalizes any power of A:

$$A^n = \underbrace{A \cdot A \cdot A \cdot A \cdots A}_{n \text{ terms}} = \underbrace{N\Lambda N^{-1} \cdot N\Lambda N^{-1} \cdots N\Lambda N^{-1}}_{n \text{ terms}} = N\Lambda^n N^{-1}$$

So we have the result that

$$A^n = N\Lambda^n N^{-1} \qquad \text{or} \qquad \Lambda^n = N^{-1}A^n N$$

Finally, we conclude that N will diagonalize any polynomial in A. Thus, any equation that is satisfied by the diagonalized form of the matrix is satisfied by the matrix itself. Now, we write the characteristic equation for matrix A as

$$g(A) = A^n + a_1 A^{n-1} + a_2 A^{n-2} + \cdots + a_{n-1}A + a_n I$$

with the same coefficients a_i that we have in the equation

$$g(\lambda) = 0$$

From our previous argument, we have that, since $g(A)$ is a polynomial in A,

$$g(A) = Ng(\Lambda)N^{-1}$$

and since

$$g(\Lambda) = \mathbf{0} \tag{1.44a}$$

we have

$$g(A) = \mathbf{0} \tag{1.44b}$$

So we see that matrix A satisfies its own characteristic equation. Also, since the eigenvalues of a matrix are roots of its characteristic equation, they are unaffected by a similarity transformation. In other words, if matrices A and B are connected by a similarity transformation, they have the same eigenvalues. Finally, observe that if matrix A has a zero eigenvalue, there will be no constant term in the characteristic equation, and the inverse matrix does not exist.

Example 5.1 Consider

$$A = \begin{bmatrix} 2 & 4 \\ 1 & 5 \end{bmatrix}$$

(a) Determine that the eigenvalues of A are 6 and 1.
(b) Find the characteristic equation for the matrix.
(c) Show that $g(A) = \mathbf{0}$. We have

$$\begin{aligned} g(A) &= A^2 - 7A + 6I \\ &= \begin{bmatrix} 8 & 28 \\ 7 & 29 \end{bmatrix} - \begin{bmatrix} 14 & 28 \\ 7 & 35 \end{bmatrix} + \begin{bmatrix} 6 & 0 \\ 0 & 6 \end{bmatrix} = \begin{bmatrix} 0 & 0 \\ 0 & 0 \end{bmatrix} = \mathbf{0} \end{aligned}$$

(d) Introduce the matrix $C = A^3 + A$. Find the eigenvalues of matrix C. We know that there exists a matrix N such that

$$N^{-1}AN = \begin{bmatrix} 6 & 0 \\ 0 & 1 \end{bmatrix}$$

We then find

$$\begin{aligned} N^{-1}CN &= N^{-1}A^3N + N^{-1}AN \\ &= \begin{bmatrix} 6^3 & 0 \\ 0 & 1^3 \end{bmatrix} + \begin{bmatrix} 6 & 0 \\ 0 & 1 \end{bmatrix} \\ &= \begin{bmatrix} 222 & 0 \\ 0 & 2 \end{bmatrix} \end{aligned}$$

Then,

$$C = N\begin{bmatrix} 222 & 0 \\ 0 & 2 \end{bmatrix}N^{-1}$$

We know that the eigenvalues of a matrix are unchanged by the similarity transformation N. Thus, we have the result that the eigenvalues of matrix C are $\lambda_1 = 222$ and $\lambda_2 = 2$. Observe that we found the eigenvalues of matrix C without a knowledge of matrix N. ■

Example 5.2 Consider the 3×3 matrix

$$B = \begin{bmatrix} 2 & 0 & 3 \\ 0 & 5 & 0 \\ 3 & 0 & 2 \end{bmatrix}$$

(a) Find its characteristic equation and show that it has roots $= 5, 5, -1$. Show that $g(B) = \mathbf{0}$.

(b) Use the CH theorem to find its inverse. The theorem tells us that $g(B)$ is of the form

$$g(B) = B^3 + b_1B^2 + b_2B + b_3I = \mathbf{0}$$

So, multiplying through by B^{-1}, we have

$$B^2 + b_1B + b_2I + b_3B^{-1} = \mathbf{0}$$

or

$$B^{-1} = -\frac{1}{b_3}\left(B^2 + b_1B + b_2I\right)$$

Thus, we can find B^{-1} from a polynomial expansion.

Other methods, using determinants, can be quite complicated and it is necessary to keep track of many signs. ■

1.5.1 Powers of a Matrix

We often want to raise a matrix to a power $m > n$, where n is the dimension of the matrix. The CH theorem tells us that we can always express any power of a matrix as a polynomial whose highest power is one less than the dimension of the matrix. To see this, consider a 2×2 matrix A. It has a characteristic polynomial of the form

$$g(A) = A^2 + a_1A + a_2I$$

Let us assume that we want to compute A^n. Before obtaining the general result, we consider an illustrative example.

Example 5.3 Consider the matrix

$$A = \begin{bmatrix} 2 & 1 \\ 3 & 4 \end{bmatrix}$$

and let us assume that we want to compute A^4. We introduce the characteristic equation

$$g(A) = A^2 - 6A + 5I$$

and use synthetic division, as follows:

$$\begin{array}{r|l}
 & A^2 + 6A + 31I \\
\hline
A^2 - 6A + 5I & A^4 \\
 & A^4 - 6A^3 + 5A^2 \\
\hline
 & \quad 6A^3 - 5A^2 \\
 & \quad 6A^3 - 36A^2 + 30A \\
\hline
 & \qquad 31A^2 - 30A \\
 & \qquad 31A^2 - 186A + 155I \\
\hline
 & \qquad\qquad 156A - 155I
\end{array}$$

We write the result in the form

$$A^4 = (A^2 - 6A + 5I)(A^2 + 6A + 31I) + (156A - 155I)$$

where the first term is the product of the characteristic polynomial $g(A)$, which equals zero, and the quotient obtained by dividing $g(A)$ into the desired power of the matrix, i.e., A^4. The second term is the remainder and clearly it is at most of degree 1. Thus,

$$A^4 = 156A - 155I$$

■

The example is general enough for us to conclude that for an arbitrary 2×2 matrix A, we can write:

$$A^n = g(A) * (\text{polynomial of degree } n - 2) + r(A)$$

where the remainder term $r(A)$ is at most of degree 1. The first factor is identically equal to 0 since the characteristic polynomial equals 0, and thus we have the result:

$$A^n = \alpha A + \beta I \tag{1.45}$$

where α and β are constants to be determined.

Example 5.4 Given the matrix

$$A = \begin{bmatrix} 1 & -3 \\ 3 & 1 \end{bmatrix}$$

with the characteristic equation

$$g(A) = A^2 - 2A + 10I$$

By the same method as in Example 5.3, verify that $g(A) = \mathbf{0}$, and

$$A^4 = -32A + 60I$$

■

In a general problem, we can solve for the quantities α and β, and note that their values are fixed regardless of whether matrix A is in its diagonal form or not. To see this, *assume* that we put A into diagonal form by

$$A^n = \alpha A + \beta I \qquad \text{and} \qquad N^{-1}A^n N = \alpha N^{-1}AN + \beta N^{-1}IN$$

Since matrix N diagonalizes any power of A, we have

$$\Lambda^n = \alpha\Lambda + \beta I$$

$$\begin{bmatrix} \lambda_1^n & 0 \\ 0 & \lambda_2^n \end{bmatrix} = \alpha\begin{bmatrix} \lambda_1 & 0 \\ 0 & \lambda_2 \end{bmatrix} + \beta\begin{bmatrix} 1 & 0 \\ 0 & 1 \end{bmatrix} \tag{1.46}$$

without a knowledge of matrix N. This is crucial since the Cayley–Hamilton theorem frees one from finding the eigenvectors of matrix A. You just have to find the eigenvalues.

We have two linear equations in two unknowns α and β.

We solve these, and obtain:

$$\alpha = \frac{\lambda_1^n - \lambda_2^n}{\lambda_1 - \lambda_2} \tag{1.47a}$$

$$\beta = \frac{\lambda_2\lambda_1^n - \lambda_1\lambda_2^n}{\lambda_2 - \lambda_1} \tag{1.47b}$$

Thus, we have

$$A^n = \frac{\lambda_1^n - \lambda_2^n}{\lambda_1 - \lambda_2}A + \frac{\lambda_2\lambda_1^n - \lambda_1\lambda_2^n}{\lambda_2 - \lambda_1}I \tag{1.48}$$

as a general result for any 2×2 matrix with distinct eigenvalues.

Example 5.5 Consider the matrix

$$A = \begin{bmatrix} 2 & 1 \\ 3 & 4 \end{bmatrix}$$

with eigenvalues $\lambda_1 = 5$ and $\lambda_2 = 1$. We then have

$$A^n = \alpha A + \beta I; \qquad \alpha = \frac{5^n - 1^n}{5 - 1}, \quad \beta = \frac{1 \cdot 5^n - 5 \cdot 1^n}{1 - 5}$$

In particular, if $n = 4$, we find

$$\left.\begin{array}{r} 5^4 = 5\alpha + \beta \\ 1^4 = \alpha + \beta \end{array}\right\} \Rightarrow \alpha = \frac{624}{4} = 156; \qquad \beta = \frac{-620}{4} = -155$$

Then

$$A^4 = \alpha A + \beta I = \left[\begin{array}{c|c} 2 \cdot 156 - 155 & 156 \\ \hline 3 \cdot 156 & 4 \cdot 156 - 155 \end{array}\right]$$
$$= \begin{bmatrix} 157 & 156 \\ 468 & 469 \end{bmatrix}$$

■

Example 5.6 Given the matrix

$$A = \begin{bmatrix} 1 & -3 \\ 3 & 1 \end{bmatrix}$$

use the CH theorem to find A^4. (This is the same matrix that we had in Example 5.4.) This problem is considerably more difficult than the preceding one because the eigenvalues are complex:

$$\lambda_1 = 1 + 3i; \qquad \lambda_2 = 1 - 3i$$

We write our basic equations for constants α and β:

$$\alpha = \frac{(1 + 3i)^4 - (1 - 3i)^4}{6i}$$

$$\beta = \frac{(1 - 3i)(1 + 3i)^4 - (1 + 3i)(1 - 3i)^4}{-6i}$$

and solve them to obtain the result.

$$(1 + 3i)^4 = 1 + 4(3i) + 6(3i)^2 + 4(3i)^3 + (3i)^4$$
$$(1 - 3i)^4 = 1 - 4(3i) + 6(3i)^2 - 4(3i)^3 + (3i)^4$$
$$\alpha = -32; \qquad \beta = 60$$

■

If our eigenvalues coincide, we could write them as λ_1 and $\lambda_1 + \varepsilon$, and then let ε go to zero. Then Equations (1.47a) and (1.47b) take the form:

$$\alpha = \frac{\lambda_1^n - (\lambda_1 + \varepsilon)^n}{\lambda_1 - (\lambda_1 + \varepsilon)}$$

$$\beta = \frac{(\lambda_1 + \varepsilon)\lambda_1^n - \lambda_1(\lambda_1 + \varepsilon)^n}{(\lambda_1 + \varepsilon) - \lambda_1}$$

or

$$\alpha = n\lambda_1^{n-1}; \qquad \beta = \lambda_1^n(1 - n); \qquad \varepsilon \to 0$$

We obtain the same result by noting that we are effectively using l'Hospital's rule by differentiating our expressions for α and β with respect to λ_2, holding λ_1 fixed and then letting $\lambda_2 \to \lambda_1$. Thus, our expression for a 2×2 matrix with degenerate roots becomes

$$A^n = n\lambda_1^{n-1}A + (1 - n)\lambda_1^n I \tag{1.49}$$

Example 5.7 Given the matrix

$$A = \begin{bmatrix} 6 & 8 \\ -2 & -2 \end{bmatrix}$$

we find the eigenvalues to be degenerate and equal to 2. We want to know A^3.
Solve for α and β and verify that they are 12 and -16, respectively.

$$A^3 = 3 \cdot 2^2 A + (1 - 3)2^3 I$$

$$A^3 = \begin{bmatrix} 56 & 96 \\ -24 & -40 \end{bmatrix}$$

■

EXERCISE 5.1 Given the matrix

$$A = \begin{bmatrix} 1 & 2 \\ 2 & 1 \end{bmatrix}$$

Use the CH theorem to find A^3 and compute it directly. ■

1.5.2 Difference Equations

We often encounter difference equations of the form

$$x_{n+1} = ax_n + bx_{n-1} \tag{1.50}$$

where a and b are constants.

Equation (1.50) can be written in the form

$$x_{n+1} = ax_n + by_n$$
$$y_{n+1} = x_n$$
$$\begin{bmatrix} x \\ y \end{bmatrix}_{n+1} = \begin{bmatrix} a & b \\ 1 & 0 \end{bmatrix} \begin{bmatrix} x \\ y \end{bmatrix}_n; \qquad n \geq 1$$

And we are generally given the initial conditions x_1, $x_0 = y_1$, and asked to find the value of x_{n+1}.

We see that this is equivalent to operating with the matrix n times or

$$\begin{bmatrix} x \\ y \end{bmatrix}_{n+1} = A^n \begin{bmatrix} x \\ y \end{bmatrix}_1; \qquad A = \begin{bmatrix} a & b \\ 1 & 0 \end{bmatrix}$$

So we can use the CH theorem to find the result without specifying the power n. We proceed as follows. We designate the two eigenvalues as λ_1 and λ_2. We know that

$$A^n = \alpha \begin{bmatrix} a & b \\ 1 & 0 \end{bmatrix} + \beta \begin{bmatrix} 1 & 0 \\ 0 & 1 \end{bmatrix}$$

Coefficients α and β are independent of whether matrix A is diagonal, so we proceed as before and find

$$A^n = \begin{bmatrix} \alpha a + \beta & \alpha b \\ \alpha & \beta \end{bmatrix} \tag{1.51}$$

Given a and b, we can solve for α and β.

EXERCISE 5.2 Consider the difference equation

$$x_{n+1} = 4x_n + 12x_{n-1}; \qquad n \geq 1, \quad x_1 = 1, \quad x_0 = 0$$

We seek an expression for x_{n+1} in terms of x_1 and x_0.

(a) Construct the associated matrix equation

$$\begin{bmatrix} x \\ y \end{bmatrix}_{n+1} = \begin{bmatrix} 4 & 12 \\ 1 & 0 \end{bmatrix} \begin{bmatrix} x \\ y \end{bmatrix}_n = \begin{bmatrix} 4 & 12 \\ 1 & 0 \end{bmatrix}^n \begin{bmatrix} x \\ y \end{bmatrix}_1$$
$$x_1 = 1; \qquad y_1 = x_0 = 0$$

(b) Find the eigenvalues 6 and -2.

(c) Substitute them in the expressions for α and β, obtaining

$$\alpha = \frac{6^n - (-2)^n}{8}; \qquad \beta = \frac{-2 \cdot (6^n) - 6 \cdot (-2)^n}{-8}$$

(d) We then have

$$A^n = \left[\begin{array}{c|c} \frac{3}{4}\cdot 6^n + \frac{1}{4}(-2)^n & \frac{3}{2}\left[6^n - (-2)^n\right] \\ \hline \frac{1}{8}\left[6^n - (-2)^n\right] & \frac{1}{4}\left[6^n + 3(-2)^n\right] \end{array}\right]$$

(e) We are interested in

$$\begin{aligned} x_{n+1} &= \left[\tfrac{3}{4}\cdot 6^n + \tfrac{1}{4}(-2)^n\right]x_1 + \tfrac{3}{2}\left[6^n - (-2)^n\right]x_0 \\ &= \tfrac{3}{4}\cdot 6^n + \tfrac{1}{4}(-2)^n \end{aligned}$$

Check the result for $n = 1, 2$ ($x_2 = 4$, $x_3 = 28$). ■

EXERCISE 5.3 Consider the difference equation that defines the Fibonacci numbers:

$$x_{n+1} = x_n + x_{n-1}; \qquad n \geq 1, \quad x_1 = 1, \quad x_0 = 0$$

(The numbers are $0, 1, 1, 2, 3, 5, 8, 13, \ldots$). We have

$$\left.\begin{aligned} x_{n+1} &= x_n + y_n \\ y_{n+1} &= x_n \end{aligned}\right\} \quad \text{or} \quad \begin{bmatrix} x \\ y \end{bmatrix}_{n+1} = \begin{bmatrix} 1 & 1 \\ 1 & 0 \end{bmatrix}^n \begin{bmatrix} x \\ y \end{bmatrix}_1$$

We solve for the eigenvalues:

$$\lambda_{1,2} = \tfrac{1}{2}(1 \pm \sqrt{5})$$

and find α and β. Our expression for A^n becomes

$$\begin{aligned} A^n = &\frac{\lambda_1^n - \lambda_2^n}{\lambda_1 - \lambda_2}\begin{bmatrix} 1 & 1 \\ 1 & 0 \end{bmatrix} \\ &+ \frac{\lambda_2\lambda_1^n - \lambda_1\lambda_2^n}{\lambda_2 - \lambda_1}\begin{bmatrix} 1 & 0 \\ 0 & 1 \end{bmatrix} \end{aligned}$$

Since we are interested in x_{n+1}, we expand our result and write:

$$x_{n+1} = \frac{\lambda_1^n(1 - \lambda_2) + \lambda_2^n(\lambda_1 - 1)}{\lambda_1 - \lambda_2}$$

Note that

$$\lambda_1 + \lambda_2 = 1; \qquad \lambda_1 - \lambda_2 = \sqrt{5}$$

so that we can simplify our result to read

$$x_n = \frac{1}{\sqrt{5}}(\lambda_1^n - \lambda_2^n)$$

(It is interesting to note how the $\sqrt{5}$'s cancel so as to give a result in integers.) Check that you get the correct result for $n = 4$. ■

EXERCISE 5.4 Given the difference equation

$$x_{n+1} = 2x_n + y_n$$
$$y_{n+1} = -x_n$$
$$n \geq 0; \qquad x_0 = y_0 = 1$$

find the value of x_n in terms of λ_1 and λ_2. Note that the eigenvalues are both equal to 1. We use the CH theorem for degenerate roots:

$$A^n = \begin{bmatrix} 2 & 1 \\ -1 & 0 \end{bmatrix}^n = \alpha A + \beta I$$
$$\alpha = n\lambda^{n-1} = n \cdot 1$$
$$\beta = (1 - n)\lambda^n = 1 - n$$

So we have

$$A^n = n\begin{bmatrix} 2 & 1 \\ -1 & 0 \end{bmatrix} + (1 - n)\begin{bmatrix} 1 & 0 \\ 0 & 1 \end{bmatrix}$$
$$\begin{bmatrix} x \\ y \end{bmatrix}_n = \begin{bmatrix} n+1 & n \\ -n & 1-n \end{bmatrix}\begin{bmatrix} x \\ y \end{bmatrix}_0$$
$$x_n = (n+1)x_0 + ny_0 = 2n + 1$$

Check this result explicitly for $n = 2, 3$. ■

1.5.3 Functions of a Matrix

We often have to find functions of a matrix, the most common being the exponential, e.g.,

$$e^A \qquad \text{or} \qquad e^{At}$$

where A is a constant matrix. (The situation is much more complicated if the elements of the matrix depend on time. See the Bibliography.) Since the Taylor-series expansion of the exponential function has an infinite radius of convergence, we define

$$e^{At} \equiv \Sigma \frac{A^n}{n!} t^n = I + \frac{At}{1!} + \frac{A^2t^2}{2!} + \frac{A^3t^3}{3!} + \cdots \tag{1.52a}$$

and then

$$\frac{d}{dt}e^{At} = A + \frac{A^2t}{1!} + \frac{A^3t^2}{2!} + \cdots = A \cdot \Sigma \frac{A^n t^n}{n!} = Ae^{At} \qquad (1.52b)$$

We know from our analysis of the CH theorem that if matrix N diagonalizes matrix A, then it diagonalizes any power of A. Thus, we write

$$N^{-1}e^{A}N = \Sigma \frac{N^{-1}A^nN}{n!} = \Sigma \frac{\Lambda^n}{n!} = e^{\Lambda}$$

The same argument shows that N diagonalizes any function of A, with the understanding that the function is represented by a power-series expansion. (The text by Joshi has many examples, including the logarithm, sine, cosh, etc.)

Let us use the CH theorem to find

$$e^{At} = \alpha At + \beta I = (\alpha t)A + \beta I$$

where we are still assuming A to be a 2×2 matrix.

It follows that

$$\exp \lambda_1 t = \alpha(\lambda_1 t) + \beta = (\alpha t)\lambda_1 + \beta$$
$$\exp \lambda_2 t = \alpha(\lambda_2 t) + \beta = (\alpha t)\lambda_2 + \beta$$

which yields

$$\alpha t = \frac{\exp \lambda_1 t - \exp \lambda_2 t}{\lambda_1 - \lambda_2}$$
$$\beta = \frac{\lambda_2 \exp \lambda_1 t - \lambda_1 \exp \lambda_2 t}{\lambda_2 - \lambda_1}$$

Notice that time t is a passive participant in the calculation of αt and β. You can think of it as attached to matrix A and hence to its eigenvalues. Substituting, we find

$$\exp At = \frac{\exp \lambda_1 t - \exp \lambda_2 t}{(\lambda_1 - \lambda_2)} A + \frac{\lambda_2 \exp \lambda_1 t - \lambda_1 \exp \lambda_2 t}{(\lambda_2 - \lambda_1)} I \qquad (1.53)$$

EXERCISE 5.5 Given the matrix

$$A = \begin{bmatrix} 3 & 2 \\ 2 & 3 \end{bmatrix}$$

we are interested in finding $\exp At$. Show that

$$\lambda_1 = 5 \qquad \text{and} \qquad \lambda_2 = 1$$

and

$$\exp\begin{bmatrix} 3 & 2 \\ 2 & 3 \end{bmatrix} t = \tfrac{1}{4}(e^{5t} - e^t)\begin{bmatrix} 3 & 2 \\ 2 & 3 \end{bmatrix} + \tfrac{1}{4}(5e^t - e^{5t})\begin{bmatrix} 1 & 0 \\ 0 & 1 \end{bmatrix}$$ ■

EXERCISE 5.6 Given the matrix

$$A = \begin{bmatrix} \theta & -\phi \\ \phi & \theta \end{bmatrix}$$

show that

$$e^{At} = e^{\theta t}\begin{bmatrix} \cos\phi t & -\sin\phi t \\ \sin\phi t & \cos\phi t \end{bmatrix}$$

where we use the exponential form for the sine and cosine. ■

EXERCISE 5.7 Given the Pauli matrix

$$\sigma = \begin{bmatrix} 0 & 1 \\ 1 & 0 \end{bmatrix}$$

show that

$$e^{\sigma t} = \begin{bmatrix} \cosh t & \sinh t \\ \sinh t & \cosh t \end{bmatrix}$$ ■

EXERCISE 5.8 Given the matrix

$$A = \begin{bmatrix} \omega & 0 \\ 1 & \omega \end{bmatrix}$$

show that

$$e^{At} = e^{\omega t}\begin{bmatrix} 1 & 0 \\ t & 1 \end{bmatrix}$$ ■

EXERCISE 5.9 (This exercise is related to Exercise 5.6.) We want to consider a rotation of the x and y axes through a finite angle ϕ, as indicated in Figure 1.1.

We know that this rotation from coordinates (x, y) to (x', y') is given by the rotation matrix:

$$\begin{bmatrix} x' \\ y' \end{bmatrix} = R(\phi)\begin{bmatrix} x \\ y \end{bmatrix}; \qquad R(\phi) = \begin{bmatrix} \cos\phi & \sin\phi \\ -\sin\phi & \cos\phi \end{bmatrix}$$

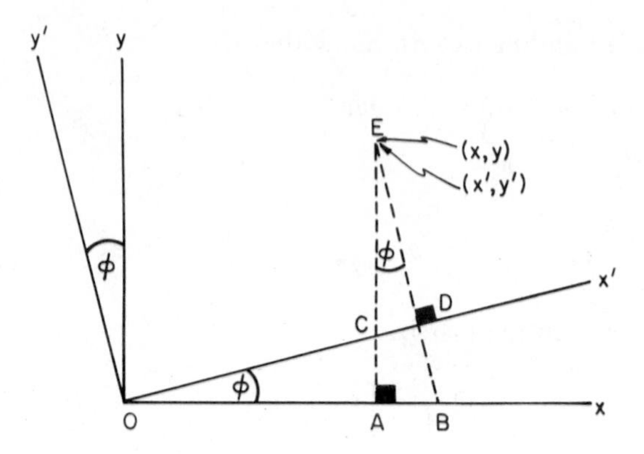

Figure 1.1 The point E is given by the coordinates (x, y) and (x', y').

We want to look at this problem from a different aspect, showing that one can generate a finite rotation through angle ϕ by an infinite series of infinitesimal rotations θ such that

$$n\theta = \phi: \theta \to 0; \qquad n \to \infty; \quad n\theta = \phi = \text{finite}$$

We then have for the infinitesimal rotation matrix:

$$\begin{bmatrix} x' \\ y' \end{bmatrix}_{\theta} = I(\theta)\begin{bmatrix} x \\ y \end{bmatrix}; \qquad I(\theta) = \begin{bmatrix} 1 & \theta \\ -\theta & 1 \end{bmatrix}$$

where we have used the small-angle approximation for the sine and cosine. We now generate the finite rotation as follows:

$$\begin{bmatrix} x' \\ y' \end{bmatrix}_{n\theta=\phi} = I^n(\theta)\begin{bmatrix} x \\ y \end{bmatrix}; \qquad n \to \infty, \quad n\theta \to \phi$$

We are interested in the limit $n \to \infty$. We use the CH theorem, noting that the eigenvalues of $I(\theta)$ are

$$\lambda_{1,2} = 1 \pm i\theta$$

Thus, we have

$$I^n(\theta) = \alpha I(\theta) + \beta I$$

$$\alpha = \frac{(1 + i\theta)^n - (1 - i\theta)^n}{2i\theta}$$

$$\beta = \frac{(1 - i\theta)(1 + i\theta)^n - (1 + i\theta)(1 - i\theta)^n}{-2i\theta}$$

We also have to observe that

$$(1+i\theta)^n = \left(1+\frac{i\phi}{n}\right)^n \to \exp(i\phi) \quad \text{as} \quad n\to\infty$$

$$(1-i\theta)^n = \left(1-\frac{i\phi}{n}\right)^n \to \exp(-i\phi) \quad \text{as} \quad n\to\infty$$

so that we can write our coefficients α and β as

$$\alpha = \frac{e^{i\phi}-e^{-i\phi}}{2i\theta} = \frac{\sin\phi}{\theta}$$

$$\beta = -\frac{e^{i\phi}-e^{-i\phi}}{2i\theta} + \frac{e^{i\phi}+e^{-i\phi}}{2} = -\frac{\sin\phi}{\theta} + \cos\phi$$

Finally, we have

$$R(\phi) = I^n_{n\to\infty}(\theta) = \begin{bmatrix} \cos\phi & \sin\phi \\ -\sin\phi & \cos\phi \end{bmatrix}$$

as required. ■

EXERCISE 5.10 We are interested in solving the following set of differential equations:

$$dx/dt = 3x + 2y$$
$$dy/dt = 2x + 3y$$
$$x(0),\ y(0) \text{ given}$$

for $x(t)$ and $y(t)$. We set up the matrix equation

$$d/dt\begin{bmatrix} x \\ y \end{bmatrix} = \begin{bmatrix} 3 & 2 \\ 2 & 3 \end{bmatrix}\begin{bmatrix} x \\ y \end{bmatrix} = A\begin{bmatrix} x \\ y \end{bmatrix}$$

and, since our matrix A is constant, we can write the solution immediately:

$$\begin{bmatrix} x \\ y \end{bmatrix} = e^{At}\begin{bmatrix} x(0) \\ y(0) \end{bmatrix}$$

Differentiate both sides to verify that this is correct. We now just have to find

$\exp(At)$. We have already done this in Exercise 5.5.

$$\exp At = \tfrac{1}{4}(\exp 5t - \exp t)\begin{bmatrix} 3 & 2 \\ 2 & 3 \end{bmatrix} + \tfrac{1}{4}(5\exp t - \exp 5t)\begin{bmatrix} 1 & 0 \\ 0 & 1 \end{bmatrix}$$

It is convenient to rearrange our solution into a better form by writing

$$\begin{bmatrix} x(t) \\ y(t) \end{bmatrix} = \frac{1}{2}\begin{bmatrix} e^{5t} + e^{t} & e^{5t} - e^{t} \\ e^{5t} - e^{t} & e^{5t} + e^{t} \end{bmatrix}\begin{bmatrix} x(0) \\ y(0) \end{bmatrix}$$

or, alternately,

$$\begin{bmatrix} x(t) \\ y(t) \end{bmatrix} = \frac{1}{2}\begin{bmatrix} e^{5t} & e^{t} \\ e^{5t} & -e^{t} \end{bmatrix}\begin{bmatrix} x(0) + y(0) \\ x(0) - y(0) \end{bmatrix}$$

■

EXERCISE 5.11 Consider positive definite matrices A and B. (A matrix is positive definite if it is Hermitian and all of its eigenvalues are positive.)

$$A = \begin{bmatrix} 2 & 0 \\ 0 & 20 \end{bmatrix}; \qquad B = \begin{bmatrix} 1 & 3 \\ 3 & 10 \end{bmatrix}$$

(a) Construct matrix $C = A - B$ and conclude that it is positive definite.

(b) Construct matrix $D = A^2 - B^2$ and conclude that it is not positive definite. This is perhaps most easily done by showing that it has a positive trace and negative determinant.

Thus, we see from this exercise, that the positive definite character of matrices is not necessarily maintained when we compute functions of them. This is in contrast with the corresponding results for scalar functions. For example, in Chapter 0, we obtained from Equations (0.29a) and (0.29b) that if $f(x) > g(x)$, then $\exp[f(x)] > \exp[g(x)]$.

I learned of this example from M. L. Mehta. It is a special case of a theorem due to Wigner and Yanase, discussed in Theorem 6.3.4 on page 75 of Mehta's book on matrix theory. ■

1.6 PERTURBATION THEORY

We often encounter matrices that are close to a matrix whose properties are known. For example, we might have a problem in which symmetry dictates a special form for a matrix and then this symmetry is broken by some interaction yielding a small change in the matrix elements. We might have a matrix whose original entries are integers. A perturbation is introduced that alters these entries so that the new ones are close to integers, e.g., 3.03 and 2.96, and we want to have an estimate of the eigenvalues and eigenvectors of the perturbed matrix.

There is a systematic perturbation theory procedure that develops the properties of the altered matrix in terms of the eigenvalues and eigenvectors of the original matrix. The most generally applicable method is reasonably complicated and involves a significant amount of algebra. Thus, it seems most expeditious to explain it by a simple example in which we find the first correction for 2×2 matrices with distinct eigenvalues.

Notation: The scalar product composed of a matrix sandwiched between a left vector and a right vector is called a matrix element. It is a number.

$$\langle \psi_2, A\phi_1 \rangle \tag{1.54}$$

Example 6.1 Given

$$A = \begin{bmatrix} 3 & 2 \\ 2 & 3 \end{bmatrix}$$

and

$$\langle \psi_2 | = [1 \quad 1]; \qquad |\phi_1\rangle = \begin{bmatrix} 1 \\ 5 \end{bmatrix}$$

We have

$$\langle \psi_2, A\phi_1 \rangle = [1 \quad 1] \begin{bmatrix} 3 & 2 \\ 2 & 3 \end{bmatrix} \begin{bmatrix} 1 \\ 5 \end{bmatrix} = [1 \quad 1] \begin{bmatrix} 13 \\ 17 \end{bmatrix} = 30$$ ■

Consider a problem in which we are given a 2×2 real-symmetric matrix A with distinct eigenvalues λ_1 and λ_2, and orthogonal eigenvectors $|\psi_1\rangle$ and $|\psi_2\rangle$.

(i) We normalize the eigenvectors to 1, i.e., we make them orthonormal.

$$A = \begin{bmatrix} a_{11} & a_{12} \\ a_{21} & a_{22} \end{bmatrix}; \qquad a_{21} = a_{12}$$

$$A|\psi_1\rangle = \lambda_1 |\psi_1\rangle; \qquad A|\psi_2\rangle = \lambda_2 |\psi_2\rangle; \qquad \lambda_1 \neq \lambda_2$$

$$\langle \psi_i, \psi_j \rangle = \delta_{i,j}$$

The Kronecker delta symbol $\delta_{i,j}$ was defined in Equation (1.20).

(ii) We introduce the perturbing matrix εB, where elements b_{ij} are of the same order of magnitude as a_{ij}. (Elements that are equal to 0 do not cause any difficulties.) We seek the eigenvalues μ_1 and μ_2 and the eigenvectors $|\phi_1\rangle$ and $|\phi_2\rangle$ of $A + \varepsilon B$.

$$(A + \varepsilon B)|\phi_i\rangle = \mu_i |\phi_i\rangle; \qquad i = 1, 2; \quad |\varepsilon| \ll 1 \tag{1.55}$$

(iii) Since the eigenvectors of A form a complete orthonormal set, we can expand the eigenvectors of $A + \varepsilon B$ as follows:

$$|\phi_1\rangle = c_1|\psi_1\rangle + c_2|\psi_2\rangle \tag{1.56}$$

$$|\phi_2\rangle = d_1|\psi_1\rangle + d_2|\psi_2\rangle \tag{1.57}$$

where c_i and d_i are constants to be determined.

Let us concentrate on the eigenvalue μ_1 and the eigenvector $|\phi_1\rangle$.

(iv) We know that as the small parameter ε goes to 0, we are to recover the eigenvalues and eigenvectors of the unperturbed system. Therefore, we have

$$|\phi_1\rangle \to |\psi_1\rangle \Rightarrow c_1 \to 1; \qquad c_2 \to 0; \qquad \mu_1 \to \lambda_1$$

$$|\phi_2\rangle \to |\psi_2\rangle \Rightarrow d_1 \to 0; \qquad d_2 \to 1; \qquad \mu_2 \to \lambda_2$$

Since the perturbation is given in terms of the first power of the parameter ε, we assume that there is a power-series expansion of the form:

$$\mu_i = \lambda_i + \varepsilon\lambda_i^{(1)} + O(\varepsilon^2); \qquad i = 1, 2$$

$$c_1 = c_1^{(0)} + \varepsilon c_1^{(1)} + O(\varepsilon^2); \qquad c_2 = \varepsilon c_2^{(1)} + O(\varepsilon^2)$$

$$d_1 = \varepsilon d_1^{(1)} + O(\varepsilon^2); \qquad d_2 = d_2^{(0)} + \varepsilon d_2^{(1)} + O(\varepsilon^2)$$

$$|\phi_1\rangle = \left(c_1^{(0)} + \varepsilon c_1^{(1)}\right)|\psi_1\rangle + \varepsilon c_2^{(1)}|\psi_2\rangle + O(\varepsilon^2)$$

$$|\phi_2\rangle = \varepsilon d_1^{(1)}|\psi_1\rangle + \left(d_2^{(0)} + \varepsilon d_2^{(1)}\right)|\psi_2\rangle + O(\varepsilon^2)$$

We have the result that

$$c_1^{(0)} = 1 \quad \text{and} \quad d_2^{(0)} = 1$$

since they are both constants that are equal to 1 in the limit $\varepsilon = 0$. Also, since we are finding only the first correction terms, we neglect all terms that have powers of ε greater than 1. (We use the symbol $O(\varepsilon^2)$ to indicate terms of this nature. The precise meaning of the symbol $O(\varepsilon^2)$ is explained in Chapter 3.)

(v) We construct the equations

$$(A + \varepsilon B)|\phi_1\rangle = \mu_1|\phi_1\rangle$$

or

$$(A + \varepsilon B)\left[\left(1 + \varepsilon c_1^{(1)}\right)|\psi_1\rangle + \varepsilon c_2^{(1)}|\psi_2\rangle\right]$$

$$= \left(\lambda_1 + \varepsilon\lambda_1^{(1)}\right)\left[\left(1 + \varepsilon c_1^{(1)}\right)|\psi_1\rangle + \varepsilon c_2^{(1)}|\psi_2\rangle\right] \tag{1.58}$$

Multiply Equation (1.58) by $\langle\psi_1|$ and use the orthogonality of the eigenvectors to obtain

$$\langle\psi_1, B\psi_1\rangle = \lambda_1^{(1)} \tag{1.59}$$

(vi) Multiply Equation (1.58) by $\langle\psi_2|$, noting orthogonality, and obtain

$$c_2^{(1)} = \frac{\langle\psi_2, B\psi_1\rangle}{\lambda_1 - \lambda_2} \tag{1.60}$$

(vii) We can obtain the second eigenvector and eigenvalue by interchanging subscripts 1 and 2 and c and d:

$$1 \to 2; \qquad 2 \to 1; \qquad c \to d$$

Our results are then

$$\mu_i = \lambda_i + \varepsilon\langle\psi_i, B\psi_i\rangle; \qquad i = 1, 2 \tag{1.61a}$$

$$|\phi_1\rangle = \left(1 + \varepsilon c_1^{(1)}\right)|\psi_1\rangle + \varepsilon\frac{\langle\psi_2, B\psi_1\rangle}{\lambda_1 - \lambda_2}|\psi_2\rangle \tag{1.61b}$$

$$|\phi_2\rangle = \varepsilon\frac{\langle\psi_1, B\psi_2\rangle}{\lambda_2 - \lambda_1}|\psi_1\rangle + \left(1 + \varepsilon d_2^{(1)}\right)|\psi_2\rangle \tag{1.61c}$$

neglecting throughout terms with quadratic and higher powers of ε as coefficients.

We require $|\phi_1\rangle$ and $|\phi_2\rangle$ to be normalized to unity. Thus, to order ε, we have $c_1^{(1)} = d_2^{(1)} = 0$.

(We want to fix the normalization of the eigenvectors $|\phi_1\rangle$ so that to each order in the small parameter ε, they are normalized to unity. Thus, for example,

$$\langle\phi_1, \phi_1\rangle = 1 + 0(\varepsilon^2) \Rightarrow c_1^{(1)} + c_1^{(1)*} = 0$$

This fixes the real part of $c_1^{(1)} = 0$. The imaginary part remains undetermined and plays no role in the computation of matrix elements. Thus, we are safe to pick $c_1^{(1)} = 0$. The argument is unchanged for the coefficient $d_1^{(1)}$, so we also set $d_1^{(1)} = 0$.) Our final expressions for the perturbed eigenvectors are

$$|\phi_1\rangle = |\psi_1\rangle + \varepsilon\frac{\langle\psi_2, B\psi_1\rangle}{\lambda_1 - \lambda_2}|\psi_2\rangle + O(\varepsilon^2) \tag{1.62a}$$

$$|\phi_2\rangle = \varepsilon\frac{\langle\psi_1, B\psi_2\rangle}{\lambda_2 - \lambda_1}|\psi_1\rangle + |\psi_2\rangle + O(\varepsilon^2) \tag{1.62b}$$

We see that the shift in the eigenvalues is computed from the matrix elements of the perturbation, εB, taken between eigenvectors of the unperturbed system. Thus, if one is seeking only this shift, it is unnecessary to compute the new eigenvectors.

Example 6.2 Given matrix A with known eigenvalues and eigenvectors

$$A = \begin{bmatrix} 3 & 2 \\ 2 & 3 \end{bmatrix}; \qquad \lambda_1 = 5; \qquad \lambda_2 = 1$$

$$|\psi_1\rangle = \frac{1}{\sqrt{2}} \begin{bmatrix} 1 \\ 1 \end{bmatrix}; \qquad |\psi_2\rangle = \frac{1}{\sqrt{2}} \begin{bmatrix} 1 \\ -1 \end{bmatrix}$$

and the matrix

$$B = \begin{bmatrix} 2 & 1 \\ 3 & 4 \end{bmatrix}; \qquad A + \varepsilon B = \begin{bmatrix} 3 + 2\varepsilon & 2 + \varepsilon \\ 2 + 3\varepsilon & 3 + 4\varepsilon \end{bmatrix}; \qquad |\varepsilon| \ll 1$$

We seek the eigenvectors and eigenvalues of matrix $A + \varepsilon B$ correct through the first order in the small parameter ε.

When we obtain the solution, we note that we can easily check the correctness of the eigenvalues by taking the trace and determinant of matrix $A + \varepsilon B$ and noting that they completely determine the eigenvalues of a 2×2 matrix. It is possible to check the correctness of the eigenvectors by solving the equation

$$(A + \varepsilon B)|\phi_i\rangle = \mu_i|\phi_i\rangle; \qquad i = 1, 2$$

and seeing that it is correct, neglecting, as before, powers of ε higher than the first.

SOLUTION

$$\langle \psi_1, B\psi_1 \rangle = \tfrac{1}{2}[1 \quad 1] \begin{bmatrix} 2 & 1 \\ 3 & 4 \end{bmatrix} \begin{bmatrix} 1 \\ 1 \end{bmatrix} = 5$$

$$\langle \psi_2, B\psi_2 \rangle = 1$$

$$\langle \psi_1, B\psi_2 \rangle = 0; \qquad \mu_1 = 5 + 5\varepsilon$$

$$\langle \psi_2, B\psi_1 \rangle = -2; \qquad \mu_2 = 1 + \varepsilon$$

$$|\phi_1\rangle = \frac{1}{\sqrt{2}} \left[\begin{bmatrix} 1 \\ 1 \end{bmatrix} - \frac{2\varepsilon}{5 - 1} \begin{bmatrix} 1 \\ -1 \end{bmatrix} \right]$$

$$= \frac{1}{\sqrt{2}} \begin{bmatrix} 1 - \varepsilon/2 \\ 1 + \varepsilon/2 \end{bmatrix}$$

$$|\phi_2\rangle = \frac{1}{\sqrt{2}} \left[\mathbf{0} + \begin{bmatrix} 1 \\ -1 \end{bmatrix} \right] = \frac{1}{\sqrt{2}} \begin{bmatrix} 1 \\ -1 \end{bmatrix}$$

Check, neglecting terms $O(\varepsilon^2)$:

$$\mathrm{Tr}(A + \varepsilon B) = 6 + 6\varepsilon = \mu_1 + \mu_2$$
$$\det(A + \varepsilon B) = 5 + 10\varepsilon = \mu_1 \cdot \mu_2$$

$$\begin{bmatrix} 3 + 2\varepsilon & 2 + \varepsilon \\ 2 + 3\varepsilon & 3 + 4\varepsilon \end{bmatrix} \begin{bmatrix} 1 - \dfrac{\varepsilon}{2} \\ 1 + \dfrac{\varepsilon}{2} \end{bmatrix} = \begin{bmatrix} 5 + \frac{5}{2}\varepsilon \\ 5 + \frac{15}{2}\varepsilon \end{bmatrix}$$

$$= (5 + 5\varepsilon) \begin{bmatrix} 1 - \dfrac{\varepsilon}{2} \\ 1 + \dfrac{\varepsilon}{2} \end{bmatrix}$$

$$\begin{bmatrix} 3 + 2\varepsilon & 2 + \varepsilon \\ 2 + 3\varepsilon & 3 + 4\varepsilon \end{bmatrix} \begin{bmatrix} 1 \\ -1 \end{bmatrix} = \begin{bmatrix} 1 + \varepsilon \\ -1 - \varepsilon \end{bmatrix} = (1 + \varepsilon) \begin{bmatrix} 1 \\ -1 \end{bmatrix} \qquad \blacksquare$$

1.7 CONCLUDING REMARKS

In this section, we collect a variety of results that relate to the stability of solutions to linear difference and differential equations. These will be particularly useful when we discuss weakly nonlinear systems in which we use the linear part of the equation as a starting point for a perturbative solution.

We confine our attention almost exclusively to matrices that have distinct eigenvalues and hence are diagonalizable by an invertible transformation. This is shown to be reasonable by the theorem, discussed in Section 1.7.1, that states that arbitrarily close in norm to a given matrix A is a matrix B with distinct eigenvalues. Thus, although one has to be careful, most of the results associated with eigenvalues remain valid when the eigenvalues coincide. On the other hand, eigenvectors are affected and special attention is needed in the case of degenerate eigenvectors. (This is made clear in Exercise 7.6 at the end of the chapter.)

1.7.1 Rayleigh's Inequality and Related Material

A Hermitian matrix H can be diagonalized by a unitary matrix U. Thus, we have

$$H = H^\dagger; \qquad U^\dagger H U = \Lambda; \qquad UU^\dagger = I \tag{1.63}$$

A Hermitian matrix has real eigenvalues and a complete set of eigenvectors. Eigenvectors associated with distinct eigenvalues are orthogonal and those associated with degenerate eigenvalues can be orthogonalized by the Gram–Schmidt method. These results are proved in standard texts on matrix analysis.

Since $U^{\dagger}U = I$, we see that the scalar product $\langle Ux, Uy \rangle = \langle x, y \rangle$. Thus, the scalar product is preserved under unitary transformations.

Rayleigh's inequality tells us that the maximum value of the scalar product

$$R = \langle x, Hx \rangle / \langle x, x \rangle$$

where H is a Hermitian matrix, and $|x\rangle$ is any vector, is equal to the largest eigenvalue. The proof depends upon the completeness and the orthogonality of the eigenvectors of the matrix H.

Proof: Expand a general vector $|x\rangle$ in terms of the eigenvectors $|y_i\rangle$ of H:

$$|x\rangle = \Sigma c_i |y_i\rangle, \quad \text{where } H|y_i\rangle = \lambda_i |y_i\rangle$$

(We have normalized vectors $|y_i\rangle$ to unity.) Then

$$R = \langle x, Hx \rangle / \langle x, x \rangle = \frac{\Sigma c_i c_i^* \lambda_i}{\Sigma c_i c_i^*}; \qquad \langle y_i, y_j \rangle = \delta_{i,j}$$

Now order the eigenvalues:

$$\lambda_{\text{min}} \leq \cdots \lambda_i \cdots \leq \lambda_{\text{max}}$$

and since

$$\lambda_{\text{min}} \cdot \Sigma c_i c_i^* \leq \Sigma c_i c_i^* \lambda_i \leq \lambda_{\text{max}} \Sigma c_i c_i^*$$

we have

$$R = \langle x, Hx \rangle / \langle x, x \rangle \leq \lambda_{\text{max}} \tag{1.64a}$$

Notice that in an application of Rayleigh's inequality, you can take as your trial vector any vector of your choice and be assured that the quantity R is at most equal to the largest eigenvalue of matrix H. It is also clear that we can work the argument backwards and obtain the result

$$R \geq \lambda_{\text{min}} \qquad \blacksquare \tag{1.64b}$$

It is important to observe that Rayleigh's inequality does not apply to matrices that do not have a complete set of orthogonal eigenvectors. For example, consider matrix A, which has real eigenvalues equal to a and c, with $a > c$:

$$A = \begin{bmatrix} a & b \\ 0 & c \end{bmatrix}$$

Evaluate $R = \langle x, Ax \rangle / \langle x, x \rangle$, where we choose as our vector $|x\rangle$:

$$|x\rangle = \begin{bmatrix} 1 \\ 1 \end{bmatrix}$$

We find that $R = (a + b + c)/2$. With an appropriate choice of b, we can make $R > a$, the largest eigenvalue of A. (The eigenvectors of A are complete but not orthogonal.)

We can maneuver around this difficulty by considering a general matrix A and forming the product $A^{\dagger}A$, which is a positive semidefinite Hermitian matrix. This means that all of its eigenvalues are real and are greater than or equal to 0. We see this by considering a general vector $|x\rangle$ and forming the scalar product $\langle x, A^{\dagger}Ax\rangle = \langle Ax, Ax\rangle \geq 0$.

Example 7.1 Consider the matrix

$$A = \begin{bmatrix} 0 & 2 \\ 0 & 0 \end{bmatrix}; \qquad A^{\dagger}A = \begin{bmatrix} 0 & 0 \\ 0 & 4 \end{bmatrix}$$

Note $\langle Ax, Ax\rangle \geq 0$. ■

We introduce the concept of a norm since we will be using it in our comments on stability theory. There are many possible definitions of a norm, and it is sufficient for most purposes to consider the Hermitian norm. The norm of vector $|x\rangle$ is defined as the quantity

$$\|x\| \equiv \sqrt{\langle x, x\rangle} = \sqrt{\Sigma x_i^* x_i} \tag{1.65}$$

It gives us a measure of the length of a vector. Notice that it is invariant under a unitary transformation:

$$\langle Ux, Ux\rangle = \langle x, U^{\dagger}Ux\rangle = \langle x, x\rangle$$

Associated with this vector norm is the norm of a matrix. We define the norm of matrix A by

$$\|A\| = \max_{x \neq 0} \frac{\|Ax\|}{\|x\|} = \frac{\sqrt{\langle Ax, Ax\rangle}}{\sqrt{\langle x, x\rangle}} = \frac{\sqrt{\langle x, A^{\dagger}Ax\rangle}}{\sqrt{\langle x, x\rangle}} \tag{1.66}$$

Thus, in order to evaluate the matrix norm, we find the maximum value of the scalar product $\langle Ax, Ax\rangle/\langle x, x\rangle$ when we vary vector $|x\rangle$ over all possible vectors in the space. We assume that vector $|x\rangle$ is normalized to unity since this does not change the result. We consider the related quantity:

$$N = \langle x, A^{\dagger}Ax\rangle \tag{1.67}$$

We then observe that we have generated, by matrix A, a related Hermitian matrix $A^{\dagger}A$ whose eigenvalues are all greater than or equal to 0. We then apply Rayleigh's inequality, given by Equation (1.64a), and note that the norm of matrix A is equal to the square root of the largest eigenvalue of matrix $A^{\dagger}A$.

Finally, if matrix A is Hermitian, we have $A^\dagger A = A^2$, and thus the norm of a Hermitian matrix is equal to the maximum, in absolute value of its eigenvalues.

We want to show that given matrix A with coincident eigenvalues, we can find matrix B arbitrarily close in norm whose eigenvalues are distinct. We proceed as follows:

(a) We know that an arbitrary matrix A can be brought to triangular form by a unitary matrix U. (See a standard text on matrix theory, e.g., Franklin or Bellman.) When a matrix is in triangular form, the diagonal elements are the eigenvalues of matrix A and all of the elements below the diagonal are equal to zero. We call "T" the triangular matrix; thus, we have

$$A = UTU^\dagger; \qquad T = \begin{bmatrix} \lambda_1 & & & \\ & \lambda_2 & & \\ & & \ddots & t_{ij} \\ & 0 & & \\ & & & \lambda_n \end{bmatrix}$$

$$t_{ij} = 0; \qquad i > j \tag{1.68}$$

(b) We now consider related matrices B and S. The elements of matrix S are chosen to be the same as those of matrix T, except that each diagonal element has a small additional part, ε_i. Thus, we have

$$B \equiv USU^\dagger; \qquad S = \begin{bmatrix} \lambda_1 + \varepsilon_1 & & & \\ & \lambda_2 + \varepsilon_2 & & s_{ij} \\ 0 & & \ddots & \\ & & & \lambda_n + \varepsilon_n \end{bmatrix}$$

$$s_{ij} = t_{ij}; \qquad i \neq j \tag{1.69}$$

We now subtract A from B and obtain:

$$B - A = U(S - T)U^\dagger = U \begin{bmatrix} \varepsilon_1 & & & \\ & \varepsilon_2 & & 0 \\ 0 & & \ddots & \\ & & & \varepsilon_n \end{bmatrix} U^\dagger \tag{1.70}$$

We see that we have a diagonal matrix ε sandwiched between matrices U and $U^\dagger$. It is clear that since the quantities ε_i are at our disposal, we can make the norm of $A - B$ as small as we please. Of course, we can choose the elements so that the eigenvalues of B are distinct.

1.7.2 Location of the Eigenvalues of a Matrix

It is often of interest to have an estimate of where the eigenvalues of a matrix lie. For example, questions regarding the stability of the solution of a matrix differential equation are connected with the sign of the real part of the eigenvalues of a matrix. Consider the equation

$$\frac{d}{dt}|y\rangle = A|y\rangle$$

If the eigenvalues of matrix A have negative real parts, then the solution $y(t)$ tends to zero as t tends to infinity. Theorem 7.1 enables us to obtain a simple bound on the location of the eigenvalues of a matrix.

Theorem 7.1 The $n \times n$ matrix A is nonsingular if

$$|a_{ii}| > \sum_{j \neq i} |a_{ij}|; \qquad i = 1, 2 \cdots, n$$

or, equivalently, if

$$|a_{ii} - \lambda| > \sum_{j \neq i} |a_{ij}|; \qquad i = 1, 2 \cdots, n$$

we conclude that λ is not an eigenvalue of matrix A.

Consequently, every eigenvalue of the $n \times n$ matrix A satisfies at least one of the inequalities

$$|a_{ii} - \lambda| \leq \sum_{j \neq i} |a_{ij}|; \qquad i = 1, 2 \cdots, n \tag{1.71}$$

Proof Consider matrix A. If it has an eigenvalue equal to λ, then matrix $A - \lambda I$ has a zero eigenvalue. We look at the components of the associated eigenvector and choose the element that is largest in absolute value, e.g., x_i.

$$(a_{ii} - \lambda)(x_i) + \sum_{k \neq i} a_{ik} x_k = 0; \qquad |x_i| = \max_k |x_k|$$

We rearrange the terms and take absolute values:

$$|(a_{ii} - \lambda) x_i| = |a_{ii} - \lambda| \cdot |x_i| = \left| \sum_{k \neq i} a_{ik} x_k \right|$$

establishing the related inequalities:

$$|a_{ii} - \lambda| \cdot |x_i| \leq \sum_{k \neq i} |a_{ik}| \cdot |x_k| \leq \sum_{k \neq i} |a_{ik}| \cdot |x_i|$$

But, the last inequality is in contradiction to our hypothesis that

$$|a_{ii} - \lambda| > \sum_{k \neq i} |a_{ik}|$$

So λ is not an eigenvalue of matrix A. ■

Example 7.2 Use Equation (1.71) to establish bounds on the eigenvalues of the matrix

$$B = \begin{bmatrix} -10 & 1 & 1 \\ 0.2 & -5 & -0.3 \\ -i & 0 & -4 \end{bmatrix}$$

Form the appropriate inequalities:

$$|-10 - \lambda| \leq |2| \quad \text{or} \quad -12 \leq \text{Re}\,\lambda \leq -8$$
$$|-5 - \lambda| \leq |0.2| + |-0.3| \quad \text{or} \quad -5.5 \leq \text{Re}\,\lambda \leq -4.5$$
$$|-4 - \lambda| \leq |-i| \quad \text{or} \quad -5 \leq \text{Re}\,\lambda \leq -3$$

Thus, we conclude that the eigenvalues lie so as to satisfy the three inequalities. Alternatively, draw a circle, in the complex plane, about each diagonal element with a radius equal to the sum of the absolute values of the off-diagonal elements in the same row. The eigenvalues of matrix B lie within these circles. See Figure 1.2. ■

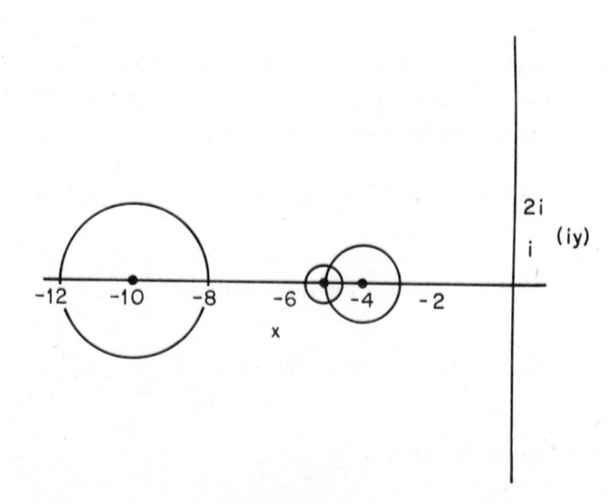

Figure 1.2

EXERCISE 7.1 Give a bound on the value of the real part of the least negative eigenvalue of matrix C (i.e., the negative eigenvalue closest to zero).

$$C = \begin{bmatrix} -11 & 1 & i \\ 2 & -8 & 0 \\ -1 & -1 & -3 \end{bmatrix}$$ ■

It is clear from this discussion that given an arbitrary matrix, it is useful to try and bring it into a form where the largest elements are on the diagonal. This will lead to a way to put bounds on the location of the roots. It also proves to be of considerable use in estimating the inverse of the matrix.

1.7.3 Stability

We are interested in discussing the stability of a solution to a linear difference equation or a differential equation under a small change in the eigenvalues. This change can be accomplished by adjoining a small perturbing matrix to our original matrix that brings about a small change in the eigenvalues. In particular, we will bring to light a situation where the small perturbation destabilizes the original matrix. Although our results are true for $n \times n$ matrices, we limit our discussion to 2×2 matrices.

Let us look at a differential equation of the form

$$\frac{d}{dt}\begin{bmatrix} x \\ y \end{bmatrix} = \begin{bmatrix} \alpha & \beta \\ 0 & \mu \end{bmatrix}\begin{bmatrix} x \\ y \end{bmatrix}$$

Elements α and μ are real and negative. We allow for the possibility that $\alpha = \mu$. We want to know how the solutions behave as t tends to infinity. It is clear that if α and μ are different, the solution will approach zero independent of the initial condition. So, we will concentrate on developing a proof that takes into consideration the possibility that $\alpha = \mu$. We write our matrix equation as two separate differential equations for components x and y:

$$\frac{dx}{dt} = \alpha x + \beta y$$

$$\frac{dy}{dt} = \mu y$$

We solve for $y(t)$ and observe that $y(t)$ goes to zero exponentially as t tends to infinity.

$$y(t) = y(0) \exp \mu t$$

We can now solve for $x(t)$ using our knowledge of $y(t)$.

$$\frac{d}{dt}x(t) = \alpha x + \beta y(0) \exp \mu t$$

$$x(t) = x(0) \exp \alpha t + \beta y(0) \int_0^t \exp[\alpha(t - t')] \exp \mu t' \, dt'$$

It is clear that $x(t)$ goes to zero as t tends to infinity as long as both α and μ are negative. This is a general result, valid for $n \times n$ matrices. Matrix A is a stability matrix if its roots have negative real part. (There is an extensive discussion of stability theory in the text by Bellman.)

Now let us show that you have to be careful. Consider the matrix equation

$$\frac{d}{dt}\begin{bmatrix} x \\ y \end{bmatrix} = A(\varepsilon)\begin{bmatrix} x \\ y \end{bmatrix} \tag{1.72a}$$

$$A(\varepsilon) = \begin{bmatrix} -\varepsilon^2 & b \\ 0 & -\varepsilon^2 \end{bmatrix}; \qquad 0 < \varepsilon \ll 1 \tag{1.72b}$$

where b takes on two different values:

$$b = \varepsilon; \qquad b = \varepsilon^p, \quad p > 1$$

Since the eigenvalues are negative, we say that matrix A is a stability matrix. We now add a small matrix D to matrix A:

$$D(\varepsilon) = \begin{bmatrix} 0 & 0 \\ c^2\varepsilon^3 & 0 \end{bmatrix}; \qquad 0 < \varepsilon \ll 1 \tag{1.73}$$

where c is a constant that will be chosen to be either > 1 or < 1.

We now look at the differential equation

$$\frac{d}{dt}\begin{bmatrix} x \\ y \end{bmatrix} = (A + D)\begin{bmatrix} x \\ y \end{bmatrix} = \begin{bmatrix} -\varepsilon^2 & b \\ c^2\varepsilon^3 & -\varepsilon^2 \end{bmatrix}\begin{bmatrix} x \\ y \end{bmatrix} \tag{1.74}$$

Since we are interested in the stability of the solution, we look at the eigenvalues. They are

$$\lambda = -\varepsilon^2 \pm \sqrt{bc^2\varepsilon^3}$$

We note that if $b = \varepsilon$ and $c > 1$, we have a positive eigenvalue equal to $(c - 1)\varepsilon^2$. In this case, the solution will grow in time. On the other hand, if $b = \varepsilon^p$, we see that the eigenvalues remain negative for any value of constant c if we choose our parameter ε to be sufficiently small. Thus, in this case, the stability of the solution is maintained under the perturbation $D(\varepsilon)$.

We now turn our attention to the related difference equation:

$$\begin{aligned} x_{n+1} &= ax_n + by_n \\ y_{n+1} &= \qquad dy_n \end{aligned}$$

$$\begin{bmatrix} x \\ y \end{bmatrix}_{n+1} = A\begin{bmatrix} x \\ y \end{bmatrix}_n; \qquad A = \begin{bmatrix} a & b \\ 0 & d \end{bmatrix} \tag{1.75}$$

We are interested in the behavior of x_n and y_n as n tends to infinity. If x_n and y_n tend to zero, we say that the solution is stable. We find, if $a \neq d$,

$$A^n = \frac{a^n - d^n}{a - d}\begin{bmatrix} a & b \\ 0 & d \end{bmatrix} + \frac{da^n - ad^n}{d - a}\begin{bmatrix} 1 & 0 \\ 0 & 1 \end{bmatrix} \tag{1.76}$$

If the eigenvalues are in absolute value less than 1, it is clear that A^n goes to zero as n tends to infinity. Hence, our solution tends to the zero vector. The result remains true if our eigenvalues coincide. This can be demonstrated by a variety of arguments. For example, we will have to consider terms of the form

$$na^n = n\exp(n \ln a)$$

that appear as off-diagonal elements. If we write $a = 1 - \mu$, where μ is small and positive, we see that

$$n\exp[n\ln(1-\mu)] \simeq n\exp(-n\mu) \to 0 \qquad \text{as} \qquad n \to \infty$$

as long as the product $n\mu > 0$. Thus, the solution of the difference equation is said to be stable if the eigenvalues of the associated matrix have moduli less than 1. We now proceed as we did in Equation (1.72b). Form the matrix

$$A(\varepsilon) = \begin{bmatrix} 1 - \varepsilon^2 & b \\ 0 & 1 - \varepsilon^2 \end{bmatrix} \tag{1.77}$$

and observe that the solution to the associated difference equation tends to 0 as n tends to ∞. [Note that $A(\varepsilon)$ given by Equation (1.77) is related to but different from $A(\varepsilon)$ given by Equation (1.72b).] If we adjoin to our matrix the small matrix

$$D(\varepsilon) = \begin{bmatrix} 0 & 0 \\ c^2\varepsilon^3 & 0 \end{bmatrix} \tag{1.78}$$

we see that the solution is no longer stable if $b = \varepsilon$ and $c > 1$.

Thus, there is a parallel between the stability of linear differential equations with constant coefficients and associated difference equations. (The particular matrices treated in this example are based on an article by Robinson entitled "Stability of Periodic Solutions from Asymptotic Expansions," pp. 173–185 in

R. L. Devaney, and Z. H. Nitecki, eds., *Classical Mechanics and Dynamical Systems*, Marcel Dekker, Inc., New York, 1981.)

We have learned that we have to be very careful when we perturb a matrix by off-diagonal terms. It is precisely because terms of this nature disrupt the connection between the norm of a matrix and its largest eigenvalue that they have such a strong effect on stability. Notice that if the off-diagonal element b of matrix A is equal to ε, it is larger than ε^2, and hence not really a small perturbation on the stability.

EXERCISE 7.2 Consider matrix $A + D$

$$A + D = \begin{bmatrix} -\varepsilon^2 & b \\ c^2\varepsilon^3 & -\varepsilon^2 \end{bmatrix}; \qquad 0 < \varepsilon \ll 1$$

with element $c > 1$. Since the eigenvalues are distinct, the matrix can be brought to diagonal form by an invertible transformation T. Find the eigenvectors of matrix $A + D$ and use them to construct matrix T. We are interested in two possibilities for element b: $b = \varepsilon$ and $b = \varepsilon^p$, where $p > 1$. ■

EXERCISE 7.3 Consider the matrix

$$A = \begin{bmatrix} 6 & 4i \\ i & 2 \end{bmatrix}$$

(a) Show that it has coincident eigenvalues = 4.

(b) Show that the matrix

$$B = \begin{bmatrix} 2 & 2 \\ i & i \end{bmatrix}$$

is such that $AB = B\Lambda$, where

$$\Lambda = \begin{bmatrix} 4 & 0 \\ 0 & 4 \end{bmatrix}$$

(c) Separate the eigenvalues slightly by considering the matrix

$$C = \begin{bmatrix} 6 + \varepsilon & 4i \\ i & 2 - \varepsilon \end{bmatrix}; \qquad 0 < \varepsilon \ll 1$$

Find the eigenvalues of matrix C and the invertible transformation T that brings it to diagonal form. Comment on what happens to the elements of matrix T when the small parameter ε goes to zero. ■

EXERCISE 7.4 Consider the Pauli Spin matrices given by

$$\sigma_x = \begin{bmatrix} 0 & 1 \\ 1 & 0 \end{bmatrix}; \qquad \sigma_y = \begin{bmatrix} 0 & -i \\ i & 0 \end{bmatrix}; \qquad \sigma_z = \begin{bmatrix} 1 & 0 \\ 0 & -1 \end{bmatrix}$$

(a) Show that

$$\sigma_x\sigma_y + \sigma_y\sigma_x = 0; \qquad \sigma_x\sigma_y = i\sigma_z; \qquad \sigma_z\sigma_x = i\sigma_y; \qquad \sigma_y\sigma_z = i\sigma_x$$

(b) Find the numerical value of the matrix element

$$\begin{bmatrix} 1 & 1 \end{bmatrix} \exp \sigma_x \begin{bmatrix} 1 \\ 1 \end{bmatrix}$$

Answer: $2e$. ■

EXERCISE 7.5 Consider the product of two matrices A and B:

$$N = AB \qquad \text{and} \qquad M = BA$$

Assume that either A or B is invertible. Show that N and M have the same eigenvalues. (*Hint*: Connect them by a similarity transformation.) The result that N and M have the same eigenvalues is still correct without the assumption of invertibility. ■

EXERCISE 7.6 Consider the matrix

$$A = \begin{bmatrix} 4 & 0 & 2 \\ 0 & 6 & 0 \\ 2 & 0 & 4 \end{bmatrix}$$

(a) Show that the eigenvalues are 6, 6, and 2.
(b) Construct a set of orthogonal eigenvectors. Normalize them to unity.

Answer:

$$\sqrt{2}\begin{bmatrix} 1 \\ 0 \\ -1 \end{bmatrix}; \qquad \begin{bmatrix} 0 \\ 1 \\ 0 \end{bmatrix}; \qquad \sqrt{2}\begin{bmatrix} 1 \\ 0 \\ 1 \end{bmatrix}$$

■

BIBLIOGRAPHY

Introductory Texts:

1. Davis, P. J., *Mathematics of Matrices*, Blaisdell, Waltham, Massachusetts, 1965.
2. Schwartz, J., *Introduction to Matrices and Vectors*, McGraw-Hill, New York, 1961.
3. Stephenson, G., *An Introduction to Matrices, Sets, and Groups for Scientists and Engineers*, Longman Group Ltd., London, 1965. Reprinted by Dover, New York, 1986.

More Advanced Texts:

4. Wilf, H. S., *Mathematics for the Physical Sciences*, Wiley, New York, 1962. Reprinted by Dover, New York, 1978.
5. Friedman, B., *Principles and Techniques of Pure and Applied Mathematics*, Wiley, New York, 1956.
6. Joshi, A. W., *Matrices and Tensors in Physics*, Halsted Press, New York, 1975.
7. Bellman, R., *Introduction to Matrix Analysis*, McGraw-Hill, New York, 1960.
8. Franklin, J., *Matrix Theory*, Prentice-Hall, Englewood Cliffs, New Jersey, 1968.
9. Mehta, M. L., *Elements of Matrix Theory*, Hindustan Publishing Co., Delhi, India, 1977.

Numerical Methods:

10. Householder, A., *Principles of Numerical Analysis*, McGraw-Hill, New York, 1953.
11. Hamming, R. W., *Numerical Methods for Scientists and Engineers*, 2d Ed., McGraw-Hill, New York, 1973. Reprinted by Dover, New York, 1988.

CHAPTER 2

THE GAMMA AND RELATED FUNCTIONS

We will soon see that the gamma function is the workhorse of mathematical physics. It enables us to either express exactly or approximately integrals that occur in diverse contexts. For example, it dominates the calculations in probability theory because it is used to evaluate Gaussian processes; it occurs in calculations in the kinetic theory of gases and in statistical mechanics because it is related to the Maxwell–Boltzmann distribution; it occurs in problems in condensed-matter physics when you encounter Fermi–Dirac and Einstein–Bose statistics; and it is used in mechanics to calculate areas, volumes, moments of inertia, centers of mass, etc.

We begin by defining the gamma function and then quickly establish a few results. We then introduce Dirichlet integrals to prepare us for the beta functions, which are expressed in terms of gamma functions. We will then take the opportunity to discuss the Riemann zeta function since it occurs in the relationship between sums and integrals. The chapter concludes with an introduction to the delta function.

Illustrative examples are used to derive a variety of useful results that, although not central to the chapter, are important for a firm grasp of the subject.

2.1 THE GAMMA FUNCTION

We choose to give a limited presentation at an elementary level, having decided that the powerful techniques of complex variables are beyond the scope of our treatment. A more comprehensive discussion of gamma and related functions can be found in the texts by Arfken and Lebedev cited in the Bibliography at the end of the chapter. We define the gamma function $\Gamma(z)$ by the integral

$$\Gamma(z) \equiv \int_0^\infty t^{z-1} \exp(-t)\, dt; \qquad z > 0 \tag{2.1}$$

We first establish that our definition of the gamma function satisfies the fundamental recursion relation:

$$\Gamma(z+1) = z\Gamma(z); \qquad z > 0 \tag{2.2}$$

We obtain this result from Equation (2.1) by writing

$$\Gamma(z+1) = \int_0^\infty t^z \exp(-t)\, dt$$

and integrating by parts:

$$\Gamma(z+1) = -t^z \exp(-t)\big|_0^\infty + \int_0^\infty z t^{z-1} \exp(-t)\, dt$$

The terms integrated by parts vanish for $z > 0$ and we obtain Equation (2.2).

The factorial function $n!$ for n integral is defined by

$$n! \equiv \int_0^\infty x^n \exp(-x)\, dx; \qquad n > 0 \tag{2.3a}$$

$$0! \equiv \int_0^\infty \exp(-x)\, dx = 1 \tag{2.3b}$$

Then, from Equation (2.3a), we have that $n! = n(n-1)!$.

From Equations (2.1), (2.2), and (2.3a) we observe that

$$n! = \Gamma(n+1) \tag{2.4}$$

Because of Equation (2.2), it is sufficient to tabulate $\Gamma(z)$ for the interval

$$1 \le z \le 2$$

See Table 2.1, which includes a list of the values of $\Gamma(z)$, and Figure 2.1, in which we plot the integrand of $\Gamma(z)$ for $z = \frac{3}{2}, \frac{5}{2}$, and 3.

We see from the table that $\Gamma(1) = 1$ and then $\Gamma(z)$ gradually decreases to, approximately, 0.8857 at $z = 1.45$, and then gradually increases and reaches 1 at $z = 2$. Thus, $\Gamma(z)$ is approximately equal to 1 in the interval [1, 2].

TABLE 2.1 Table of $\Gamma(z)$ for $1 \le z \le 2$

z	$\Gamma(z)$
1.00	1.000
1.05	0.9735
1.10	0.9514
1.15	0.9330
1.20	0.9182
1.25	0.9064
1.30	0.8975
1.35	0.8912
1.40	0.8873
1.45	0.8857
1.50	0.8862
1.55	0.8889
1.60	0.8935
1.65	0.9001
1.70	0.9086
1.75	0.9191
1.80	0.9314
1.85	0.9456
1.90	0.9618
1.95	0.9799
2.00	1.000

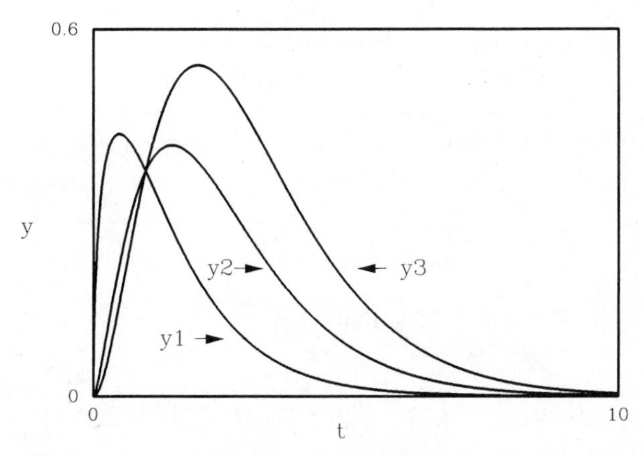

Figure 2.1 $y1 = \sqrt{t}\,\exp(-t)$; $y2 = \sqrt{t^3}\,\exp(-t)$; $y3 = t^2\exp(-t)$. Try to estimate the area under the curves. Does it agree with the value of the appropriate gamma function?

For example, if you wish to know the value of $\Gamma(3.1)$, you would write:

$$\Gamma(3.1) = (2.1)\Gamma(2.1) = (2.1)(1.1)\Gamma(1.1)$$

and then look up the value of $\Gamma(1.1)$ in Table 2.1 to obtain $\Gamma(3.1) = 2.2$.

As noted, we usually arrange our calculations to have the argument of the gamma function fall in the range $1 \le z \le 2$, except when discussing general formulas, relationships, or expressing our results using special values of $\Gamma(z)$. In this regard, we will shortly show, in Equation (2.20), that

$$\Gamma(\tfrac{1}{2}) = \sqrt{\pi}$$

If we work with our integral definition, given by Equation (2.1), we see that we are immediately able to express the following integral in terms of gamma functions by making a simple transformation of variables as indicated:

$$I(p, \lambda) = \int_0^\infty t^p \exp(-\lambda t^2)\, dt; \qquad \lambda > 0 \tag{2.5a}$$

Let $z = \lambda t^2$; then

$$I(p, \lambda) = \tfrac{1}{2}\lambda^{-(p+1)/2} \int_0^\infty z^{(p-1)/2} \exp(-z)\, dz \tag{2.5b}$$

$$I(p, \lambda) = \tfrac{1}{2}\lambda^{-(p+1)/2}\Gamma\left(\frac{p+1}{2}\right) \tag{2.5c}$$

Check the result for $p = 1$.

$$I(1, \lambda) = \frac{1}{2}\frac{1}{\lambda}\Gamma(1) = \frac{1}{2\lambda}$$

It follows from Equation (2.5a) that

$$I(2n, 1) = \int_0^\infty t^{2n} \exp(-t^2)\, dt = \tfrac{1}{2}\Gamma\left(\frac{2n+1}{2}\right) \tag{2.6a}$$

$$I(2n+1, 1) = \int_0^\infty t^{2n+1} \exp(-t^2)\, dt = \tfrac{1}{2}\Gamma(n+1) = \tfrac{1}{2}n! \tag{2.6b}$$

We will return to this line of development in Section 2.3 since it is very rich and will help us learn to manipulate various results.

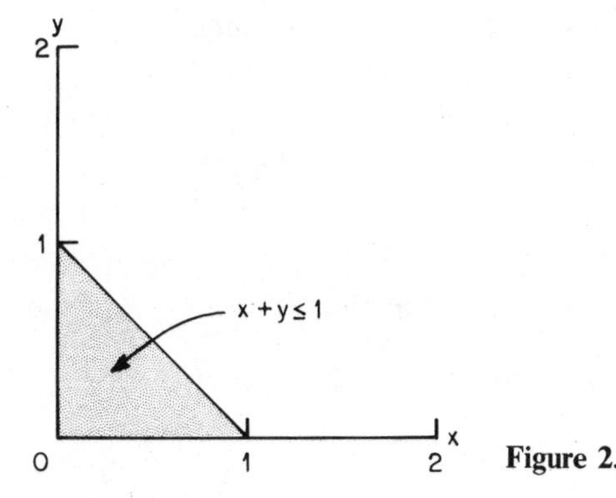

Figure 2.2

2.2 DIRICHLET INTEGRALS

We begin with two obvious and simple examples: finding the area of a triangle and of a sector of a circle.

(1) Find the area between the coordinate axes of the first quadrant, x and y, and the straight line $x + y = 1$.

First, draw the area, as indicated in Figure 2.2. It is clear that the area is equal to $\frac{1}{2}$, but let us find it by integration:

$$\text{Area} = \iint_{\substack{x+y\le 1\\ x,\, y>0}} dx\, dy = \int_0^1 dx \int_0^{1-x} dy = \int_0^1 (1-x)\, dx = \tfrac{1}{2} \tag{2.7}$$

(2) We now would like to find the area between the same coordinate axes and the circular arc given by

$$x^2 + y^2 = 1$$

Again, we know that the answer is $\pi/4$, but let us find this result by integration using polar coordinates:

$$\begin{aligned}\text{Area} &= \iint_{\substack{x^2+y^2\le 1\\ x,\, y>0}} dx\, dy = \iint_{\substack{r\le 1\\ 0\le\theta\le\pi/2}} r\, dr\, d\theta \\ &= \int_0^{\pi/2} d\theta \int_0^1 r\, dr = \frac{1}{2}\frac{\pi}{2} = \frac{\pi}{4}\end{aligned} \tag{2.8}$$

We now do two related but slightly more complicated examples; let us call them 3a and 3b.

(3a) Instead of finding the area, let us evaluate the related integral:

$$I(A, B) = \iint_{\substack{x+y \leq 1 \\ x,\, y>0}} x^A y^B \, dx \, dy; \qquad A, B > -1 \tag{2.9}$$

that is, a weighted area. You can think of each point in the area as having a mass, where the density ρ is a function of x and y in the form

$$\rho(x, y) = x^A y^B$$

Integrals of the form $I(A, B)$ are called Dirichlet integrals. [We have used the symbol $I(p, \lambda)$ in Equations (2.5a)–(2.5c); we hope that this will not cause any confusion.] We proceed in the same manner as we did in finding the area of the triangle, but now we leave our final result in the form of an integral over the variable x.

$$I(A, B) = \int_0^1 x^A \left[\int_0^{1-x} y^B \, dy \right] dx = \int_0^1 x^A \left[(B+1)^{-1} (1-x)^{B+1} \right] dx \tag{2.10}$$

(3b) We now do the corresponding problem for the circle; namely, we consider the integral

$$J(\alpha, \beta) = \iint_{\substack{x^2+y^2 \leq 1 \\ x,\, y>0}} x^\alpha y^\beta \, dx \, dy; \qquad \alpha, \beta > -1 \tag{2.11}$$

Using polar coordinates, we perform the r integration and leave our result in the form of an integral over the angle θ:

$$x = r\cos\theta; \qquad y = r\sin\theta$$

$$J(\alpha, \beta) = \int_0^{\pi/2} \int_0^1 (r\cos\theta)^\alpha (r\sin\theta)^\beta r \, dr \, d\theta$$

$$J(\alpha, \beta) = \int_0^{\pi/2} \frac{\cos^\alpha\theta \sin^\beta\theta \, d\theta}{\alpha + \beta + 2} \tag{2.12}$$

We now connect these two examples by transforming variables as follows. In example (3a), let $x = u^2$, $y = v^2$, $dx = 2u\,du$, $dy = 2v\,dv$, and

$$x + y \leq 1 \rightarrow u^2 + v^2 \leq 1$$

Then, we have

$$I(A, B) = \iint_{\substack{x+y\leq 1 \\ x,\, y>0}} x^A y^B \, dx\, dy = \iint_{\substack{u^2+v^2\leq 1 \\ u,\, v>0}} (u^{2A} 2u\, du)(v^{2B} 2v\, dv)$$

$$I(A, B) = 4\iint_{\substack{u^2+v^2\leq 1 \\ u,\, v>0}} u^{2A+1} v^{2B+1}\, du\, dv \tag{2.13}$$

Thus, we see that if we adjust the exponents appropriately, we can establish the identity

$$I(A, B) = 4J(\alpha = 2A + 1, \beta = 2B + 1)$$

or

$$\begin{aligned} I(A, B) &= \iint_{\substack{x+y\leq 1 \\ x,\, y>0}} x^{(\alpha-1)/2} y^{(\beta-1)/2}\, dx\, dy \\ &= \frac{4}{\alpha + \beta + 2} \int_0^{\pi/2} \cos^\alpha\theta \sin^\beta\theta \, d\theta \end{aligned} \tag{2.14}$$

Let us check this result for $\alpha = \beta = 1$ and $A = B = 0$.

$$I(0, 0) = \iint_{\substack{x+y\leq 1 \\ x,\, y>0}} dx\, dy = \frac{4}{1 + 1 + 2} \int_0^{\pi/2} \cos\theta \sin\theta \, d\theta = \frac{1}{2}$$

Consider $\alpha = \beta = 0$ and $A = B = -\frac{1}{2}$. We have

$$\begin{aligned} I\left[\left(\frac{-1}{2}\right), \left(\frac{-1}{2}\right)\right] &= \iint_{\substack{x+y\leq 1 \\ x,\, y>0}} \frac{1}{\sqrt{x}\sqrt{y}}\, dx\, dy \\ &= \frac{4}{0 + 0 + 2} \int_0^{\pi/2} 1 \cdot d\theta = \pi \end{aligned}$$

So,

$$\int_0^1 \frac{1}{\sqrt{x}} 2\sqrt{1 - x}\, dx = \pi$$

or

$$\int_0^1 \frac{1}{\sqrt{x}} \sqrt{(1 - x)}\, dx = \frac{\pi}{2} \tag{2.15}$$

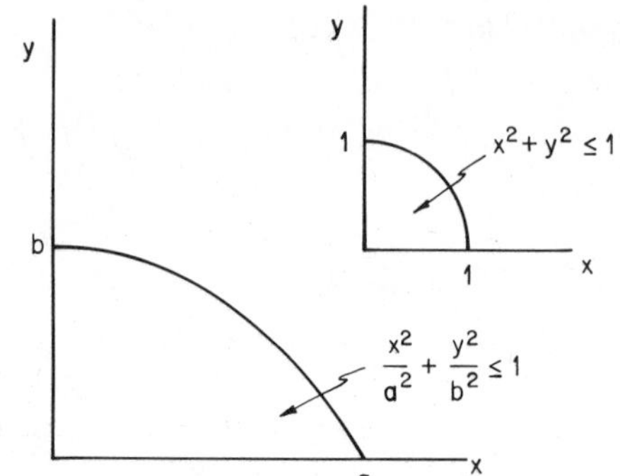

Figure 2.3 We use a simple transformation of variables to find the area of the ellipse in terms of the area of the related circle.

At first glance, this is a curious result because it shows that you can get the constant π while working with straight-line problems if you assign appropriate weights. We were always taught that π is the ratio of the circumference of a circle to its diameter. Somehow it had to do with circles. Here we see it emerge in a different context. We will see that it comes up in many unexpected places, for example, in probability theory.

In principle, we can find the value of any weighted area in the first quadrant by performing a transformation of variables from cartesian to polar coordinates and vice versa, as we choose. We have to understand only how to perform the last integration over variable x or over variable θ, as the case may be. We will see that we have to learn about beta functions and their relationship to gamma functions. Before we do so, let us find the area of the quarter of an ellipse bounded by the coordinate axes and a curve given by the equation

$$\frac{x^2}{a^2}+\frac{y^2}{b^2}=1; \qquad x,\, y \ge 0$$

$$E=\iint_{\substack{x^2/a^2+y^2/b^2\le 1\\ x,\,y>0}} dx\,dy$$

See Figure 2.3.

We recognize this as a Dirichlet integral. In order to put it in standard form, we change the scales of the coordinates and transform the problem to a circle. We then let $x = au$ and $y = bv$; then

$$E = ab\iint_{\substack{u^2+v^2\le 1\\ u,\,v>0}} du\,dv = abJ(0,0) = ab\frac{\pi}{4} \tag{2.16}$$

2.3 BETA FUNCTIONS

We now consider $J(\alpha, \beta)$ given by Equation (2.12) and perform the integration over the angle θ. We use a trick identity to do all of our work. Consider two forms of the gamma function related to those we introduced in Equations (2.1) and (2.5b):

$$\Gamma(p) = \int_0^\infty t^{p-1} \exp(-t)\, dt$$

Let $t = x^2$, $p > 0$:

$$\Gamma(p) = 2\int_0^\infty x^{2p-1} \exp(-x^2)\, dx \tag{2.17a}$$

Similarly, relabeling p by q and x by y:

$$\Gamma(q) = \int_0^\infty t^{q-1} \exp(-t)\, dt$$

$$\Gamma(q) = 2\int_0^\infty y^{2q-1} \exp(-y^2)\, dy \tag{2.17b}$$

We can think of the x and y variables as the coordinates of the axes in the first quadrant, and then write the two integrals given by Equations (2.17a) and (2.17b) as a product, in which the region of integration is the first quadrant. We then transform the variables to polar coordinates and perform the radial integration to obtain an integral over the angle θ.

$$\Gamma(p)\Gamma(q) = 4\int_0^\infty \int_0^\infty x^{2p-1} y^{2q-1} \exp\left[-(x^2 + y^2)\right]\, dx\, dy$$

Let $x = r\cos\theta$ and $y = r\sin\theta$.

$$\Gamma(p)\Gamma(q) = 4\int_0^{\pi/2} \cos^{2p-1}\theta \sin^{2q-1}\theta\, d\theta \int_0^\infty r^{2p+2q-1} \exp(-r^2)\, dr$$

But, we know

$$\Gamma(p+q) = \int_0^\infty t^{p+q-1} \exp(-t)\, dt = 2\int_0^\infty r^{2p+2q-1} \exp(-r^2)\, dr$$

Therefore,

$$\frac{\Gamma(p)\Gamma(q)}{\Gamma(p+q)} = 2\int_0^{\pi/2} \cos^{2p-1}\theta \sin^{2q-1}\theta\, d\theta; \qquad p, q > 0$$

We know the value of the l.h.s. if we are given the values of p and q. Thus, we have obtained the value of the angular integrals, as follows:

$$\beta(p,q) \equiv 2\int_0^{\pi/2} \cos^{2p-1}\theta \sin^{2q-1}\theta \, d\theta$$
$$= \frac{\Gamma(p)\Gamma(q)}{\Gamma(p+q)}; \qquad p, q > 0 \tag{2.18a}$$

We call $\beta(p,q)$ a beta function and we see that it is related to the gamma function. Note the factor 2 multiplying the integral. A common alternative form of the beta function is obtained by the substitution in Equation (2.18a) of $t = \sin^2\theta$. We then have

$$\beta(p,q) = \int_0^1 t^{p-1}(1-t)^{q-1}\, dt; \qquad p, q > 0 \tag{2.18b}$$

Notice that the factor 2 is not present as a multiplier in Equation (2.18b).

EXERCISE 3.1 Verify that the beta function is symmetric in its arguments, i.e., $\beta(p,q) = \beta(q,p)$. ■

We also have from Equation (2.12) that

$$\iint_{\substack{x^2+y^2\le 1\\ x,\,y>0}} x^\alpha y^\beta \, dx\, dy = \frac{1}{\alpha+\beta+2}\int_0^{\pi/2} \cos^\alpha\theta \sin^\beta\theta \, d\theta; \qquad \alpha, \beta > -1$$

So, referring to Equation (2.18a), we have

$$2p - 1 = \alpha; \qquad 2q - 1 = \beta \quad \text{or} \quad p = \frac{\alpha+1}{2}; \qquad q = \frac{\beta+1}{2}$$

and then

$$\iint_{\substack{x^2+y^2\le 1\\ x,\,y>0}} x^\alpha y^\beta \, dx\, dy = \frac{1}{\alpha+\beta+2}\,\frac{1}{2}\,\frac{\Gamma\left(\frac{\alpha+1}{2}\right)\Gamma\left(\frac{\beta+1}{2}\right)}{\Gamma\left(\frac{\alpha+\beta+2}{2}\right)}$$

But we can simplify the denominator by writing

$$2(\alpha+\beta+2)\Gamma\left(\frac{\alpha+\beta+2}{2}\right) = 4\Gamma\left(\frac{\alpha+\beta+4}{2}\right)$$

So,

$$\iint_{\substack{x^2+y^2\le 1\\ x,\,y>0}} x^\alpha y^\beta \,dx\,dy = \frac{1}{4}\frac{\Gamma\left(\dfrac{\alpha+1}{2}\right)\Gamma\left(\dfrac{\beta+1}{2}\right)}{\Gamma\left(\dfrac{\alpha+\beta+4}{2}\right)}; \qquad \alpha,\beta > -1 \quad (2.19)$$

For $\alpha = \beta = 0$, we know

$$\iint_{\substack{x^2+y^2\le 1\\ x,\,y>0}} dx\,dy = \frac{\pi}{4}$$

so we have

$$\frac{\pi}{4} = \frac{1}{4}\Gamma\left(\frac{1}{2}\right)\Gamma\left(\frac{1}{2}\right)$$

or

$$\Gamma\left(\frac{1}{2}\right) = \sqrt{\pi} \qquad (2.20)$$

This result is used very frequently in evaluating integrals.

We also have

$$I(A,B) = \iint_{\substack{x+y\le 1\\ x,\,y>0}} x^A y^B \,dx\,dy = 4J(\alpha = 2A+1, \beta = 2B+1)$$

$$I(A,B) = \frac{\Gamma(A+1)\Gamma(B+1)}{\Gamma(A+B+3)} \qquad (2.21)$$

Check $A = B = 0$. We should get $I(0,0) = \frac{1}{2}$.

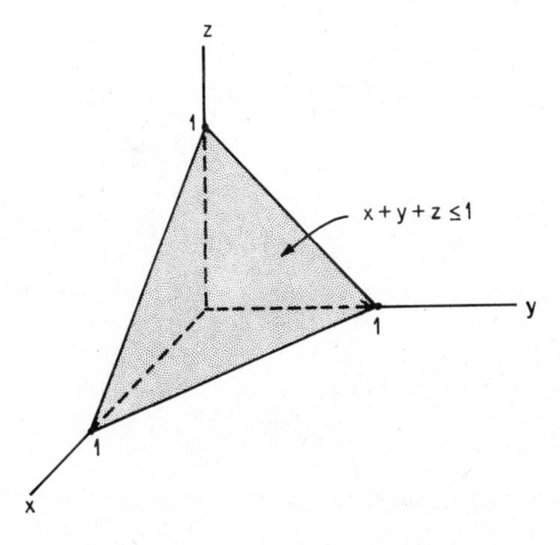

Figure 2.4 It is interesting that we can conclude from Equation (2.22) that the enclosed volume is equal to $\frac{1}{6}$. It takes some effort to be convinced that this is correct by simply studying the geometry of the figure.

The Dirichlet integral $I(A, B)$ is most often written in the form given by Equation (2.21).

We now extend these results to volume integrals over the first octant. The algebra is a little difficult, but the principle is the same as with the area integrals. You transform the given region to one that is confined to the first octant and bounded by the plane

$$x + y + z = 1$$

See Figure 2.4.

This gives a weighted integral. The value of the integral is given by Equation (2.22), which follows. We establish this result by transforming our region to an octant of a sphere and then evaluating the integral using spherical coordinates, performing the radial integration, and then expressing the angular integrations in terms of beta functions.

Check that your result properly reflects the symmetry of the original problem. Let $x = u^2$, $y = v^2$, and $z = w^2$. And let $u = r \sin\theta \cos\phi$; $v = r \sin\theta \sin\phi$, and $w = r\cos\theta$; $du\,dv\,dw = r^2 \sin\theta\, dr\, d\theta\, d\phi$. Then

$$I(A, B, C) = \iiint_{\substack{x+y+z\le 1\\ x,\,y,\,z>0}} x^A y^B z^C \, dx\, dy\, dz; \qquad A, B, C > -1$$

$$\begin{aligned} I(A, B, C) &= 8 \iiint_{\substack{u^2+v^2+w^2\le 1\\ u,\,v,\,w>0}} u^{2A+1} v^{2B+1} w^{2C+1}\, du\, dv\, dw \\ &= 2\int_0^{\pi/2} \sin^{2A+2B+3}\theta \cos^{2C+1}\theta\, d\theta \\ &\quad \cdot 2\int_0^{\pi/2} \cos^{2A+1}\phi \sin^{2B+1}\phi\, d\phi \cdot 2\int_0^1 r^{2A+2B+2C+5}\, dr \\ &= \frac{2}{2A+2B+2C+6} \cdot \frac{\Gamma(A+B+2)\Gamma(C+1)}{\Gamma(A+B+C+3)} \\ &\quad \cdot \frac{\Gamma(A+1)\Gamma(B+1)}{\Gamma(A+B+2)} \end{aligned}$$

or

$$I(A, B, C) = \frac{\Gamma(A+1)\Gamma(B+1)\Gamma(C+1)}{\Gamma(A+B+C+4)} \tag{2.22}$$

EXERCISE 3.2 Show that

$$J(\alpha,\beta,\gamma)=\iiint_{\substack{x^2+y^2+z^2\le 1\\ x,\,y,\,z>0}} x^{\alpha}y^{\beta}z^{\gamma}\,dx\,dy\,dz$$

$$=\left(\frac{1}{8}\right)\frac{\Gamma\left(\frac{\alpha+1}{2}\right)\Gamma\left(\frac{\beta+1}{2}\right)\Gamma\left(\frac{\gamma+1}{2}\right)}{\Gamma\left(\frac{\alpha+\beta+\gamma+5}{2}\right)} \quad \blacksquare \quad (2.23)$$

Notice that Equations (2.22) and (2.23) are indeed symmetric in the variables A, B, C and α, β, γ, respectively.

We sometimes encounter the double factorial notation in the literature and we introduce it at this time. We begin with $(2n)!$ and write it by separating the factors into even and odd terms, as follows:

$$\begin{aligned}(2n)! &= (2n)(2n-1)(2n-2)(2n-3)\cdots 3\cdot 2\cdot 1\\ &= [(2n)(2n-2)(2n-4)\cdots 4\cdot 2][(2n-1)(2n-3)\cdots 3\cdot 1]\end{aligned}$$

The even terms become $2^n n!$.

We then introduce the double-factorial notation by writing

$$(2n-1)!! \equiv (2n-1)(2n-3)\cdot\cdots\cdot 5\cdot 3\cdot 1$$

Remember that $(2n)!$ is defined for all integers greater than or equal to 0. So we have

$$(2n)! = 2^n n!(2n-1)!! \qquad (2.24)$$

Finally, note that the double factorial gives us a connection with $\Gamma(n+\frac{1}{2})$ since

$$\begin{aligned}(2n-1)!! &= (2n-1)(2n-3)(2n-5)\cdot\cdots\cdot 5\cdot 3\cdot 1\\ &= 2^n\left(n-\tfrac{1}{2}\right)\left(n-\tfrac{3}{2}\right)\cdots\tfrac{5}{2}\cdot\tfrac{3}{2}\cdot\tfrac{1}{2}\\ &= \frac{2^n\Gamma\left(n+\frac{1}{2}\right)}{\sqrt{\pi}}\end{aligned}$$

or

$$\Gamma\left(n+\tfrac{1}{2}\right)=\frac{(2n)!\sqrt{\pi}}{2^{2n}n!} \qquad (2.25)$$

2.4 APPLICATIONS AND EXERCISES

EXERCISE 4.1 Express the following integral in terms of gamma functions. Draw the integrand for $\lambda = 1$ and conclude that $I(1) \approx 1$.

$$I(\lambda) = \int_0^\infty \exp(-\lambda x^3)\, dx; \qquad \lambda > 0$$

Let $\lambda x^3 = y$.

$$\begin{aligned} I(\lambda) &= \tfrac{1}{3}\lambda^{-1/3} \int_0^\infty y^{-2/3} \exp(-y)\, dy \\ &= \lambda^{-1/3}\Gamma\left(\tfrac{4}{3}\right) \end{aligned}$$

■

EXERCISE 4.2 We want to find the moment of inertia about the z axis of a homogeneous sphere of radius R and constant density ρ.

$$J = \iiint_{\substack{x^2+y^2+z^2 \le R^2 \\ x,\, y,\, z > 0}} (x^2 + y^2)\rho\, dx\, dy\, dz = 2\iiint x^2 \rho\, dx\, dy\, dz$$

Let $x^2 = R^2u^2$, $y^2 = R^2v^2$, and $z^2 = R^2w^2$.

$$J = 2\rho R^5 \iiint_{\substack{u^2+v^2+w^2 \le 1 \\ u,\, v,\, w \ge 0}} u^2\, du\, dv\, dw = \frac{2\rho R^5}{8} \frac{\Gamma\left(\frac{3}{2}\right)\Gamma\left(\frac{1}{2}\right)\Gamma\left(\frac{1}{2}\right)}{\Gamma\left(\frac{7}{2}\right)}$$

$$J = \frac{\pi}{15}\rho R^5$$

The mass M of an octant of the material is

$$M = \left(\frac{1}{8}\right)\left(\frac{4}{3}\right)\pi\rho R^3$$

So,

$$J = \tfrac{2}{5}MR^2$$

■

EXERCISE 4.3 Consider a lamina described by the conditions

$$c(m, n) = x^n + y^m \le 1; \qquad x,\, y,\, \ge 0; \qquad n, m \ge 0$$

Find its area.

$$\text{Area} = \iint_{\substack{x^n + y^m \le 1 \\ x,\, y \ge 0}} dx\, dy$$

Let $u = x^n$ and $v = y^m$.

$$\text{Area} = \iint_{\substack{u+v\leq 1\\ u,v\geq 0}} \left(\frac{1}{n}\right)\left(\frac{1}{m}\right) u^{1/n-1} v^{1/m-1}\, du\, dv$$

Using Equation (2.21),

$$\text{Area} = \frac{\Gamma\left(1+\dfrac{1}{n}\right)\Gamma\left(1+\dfrac{1}{m}\right)}{\Gamma\left(1+\dfrac{1}{n}+\dfrac{1}{m}\right)}$$

In arriving at this result, we used the relationship

$$\frac{1}{n}\Gamma\left(\frac{1}{n}\right) = \Gamma\left(1+\frac{1}{n}\right) \quad \text{and} \quad \frac{1}{m}\Gamma\left(\frac{1}{m}\right) = \Gamma\left(1+\frac{1}{m}\right)$$

Draw $c(2,2)$, $c(4,4)$, $c(\frac{1}{2},\frac{1}{2})$ on a single plot. Estimate the area and compare with the results given by the formula.

Note that if n, m, $\to \infty$, the region goes to a square. We get this result by observing that

$$\Gamma\left(1+\frac{1}{n}\right) \to \Gamma(1) = 1 \quad \text{as} \quad n \to \infty$$

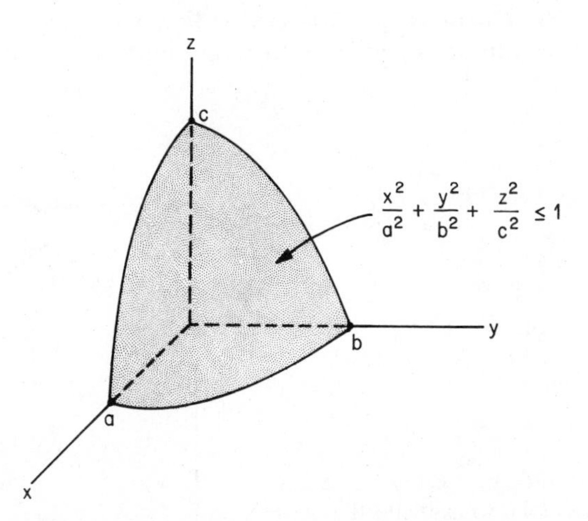

Figure 2.5 We use Dirichlet integrals to transform the volume of the octant of the ellipsoid to that of a sphere.

This also tells us how $\Gamma(z)$ behaves as $z \to 0$. We have $\Gamma(z+1) = z\Gamma(z)$ and $\Gamma(1) = 1$, so $\Gamma(z)$ must go to infinity like $1/z$ as $z \to 0$. ■

EXERCISE 4.4 Find the volume of that part of an ellipsoid contained in the first octant.

$$E = \iiint_{\substack{(x/a)^2+(y/b)^2+(z/c)^2 \le 1 \\ x,\, y,\, z \ge 0}} dx\, dy\, dz$$

See Figure 2.5. Let $u = x/a$, $v = y/b$, and $w = z/c$.

$$\begin{aligned} E &= abc \iiint_{\substack{u^2+v^2+w^2 \le 1 \\ u,\, v,\, w \ge 0}} du\, dv\, dw = abcJ(0,0,0) \\ &= \tfrac{1}{8}abc \frac{\Gamma(\frac{1}{2})\Gamma(\frac{1}{2})\Gamma(\frac{1}{2})}{\Gamma(\frac{5}{2})} \\ &= \frac{\frac{1}{8}abc\sqrt{\pi}^3}{\left[(\frac{3}{2})(\frac{1}{2})\sqrt{\pi}\right]} = abc\left(\frac{\pi}{6}\right) \end{aligned}$$

■

EXERCISE 4.5 Find the value of the integral

$$I = \int_0^\infty \frac{1}{\sqrt{t^3}}[1 - \exp(-2t)]\, dt$$

correct to three significant figures.

Answer: $2\sqrt{2\pi} = 5.01$. ■

EXERCISE 4.6 A thin metal plate is cut so that its shape is given by the region in the first quadrant bounded by the coordinate axes and the curve

$$x^4 + y^4 = 16$$

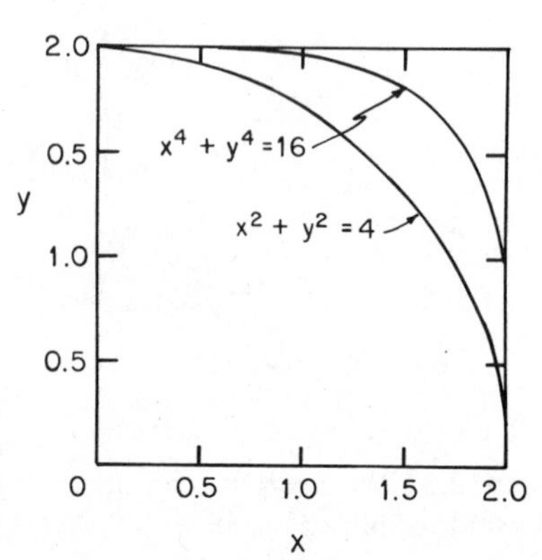

Figure 2.6 See if you can use the figure and the result of Exercise 4.6 to estimate the mass of a plate made of the same material that is confined to the region $x^2 + y^2 \le 4$; $x, y \ge 0$.

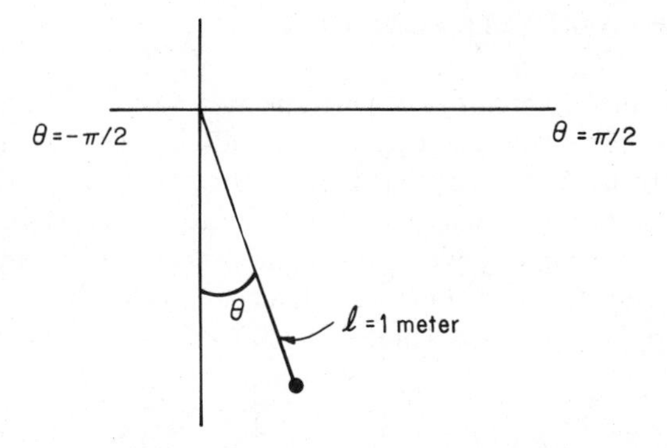

Figure 2.7 We are considering a pendulum that oscillates through an angle of π. Note that if it oscillated through a small angle, $\theta \ll \pi$, its period would be approximately 20% less.

It has a density ρ expressed in arbitrary units:

$$\rho = \tfrac{1}{20}\left(x^4 + y^4\right)^2$$

Find the mass of the plate and the coordinates of the center of mass correct to two significant figures. See Figure 2.6.

Answer: $M = 9.49$; $\quad \bar{x} = \bar{y} = 1.29$. ■

EXERCISE 4.7 A simple pendulum of 1-meter length oscillates through an angle of π radians. Find the period of oscillation of the pendulum correct to two significant figures. Take the acceleration of gravity, g, as 9.8 m/s^2. Does your answer make sense? See Figure 2.7. *Hint*: The equation of motion of the pendulum is given by

$$l\frac{d^2\theta}{dt^2} + g\sin\theta = 0$$

where θ is the angle of oscillation. We can multiply the equation by $d\theta/dt$ and obtain a perfect differential

$$\frac{1}{2}\frac{d}{dt}\left[\left(\frac{d\theta}{dt}\right)^2 - \frac{2g}{l}\cos\theta\right] = 0$$

We can then solve for period T.

Answer: $T = 2.4$ s. The small angle approximation gives $T = 2.0$ s. ■

2.5 THE RIEMANN ZETA FUNCTION

We encounter the Riemann zeta function in a variety of areas of pure and applied mathematics. We use it extensively in Chapter 4 when we discuss the Euler–MacLaurin sum expansion. It is important in evaluating a variety of sums and integrals associated with Fermi–Dirac and Einstein–Bose statistics. We concentrate, as we did with the gamma function, on developing a facility for using the zeta function and cite references to the literature for theorems and further applications. We define

$$\zeta(s) \equiv \sum_{n=1}^{\infty} n^{-s} = 1 + 2^{-s} + 3^{-s} + 4^{-s} + \cdots; \qquad s > 1 \tag{2.26}$$

We observe that the terms in the sum are alternately powers of even and odd integers, so we can write our definition as

$$\zeta(s) = \sum_{n=1}^{\infty} (2n)^{-s} + \sum_{n=1}^{\infty} (2n-1)^{-s}$$

This permits us to express other related sums in terms of the zeta function. Thus, consider

$$\zeta(s; -) \equiv \sum_{n=1}^{\infty} (-1)^{n+1} n^{-s} = 1 - 2^{-s} + 3^{-s} - 4^{-s} + \cdots$$

We separate the sum into odd and even terms:

$$\zeta(s; -) = \sum_{n=1}^{\infty} (2n-1)^{-s} - \sum_{n=1}^{\infty} (2n)^{-s}$$

Then, by adding and subtracting the sum of the even terms to the r.h.s. of our equation, we have

$$\zeta(s; -) \equiv \sum_{n=1}^{\infty} n^{-s} - 2\sum_{n=1}^{\infty} (2n)^{-s} = \zeta(s)(1 - 2^{-s+1}) \tag{2.27}$$

Also, we have

$$\zeta(s; \text{odd}) \equiv \sum_{n=1}^{\infty} (2n-1)^{-s} = \sum_{n=1}^{\infty} \left[n^{-s} - (2n)^{-s}\right] = \zeta(s)[1 - 2^{-s}] \tag{2.28}$$

The zeta function is tabulated and, for even integer values, has a simple closed form. For example,

$$\zeta(2) = \frac{\pi^2}{6}; \qquad \zeta(4) = \frac{\pi^4}{90}; \qquad \zeta(6) = \frac{\pi^6}{945}$$

[These results can be established by evaluating an appropriate Fourier series. See Chapter 4, Equation (4.18).]

In addition, we note that $\zeta(s)$ rapidly approaches 1 as s increases. We also observe the following inequalities:

$$\zeta(s+1) < \zeta(s) < \zeta(s-1); \qquad s > 2 \tag{2.29}$$

Then, we have

$$1 < \zeta(s) < 2 \qquad \text{for} \quad s > 2$$

$$\zeta(s) \approx 1 \qquad \text{for} \quad s > 4 \quad \text{since} \quad \zeta(4) \approx 1.1$$

EXERCISE 5.1 Consider the integral that arises in problems involving Einstein–Bose statistics.

$$I(s) = \int_0^\infty \frac{x^s}{\exp x - 1}\, dx; \qquad s > 0 \tag{2.30a}$$

First, multiply numerator and denominator by $\exp(-x)$. We know that

$$\frac{1}{1-x} = \sum_{n=0}^{\infty} x^n; \qquad |x| < 1$$

and we use this to express the denominator as a sum. We perform the integral after noting that it is a gamma function. We then have a simple result in terms of the gamma and zeta functions.

$$I(s) = \sum_{n=0}^{\infty} \int_0^\infty x^s \exp[-(n+1)x]\, dx$$

Let $y = (n+1)x$.

$$I(s) = \sum_{n=0}^{\infty} (n+1)^{-(s+1)} \int_0^\infty y^s \exp(-y)\, dy$$

$$I(s) = \Gamma(s+1)\zeta(s+1) \qquad \blacksquare \tag{2.30b}$$

EXERCISE 5.2 Consider the following integral that occurs in problems associated with Fermi–Dirac statistics:

$$J(s) = \int_0^\infty \frac{x^s}{\exp x + 1}\, dx; \qquad s > 0$$
$$= \sum_{n=0}^{\infty} (-1)^n \int_0^\infty x^s \exp[-(n+1)x]\, dx \tag{2.31a}$$

We then have, using Equation (2.27),

$$J(s) = \Gamma(s+1)\zeta(s+1)[1 - 2^{-s}] \qquad \blacksquare \tag{2.31b}$$

EXERCISE 5.3 Show that

$$K = \int_0^\infty \frac{1}{\exp x + 1}\, dx = \int_1^\infty \frac{1}{[t(1+t)]}\, dt = \ln 2 \tag{2.32}$$

[*Hint*: Let $t = \exp x$. This exercise can be used to study the behavior of $\zeta(1+s)$ as $s \to 0$. We have $K = \lim_{s \to 0} J(s)$ given by Equation (2.31b). Then, from Equation (2.31b), and writing 2^{-s} as $\exp(-s \ln 2)$,

$$J(s) \approx \zeta(s+1)[1 - \exp(-s \ln 2)]; \quad s \to 0$$
$$\approx \zeta(s+1) \cdot s \ln 2; \quad s \to 0$$

Combining this result with Equation (2.32), we obtain

$$s\zeta(s+1) \to 1 \qquad \text{as} \quad s \to 0$$

Hence, $\zeta(s+1)$ goes like $1/s$ for small s. ■

EXERCISE 5.4 Show that

$$M = \int_0^\infty \frac{1}{\exp x + a}\, dx = \frac{1}{a} \ln(1+a); \qquad a \geq 0 \tag{2.33}$$

This leads us to observe that

$$\frac{1}{a} \ln(1+a) \to 1 \qquad \text{when } a \to 0$$

[This result is also obtained from a Taylor-series expansion of $\ln(1+a)$ for small a.]

■

EXERCISE 5.5 Consider the sum $\zeta(N = 2M, s; -)$ related to the sum $\zeta(s; -)$ given by Equation (2.27):

$$\zeta(2M, s; -) = \sum_{n=1}^{2M} (-1)^{n+1} n^{-s} \tag{2.34a}$$

Note that the last term in the sum is $-(2M)^{-s}$. We can write our sum as

$$\zeta(2M, s; -) = \sum_{n=1}^{M} (2n-1)^{-s} - \sum_{n=1}^{M} (2n)^{-s}$$

We can, using the same procedures that we followed in obtaining Equation (2.27), write the sum $\zeta(2M, s; -)$ as

$$\zeta(2M, s; -) = \zeta(2M, s) - 2^{-s+1}\zeta(M, s) \tag{2.34b}$$

where we define $\zeta(M, s)$ as

$$\zeta(M, s) \equiv \sum_{n=1}^{M} n^{-s} \tag{2.34c}$$

Note that the limits on the sums are equal to M and $2M$. Clearly, if M equals infinity, this is of no consequence. But, if M is finite, there is the possibility of confusion. Let us consider a simple example to make certain that we understand this result. Consider the sum

$$S = \sum_{n=1}^{10} (-1)^{n+1} n^{-2}$$

We can evaluate this sum by just adding the terms using a hand calculator. Verify that Equation (2.34b) gives the correct result. Notice that it was important to get the limits correctly. It is also easy to obtain a similar formula when the last term in the sum is odd; $N = 2M + 1$. This just involves adding one more term to the sum that we already had. We return to this problem when we consider divergent sums in Section 4.3 of Chapter 4. ■

2.6 THE DIRAC DELTA FUNCTION

The Dirac delta function $\delta(x)$ is a curiosity in that it is not a function, and for quite some time, it was not regarded as a useful and powerful tool. We see clearly from its definition (given in what follows) that it has properties that are, to say the least, unusual. It serves two purposes for us. Firstly, it enables us to interchange orders of integration, thereby greatly simplifying some proofs. Secondly, it enables us to easily evaluate some integrals that would otherwise be difficult. The book by Lighthill provides an excellent discussion of the delta function as well as an extensive treatment of generalized function theory. (Our simplified treatment of the delta function is based on the one given in Lighthill's book.)

We list three properties that the delta function should possess. It is important to observe that any function that satisfies these properties is called a delta function. We give a few representations and show that each of them satisfies the definitions.

We say that $\delta(x)$ is a delta function if

$$1.\ \int_{-\infty}^{\infty} \delta(x)\, dx = 1 \tag{2.35a}$$

$$2.\ \delta(x) = 0; \qquad x \neq 0 \tag{2.35b}$$

$$3.\ \int_{-\varepsilon}^{\varepsilon} \delta(x) f(x)\, dx = f(0); \qquad \varepsilon > 0 \tag{2.35c}$$

Let us look at these three properties.

1. This states that the delta function is normalized to unity.
2. This states that it is not a function in the usual sense since the integral in property 1 should not depend upon the value of the function at a single point. So, we choose to think of the delta function as the limit of a sequence of functions such that it has the property that it is nonzero only at $x = 0$ in the limit. This will be clearer when we give examples of the delta function.
3. This is required to hold for any function $f(x)$ that has a bounded derivative. This condition enables us to obtain a simple bound for integrals. Consider $F'(x) = f(x)$, then

$$\int_a^b F'(x)\, dx = F(b) - F(a)$$

$$= (b - a) F'(\xi) \quad \text{for some } \xi; \qquad a \leq \xi \leq b$$

Let $a = 0$ and $b = x$; then we have

$$F(x) - F(0) = (x - 0) F'(\xi) = xF'(\xi)$$

and, finally, we choose to use this result in the form

$$|F(x) - F(0)| \leq |x|\,|F'(x) \max| \tag{2.36}$$

where $F'(x)$max is the maximum value of $F'(x)$ on $[a, b]$. (We will use this result in establishing that various sequences become delta functions.)

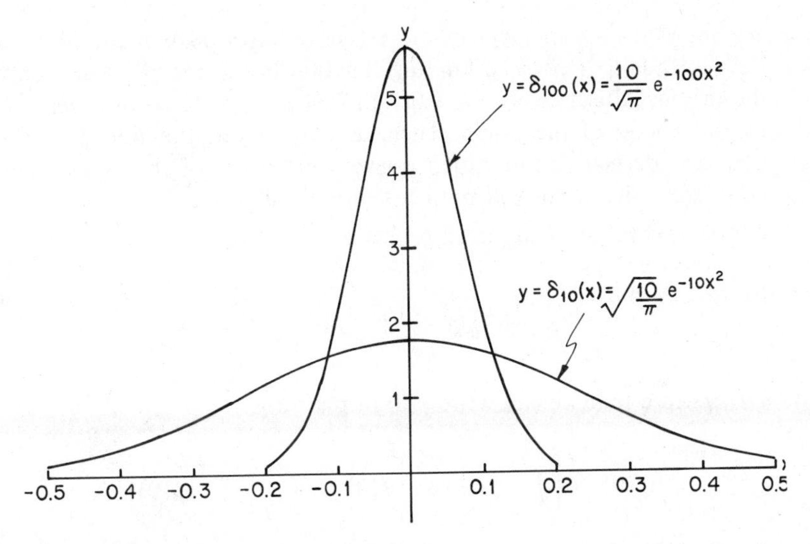

Figure 2.8 Observe that $\delta_{10}(x)$ is much broader than $\delta_{100}(x)$. Conclude that the area under each curve is approximately equal to 1.

We are now ready to give an example of a function that has the properties we require of a delta function. Consider the sequence of functions:

$$\delta_n(x) = \sqrt{\frac{n}{\pi}} \exp(-nx^2); \qquad n = 1, 2, 3, \ldots \tag{2.37}$$

See Figure 2.8.

Note that each member of the sequence is an ordinary function in that it satisfies the mean-value theorem. We identify the delta function with the limiting function:

$$\delta(x) = \operatorname*{limit}_{n \to \infty} \delta_n(x) \tag{2.38}$$

We show that this limiting function $\delta(x)$ satisfies our three requirements.

1. Normalization.

$$\int_{-\infty}^{\infty} \exp(-nx^2)\, dx = \sqrt{\frac{\pi}{n}}$$

So

$$\int_{-\infty}^{\infty} \delta_n(x)\, dx = 1; \qquad \text{every } n > 0$$

2. We see that each member of the sequence has a peak at the origin and rapidly falls to zero. The limiting function has a nonzero value only at the origin. Alternatively, we can think of it as nonzero in a very small neighborhood of the origin. Then, as long as our function $f(x)$ has a bounded derivative, this small nonzero extension of the delta function away from the origin will not cause any trouble.
3. Now let us examine the third property.

$$\int_{-\varepsilon}^{\varepsilon} \delta(x) f(x)\, dx = f(0); \qquad \varepsilon > 0$$

We observe that we can write $f(0)$ in the form

$$\begin{aligned} f(0) &= \left[\int_{-\infty}^{\infty} \delta(x)\, dx\right] f(0) = \int_{-\infty}^{\infty} [\delta(x) f(0)]\, dx \\ &= \int_{-\varepsilon}^{\varepsilon} \delta(x) f(0)\, dx \\ &= \operatorname*{limit}_{n\to\infty} \int_{-\varepsilon}^{\varepsilon} \sqrt{\frac{n}{\pi}} \exp(-nx^2) f(0)\, dx \end{aligned}$$

We use property 2 to choose our interval of integration to be from $-\varepsilon$ to $+\varepsilon$ rather than from $-\infty$ to $+\infty$. We then consider the integral $I(n)$ (defined in what follows) and show that it goes to zero in the limit n goes to infinity. Note that we are interchanging limits. This is typical of situations in which the delta function is used.

$$I(n) \equiv \int_{-\varepsilon}^{\varepsilon} \sqrt{\frac{n}{\pi}} \exp(-nx^2)[f(x) - f(0)]\, dx$$

We use Equation (2.36) to obtain:

$$\begin{aligned} I(n) &\leq \sqrt{\frac{n}{\pi}} |f'(x)\text{max}| \int_{-\varepsilon}^{\varepsilon} \exp(-nx^2)|x|\, dx \\ &= \sqrt{\frac{n}{\pi}} |f'(x)\text{max}| \frac{1 - \exp(-n\varepsilon^2)}{n} \end{aligned}$$

Thus, $I(n) \to 0$ like $1/\sqrt{n}$ as $n \to \infty$, for each $\varepsilon \to 0$ as long as $n\varepsilon^2 > 0$. To conclude, we see that the function satisfies the definition of a delta function.

At this point, we give two other examples of delta functions and an exercise; it will then become clear that it is easy to generate them.

Example 6.1 Show that

$$\delta_n(x) \equiv \frac{n}{\pi} \frac{1}{1+(nx)^2}; \qquad n \to \infty \tag{2.39a}$$

is a delta function. Assume that you know the integral

$$\int_{-\infty}^{\infty} \frac{dx}{1+x^2}\, dx = \pi \tag{2.39b}$$

If $x \neq 0$, $\delta_n(x) \to 1/\pi n x^2$ as $n \to \infty$. Thus, we need consider only

$$\begin{aligned} I(n) &= \frac{n}{\pi} \int_{-\varepsilon}^{\varepsilon} \frac{[f(x) - f(0)]}{1+(nx)^2}\, dx \\ &\leq \frac{n}{\pi} |f'(x)\max| \cdot 2 \int_0^{\varepsilon} \frac{x\, dx}{1+(nx)^2} \\ &= \frac{|f'(x)\max| \ln\left[1+(n\varepsilon)^2\right]}{n\pi} \end{aligned}$$

Thus, $I(n) \to 0$ if $(n\varepsilon)^2 > 0$; $n \to \infty$. ■

EXERCISE 6.1 Consider the sequence of functions:

$$\delta_n(x) \equiv A(n, p)\exp(-nx^{2p}) \tag{2.40}$$

Find the normalization constants $A(n, p)$ so that $\delta_n(x)$ is normalized to unity. Show that, for each integer p, each sequence becomes a delta function in the limit n goes to infinity. This demonstrates how you can generate a whole class of delta functions. Notice that as long as n is finite, the various delta functions are flat in the neighborhood of the origin, the more so as p increases. This has some relevance in problems in which n is indeed finite. (Show the flatness property by drawing the behavior of a few of these potential delta functions in the neighborhood of the origin.) ■

Example 6.2 We want to show that

$$\begin{aligned} \theta(x) &\equiv 1; \qquad x > 0 \\ \theta(x) &\equiv 0; \qquad x < 0 \end{aligned} \tag{2.41a}$$

is such that

$$\theta'(x) = \delta(x) \tag{2.41b}$$

We have to check that it has the desired properties as given by Equations (2.35a) to (2.35c).

1. $\int_{-\infty}^{\infty} \theta'(x)\, dx = \theta(x)\Big|_{-\infty}^{\infty} = 1$
2. $\theta'(x) = 0; \qquad x \neq 0$
3. $\int_{-\varepsilon}^{\varepsilon} \theta'(x) f(x)\, dx = \theta(x) f(x)\Big|_{-\varepsilon}^{\varepsilon} - \int_{-\varepsilon}^{\varepsilon} \theta(x) f'(x)\, dx = f(0)$ ■

We can use the θ function in problems in which we turn something on at, say, $t = 0$. Also, it can be used inside an integral to specify a restriction on the range of the variables. Thus, for example, in the Dirichlet integrals, it can be used as follows:

$$V(R) = \text{Volume}\,(R) = \iiint_{x^2+y^2+z^2 \leq R^2} dx\, dy\, dz$$

$$= \iiint_{\text{all space}} \theta\left[R^2 - \left(x^2 + y^2 + z^2\right)\right] dx\, dy\, dz$$

Then, we can find the surface area by differentiating the θ function to obtain a delta function:

$$\frac{dV(R)}{dR} = \text{Surface area} = \iiint 2R\delta\left[R^2 - \left(x^2 + y^2 + z^2\right)\right] dx\, dy\, dz$$

Delta functions are often used to solve inhomogeneous differential equations. The general method is called the method of Green's functions and it is too involved and difficult for us to explain in detail. (A good introduction is found, for example, in the book by Friedman.) We consider a second-order inhomogeneous differential equation of the form:

$$Dy = f(x) \tag{2.42}$$

where D is a differential operator. For example, imagine it to be $d^2/dx^2 + 1$.

We also have an integral operator I that is the inverse of operator D in the sense that

$$DI = ID = 1$$

This means that if you operate on some function with the differential operator and then with the integral operator, you recover the original function. There is a constant of integration left over that we will ignore. (It is important for actual problems, but unimportant for purposes of illustration.) Operator DI is the identity operator with no constants left over. With these comments, we can formally solve Equation (2.42) by operating on both sides with the integral operator I and writing our solution as

$$IDy(x) = y(x) = If(t) = \int G(x, t) f(t)\, dt \tag{2.43a}$$

The quantity $G(x, t)$ is called Green's function for the associated operator and boundary conditions. We have to find the equation it satisfies. This is done by operating on both sides with differential operator D:

$$Dy(x) = D\int G(x, t) f(t)\, dt = \int [DG(x, t)]\, f(t)\, dt \tag{2.43b}$$

Returning to Equation (2.42), we observe that Green's function is required to satisfy the equation

$$DG(x, t) = \delta(x - t)$$

since then we have

$$Dy = \int \delta(x - t) f(t)\, dt = f(x)$$

It is often difficult to find and use Green's function, but it is an important and valuable technique that uses the delta function as an essential element. For example, one of the difficulties in using Green's function is obtaining the correct boundary conditions.

Example 6.3 We want to establish the following four properties of the delta function that follow from defining Equations (2.35a) to (2.35c):

1. $\displaystyle\int_{-\infty}^{\infty} \delta(x) f(x)\, dx = \int_{-\infty}^{\infty} \delta(-x) f(x)\, dx$

$$\delta(x) = \delta(-x) \tag{2.44a}$$

2. $\displaystyle\int_{-\infty}^{\infty} \delta(ax) f(x)\, dx = \frac{1}{|a|} f(0)$

$$\delta(ax) = \frac{1}{|a|}\delta(x) \tag{2.44b}$$

3. $$\int_{-\infty}^{\infty} \delta(x - a) f(x)\, dx = f(a) \tag{2.44c}$$

4. $\displaystyle\int_{-\infty}^{\infty} \delta(x^2 - a^2) f(x)\, dx = \frac{f(a) + f(-a)}{2|a|}$

$$\delta(x^2 - a^2) = \frac{\delta(x - a) + \delta(x + a)}{2|a|} \qquad \blacksquare \tag{2.44d}$$

1. Consider the integrals

$$\int_{-\infty}^{\infty} \delta(x) f(x)\, dx = f(0) \qquad \text{and} \qquad \int_{-\infty}^{\infty} \delta(-x) f(x)\, dx$$

where $f(x)$ is any function with a bounded derivative. Introduce the change of variables $y = -x$ in the second integral and write

$$\int_{-\infty}^{\infty} \delta(-x) f(x)\, dx = \int_{-\infty}^{\infty} \delta(y) f(-y)\, dy = f(0)$$

Since $f(x)$ was an arbitrary function, we conclude that $\delta(x) = \delta(-x)$.

2. Assume $a > 0$. Introduce the change of variables $y = ax$ and obtain

$$\int_{-\infty}^{\infty} \delta(ax) f(x)\, dx = \frac{1}{a}\int_{-\infty}^{\infty} \delta(y) f(y/a)\, dy = \frac{f(0)}{a}$$

Assume $a < 0$. Introduce the change of variables $y = |a|x$ and use Equation (2.44a) to obtain the desired result.

3. Make the change of variables $y = x - a$ and obtain

$$\int_{-\infty}^{\infty} \delta(x-a) f(x)\, dx = \int_{-\infty}^{\infty} \delta(y) f(y+a)\, dy = f(a)$$

4. Write the quantity $x^2 - a^2$ as $(x-a)(x+a)$. We divide the interval of integration into several parts:

$$\int_{-\infty}^{\infty} \to \int_{-\infty}^{-a-\varepsilon} + \int_{-a-\varepsilon}^{-a+\varepsilon} + \int_{-a+\varepsilon}^{a-\varepsilon} + \int_{a-\varepsilon}^{a+\varepsilon} + \int_{a+\varepsilon}^{\infty}$$

Consider the term

$$\int_{-a-\varepsilon}^{-a+\varepsilon} \delta[(x-a)(x+a)]\, f(x)\, dx$$

Since the delta function yields a contribution only in the immediate neighborhood where its argument is equal to 0, the first factor can be taken as a constant in this neighborhood. Thus, we write

$$\int_{-a-\varepsilon}^{-a+\varepsilon} \delta[(-2a)(x+a)]\, f(x)\, dx$$

We then use Equations (2.44a) to (2.44c) to obtain

$$\int_{-a-\varepsilon}^{-a+\varepsilon} \delta[(-2a)(x+a)]\, f(x)\, dx = \frac{f(-a)}{2|a|}$$

The next term to consider is

$$\int_{a-\varepsilon}^{a+\varepsilon} \delta[(x-a)(x+a)]\, f(x)\, dx$$

We proceed as before and obtain

$$\int_{a-\varepsilon}^{a+\varepsilon} \delta[(x-a)(x+a)] f(x)\, dx = \int_{a-\varepsilon}^{a+\varepsilon} \delta[(x-a)(2a)] f(x)\, dx$$
$$= \frac{f(a)}{2|a|}$$

The other terms yield zero and thus we obtain our desired result.

EXERCISE 6.2 Use Equations (2.44a) to (2.44d) to show that if we write the surface area of a sphere of radius R as

$$\text{Surface area} = \iiint 2R\, \delta\left[R^2 - \left(x^2 + y^2 + z^2\right)\right] dx\, dy\, dz$$

we can obtain the value $4\pi R^2$ by performing the indicated integration. *Hint*: Use polar coordinates and note that since the radius is an intrinsically positive quantity, there is only one contribution to the integral. ■

BIBLIOGRAPHY

Gamma and Beta Functions

1. Arkfen, G., *Mathematical Methods for Physicists*, Third Edition, Academic Press, New York, 1985. This text discusses a broad range of topics in mathematical physics at the advanced undergraduate and beginning graduate student level. It is clearly written and has an extensive discussion of the special functions with many interesting applications developed in the exercises.
2. Lebedev, N. N., *Special Functions and their Applications*, Prentice-Hall, Englewood Cliffs, New Jersey, 1965. Reprinted by Dover, New York, 1972. This slim volume treats most of the special functions and discusses applications.

Delta Functions

3. Lighthill, M. J., *Introduction to Fourier Series and Generalized Functions*, Cambridge University Press, London, 1958. We find here an extensive discussion of generalized functions with an exceptionally clear treatment of the delta function.
4. Friedman, B., *Principles and Techniques of Pure and Applied Mathematics*, Wiley, New York, 1956. There is substantial discussion of the delta function and the concept of test functions and their role in finding Green's functions.

Riemann Zeta Function

5. Titchmarsh, E. C., *The Theory of the Riemann Zeta Function*, Oxford University Press, Oxford, 1951. This is the classic work on the subject. It is not easy to read, but the beginning sections are particularly accessible and interesting.

CHAPTER 3

ELEMENTS OF ASYMPTOTICS

3.1 INTRODUCTION

We will be studying problems for which we do not want to find the exact answer or the answer is just too difficult to obtain. Rather, we concentrate on finding an approximate solution or a qualitative characterization. This approach is used in the study of differential equations when we use the Frobenius, or power-series, method to generate the solution to an equation in the neighborhood of a regular or singular point. Most students are familiar with this technique. However, few courses discuss the development of approximate solutions that are valid for large values of the argument, i.e., the asymptotic region. Also, little attention is given to keeping track of the approximate solution over a long period of time and the related issue of the suppression of secular terms. We will be discussing these questions in Chapters 8 through 15.

In this chapter, we have two interrelated goals. First, we introduce the concept of approximating or describing a function by another function that is close to it. This concept arises naturally when we study the long-time behavior of a system or ask how a function behaves for large or small values of some parameter.

Our second goal is to cast this comparison in terms of the so-called asymptotic symbols. Thus, we learn the meaning of the " $\sim$ " symbol, "O or the big O" symbol, and "o or the little o" symbol. These symbols describe three important ways functions behave as a variable tends to some limiting value. In addition, they enable us to characterize the qualitative behavior of the solution of a problem. Thus, they allow us to gain a feel for what is happening without actually giving the precise answer to some question. In many circumstances, this is sufficient.

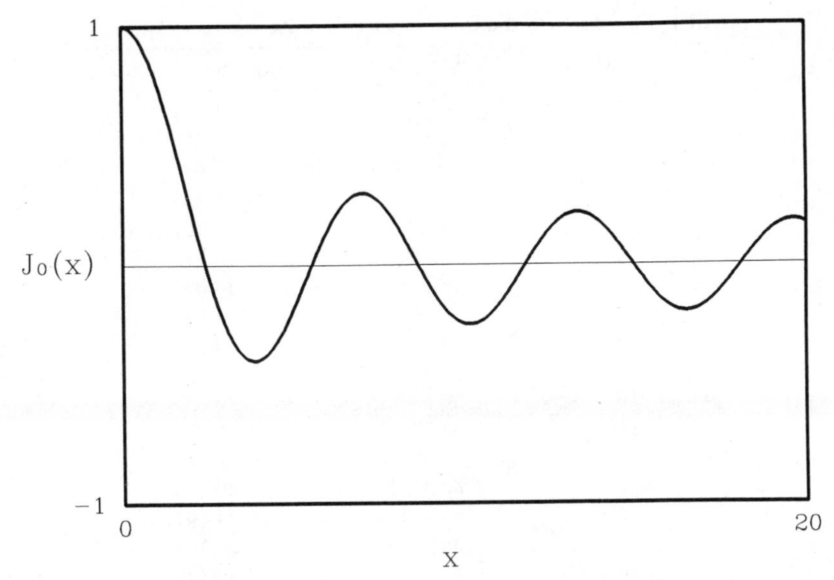

Figure 3.1 Study the plot of $J_0(x)$ vs. x and conclude that the zeros are almost equally spaced and that the envelope does indeed fall like $1/\sqrt{x}$.

As an example, let us say that we are interested in the behavior of the Bessel function $J_0(x)$. We can define it though its differential equation:

$$x^2 \frac{d^2y}{dx^2} + x \frac{dy}{dx} + x^2 y = 0 \tag{3.1}$$

It is also possible to find the power-series expansion that is useful for small values of x and the asymptotic expansion for large values of x.

$$J_0(x) = \sum_{k=0}^{\infty} \frac{(-1)^k (x/2)^{2k}}{k!k!} \tag{3.2a}$$

$$J_0(x) \quad \text{“looks like”} \quad \sqrt{\frac{2}{\pi x}} \cos\left(x - \frac{\pi}{4}\right) \qquad \text{as} \quad x \to \infty \tag{3.2b}$$

See Figure 3.1 and Table 3.1.

Let us draw our attention to the following properties:

1. $J_0(0) = 1$. Thus, we say that $J_0(x)$ looks like 1 as $x \to 0$.
2. $J_0(x)$ oscillates in an almost periodic manner so that its zeros are spaced at intervals of approximate length π.
3. The envelope of its amplitude decreases as $1/\sqrt{x}$. We will shortly learn that the asymptotic symbols provide us with a shorthand way to express

TABLE 3.1 Zeros of the Bessel Function of Zero Order, $J_0(x)$

Number of Zero s	Exact Zero x	Approximate Zero $x = (s - \frac{1}{4})\pi$
1	2.40483	2.356
2	5.52008	5.498
3	8.65373	8.639
4	11.79153	11.781
5	14.93092	14.923
6	18.07106	18.064
7	21.21164	21.206
8	24.35247	24.347
9	27.49348	27.489
10	30.63461	30.631
11	33.77582	33.772
12	36.91710	36.914
13	40.05843	40.055
14	43.19979	43.197
15	46.34119	46.338
16	49.48261	49.480
17	52.62405	52.622
18	55.76551	55.763
19	58.90698	58.905
20	62.04847	62.046

these qualities. But, since the Bessel function presents some conceptual and calculational difficulties, we turn our attention to some simple examples to gain experience with the basic ideas of asymptotics.

3.2 THE ASYMPTOTIC SYMBOLS

3.2.1 The ~ Symbol

We introduce the asymptotic symbol ~ through an example that involves the comparison of three functions.

Let us examine three curves given in Figures 3.2*a*, *b*, and *c*. We are interested both in the region close to the origin and the region of large values of x.

Let us describe each of the curves qualitatively. In Figure 3.2*a*, $f(x)$ looks like βx for small x and then grows to a limiting value β/α for large x. In Figure 3.2*b*, $g(x)$ looks like βx for small x. It reaches a maximum at $x = 1/\sqrt{\alpha}$ and then falls gradually until it looks like $\beta/\alpha x$ for large x.

In Figure 3.2*c*, $h(x)$ grows quadratically near the origin and then, while maintaining a positive slope, it approaches the line $y = (\beta/\alpha)(x - 1/\alpha)$ for large values of x.

Now let us go through the same description using the symbol ~ .

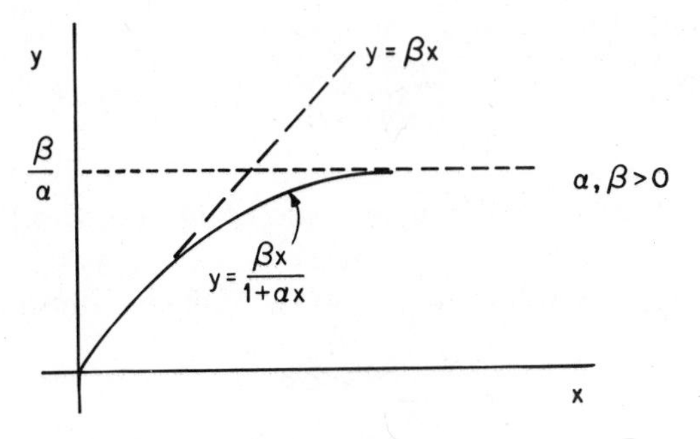

Figure 3.2*a*

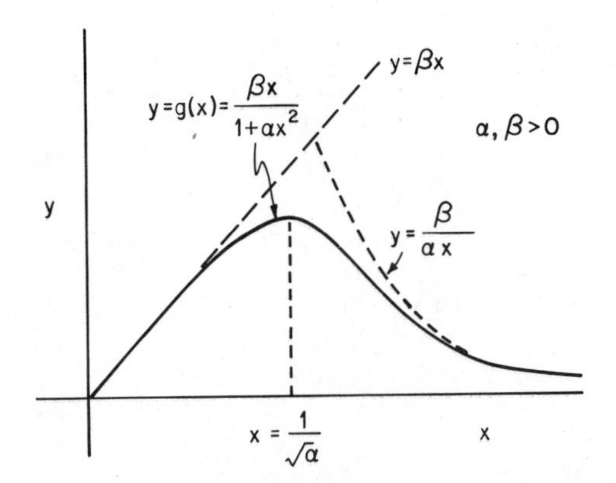

Figure 3.2*b*

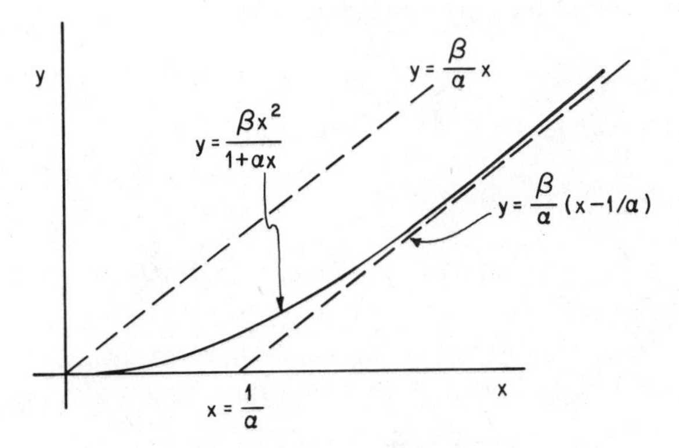

Figure 3.2*c*

Definition We say that

$$F(x) \sim G(x) \tag{3.3}$$

as x tends to some value, $x = a$, if the ratio of $F(x)$ to $G(x)$ tends to 1 as x goes to a.

Let us see how this works in the three examples illustrated in Figures 3.2a–c.

We have

$$f(x) = \frac{\beta x}{1 + \alpha x} \tag{3.4}$$

and we say that $f(x) \sim \beta x$ as $x \to 0$ because

$$\frac{f(x)}{\beta x} = \frac{1}{\beta x}\frac{\beta x}{1 + \alpha x} \to 1 \qquad \text{as} \quad x \to 0$$

Also, we have

$$f(x) \sim \frac{\beta}{\alpha} \qquad \text{as} \quad x \to \infty$$

since

$$f(x)\frac{\alpha}{\beta} = \frac{\alpha}{\beta}\frac{\beta x}{1 + \alpha x} \to 1 \qquad \text{as} \quad x \to \infty$$

We use the word "approaches" to indicate the $\sim$ symbol. It is called an asymptotic symbol because it is often used for x tending to infinity.

Now let us look at the curve in Figure 3.2b.

$$g(x) = \frac{\beta x}{1 + \alpha x^2} \tag{3.5}$$

We say that

$$g(x) \sim \beta x \qquad \text{as} \quad x \to 0$$

and

$$g(x) \sim \frac{\beta}{\alpha x} \qquad \text{as} \quad x \to \infty$$

Similarly, we have, in Figure 3.2c:

$$h(x) = \frac{\beta x^2}{1 + \alpha x} \tag{3.6}$$

Then

$$h(x) \sim \beta x^2 \quad \text{as} \quad x \to 0$$

$$h(x) \sim \frac{\beta x}{\alpha} \quad \text{as} \quad x \to \infty$$

Look at the graphs and see if the analysis given describes the behavior of the curves for small and large values of x. ■

Example 2.1 How does $\sin x$ behave as x tends to zero? We know the Taylor-series expansion begins like

$$\sin x = x - \frac{x^3}{3!} + \text{h.o.t.}$$

so we can write

$$\sin x \sim x \quad \text{as} \quad x \to 0$$

We could have equivalently written

$$\frac{\sin x}{x} \sim 1 \quad \text{as} \quad x \to 0 \tag{3.7}$$

This means that when one plots $y = \sin x$ and $y = x$, they are indistinguishable in the region close to the origin. See Figure 3.3. ■

EXERCISE 2.1 Show that $\cos x \sim 1$ as $x \to 0$. ■

EXERCISE 2.2 Show that $\tan x \sim x$ as $x \to 0$. ■

EXERCISE 2.3 Show that $\exp x \sim 1$ as $x \to 0$. ■

EXERCISE 2.4 Show that $\ln(1 + x) \sim x$ as $x \to 0$. ■

We can determine the limiting behavior of each of the functions in Example 2.1 and Exercises 2.1 to 2.4 by expanding in a Taylor series about $x = 0$ and keeping the first term. We can improve our correspondence by computing

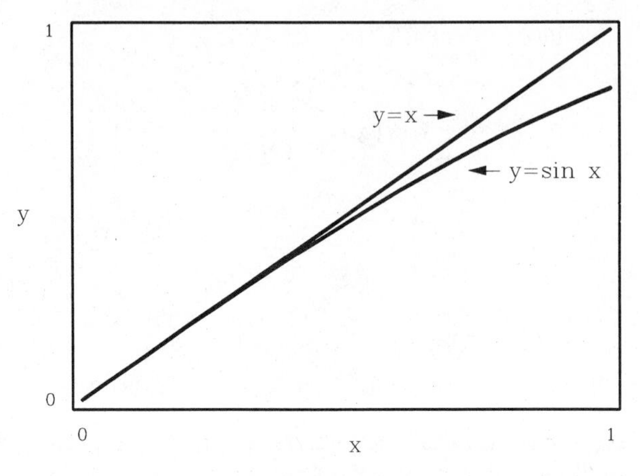

Figure 3.3

more terms in the Taylor series and using them in our relations. For instance,

$$\sin x \sim x - \frac{x^3}{3!} \qquad \text{as} \quad x \to 0$$

or

$$\frac{\sin x}{x - x^3/3!} \to 1 \qquad \text{as} \quad x \to 0$$

This is clear if we evaluate the first three terms in the Taylor-series expansion for sin x:

$$\sin x = x - \frac{x^3}{3!} + \frac{x^5}{5!} + \text{h.o.t.} \qquad \text{as} \quad x \to 0$$

or

$$\frac{\sin x}{x - x^3/3!} \sim 1 + \frac{x^5/5!}{x - x^3/3!} \to 1 \qquad \text{as} \quad x \to 0$$

Example 2.2 How does $f(x) = x + \sqrt{x}$ behave as x goes to 0? It is clear that $f(x) \sim \sqrt{x}$ since the square-root term dominates for small values of x. How does

$$f(x) = \sqrt{a + bx}\,; \qquad a, b > 0$$

behave as x goes to 0? If a and b have fixed values, we can write

$$f(x) = \sqrt{a}\sqrt{1 + \frac{bx}{a}}$$

Then it is clear that

$$f(x) \sim \sqrt{a}\left(1 + \frac{bx}{2a}\right) \qquad \text{as} \quad x \to 0$$

We will not encounter any trouble with this approach as long as the quantities a and b are fixed in value. However, let us say that we have a problem in which a small parameter is multiplied by or intertwined with the dependent variable, e.g., with time. Then, as we track the behavior of the solution for a long time, we encounter a class of concepts that are new to us, namely, secular terms and uniform and nonuniform expansions. We discuss this point at some length in Chapter 8, but conclude this section with a glimpse of the new class of difficulties to which we are referring. ■

Example 2.3 Consider

$$F(t) = \sqrt{1 + \varepsilon t} \tag{3.8}$$

If the product $|\varepsilon t| \ll 1$, we know that

$$F(t) \sim 1 + \frac{\varepsilon t}{2}$$

But if the product εt is not small, even though $|\varepsilon| \ll 1$, because t is so large, $F(t)$ does not look like $1 + \varepsilon t/2$ for small ε. Another way to say this is to point out that if we fix ε at any small value and wait a sufficiently long time, the quantity $|\varepsilon t|$ is no longer small. ■

Example 2.4 We will try and make this same point clearer through consideration of a more difficult example. Let

$$F(t) = \cos[(1 + \varepsilon)t]; \qquad 0 < \varepsilon \ll 1, \quad t = \text{time} \tag{3.9}$$

Function $F(t)$ is periodic. If $\varepsilon t \ll 1$, $F(t) \sim \cos t$ and it appears to be periodic with a period equal to 2π. But remember that we used the product $\varepsilon t \ll 1$, so that we are permitted only to follow the progress of $F(t)$ for a short time. If we observe $F(t)$ for a sufficiently long time, we will see that our approximation falls out of synchronization with the exact expression since the true period is

$$T = \frac{2\pi}{1 + \varepsilon} \tag{3.10}$$

See Figure 3.4.

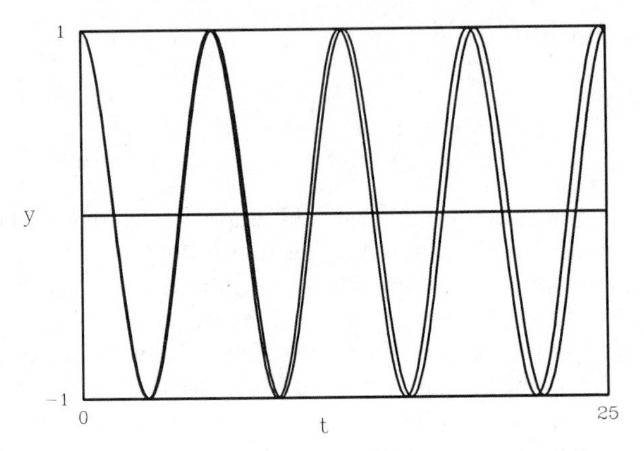

Figure 3.4 This is our first taste of time-limited approximations. Observe that no matter how small the parameter ϵ is, the two curves, $y = \cos t$ and $y = \cos[(61/60)t]$ eventually always separate in phase.

We will use the $\sim$ symbol to try and understand what is happening. We say that $F(t) \sim \cos t$ as t goes to 0 for small values of εt because

$$\frac{\cos[(1+\varepsilon)t]}{\cos t} = \frac{\cos t \cos \varepsilon t - \sin t \sin \varepsilon t}{\cos t}$$

$$= \cos \varepsilon t - \frac{\sin t}{\cos t} \sin \varepsilon t$$

or

$$F(t) \sim \cos t - (\varepsilon t) \sin t \qquad \text{as} \quad \varepsilon t \to 0 \tag{3.11}$$

We used the facts that

$$\cos \varepsilon t \sim 1 \quad \text{and} \quad \sin \varepsilon t \sim \varepsilon t \qquad \text{as} \quad \varepsilon t \to 0$$

Finally, we return to our original expression, given by Equation (3.9), and note that it is periodic, whereas the asymptotic approximation given by Equation (3.11) is not periodic. Rather, Equation (3.11) expresses the solution as a sum of a function that is periodic, with the original period ($\varepsilon = 0$) and a term that is the product of a linear function and a periodic function, $\varepsilon t \sin t$. We have no difficulties as long as we keep firmly in mind that we have imposed the restriction that εt remains small. But we often want to follow the behavior of the system for long times and in doing so lose sight of this restriction. We then find that this representation of the behavior of $F(t)$ leads us astray.

This is an important point to stress, so we will explain it again from a slightly different perspective. ■

Example 2.5 We are interested in disentangling the small parameter ε and the variable t. Let us say that we have a differential equation of the form

$$d^2x/dt^2 + f(\varepsilon)x = 0; \qquad x(0) = A, \quad \dot{x}(0) = 0$$
$$f(\varepsilon) > 0 \qquad \text{and} \qquad |\varepsilon| \ll 1 \tag{3.12}$$

The motion is periodic with period

$$T = \frac{2\pi}{\sqrt{f(\varepsilon)}}$$

and the solution is

$$x = A\cos\left(\sqrt{f(\varepsilon)}\,t\right) \tag{3.13}$$

Assume that you cannot solve this problem but are content with a power-series expansion in the small parameter ε. Develop $f(\varepsilon)$ in a Taylor series:

$$f(\varepsilon) = f(0) + \varepsilon f'(0) + \text{h.o.t.} \tag{3.14}$$

and then expand the solution in a power series in ε. However, if we are not careful, we will construct an expansion that couples parameter ε with time t. Let us consider an example when $f(\varepsilon) = 1 + \varepsilon$. We then have

$$\sqrt{f(\varepsilon)} = \sqrt{1+\varepsilon} = 1 + \frac{\varepsilon}{2} + \text{h.o.t.} \tag{3.15}$$

We know that the solution is periodic, but we don't know the exact period unless we can solve the equation exactly. A straightforward expansion yields a nonperiodic (or aperiodic) solution of the form

$$x(t) = A\cos t - \frac{\varepsilon A t}{2}\sin t + \text{h.o.t.} \tag{3.16}$$

This expansion is valid in that it faithfully characterizes the solution as long as the product εt is $\ll 1$. However, it masks the underlying periodicity of the solution. This defect is serious in that a plot of the velocity dx/dt versus the displacement x rather than being a closed trajectory is a spiral due to the terms of the form $\varepsilon t \sin t$. (This behavior is discussed at some length in Chapter 8.) We conclude that this form of the solution is unsatisfactory. It is possible to develop a solution that maintains the periodicity of the solution in each order of the perturbation expansion. In our example, this expansion would be

$$x(t) = A\cos\left[\left(1 + \frac{\varepsilon}{2} + \text{h.o.t.}\right)t\right] \tag{3.17a}$$

$$T \approx \frac{2\pi}{1 + \varepsilon/2} \tag{3.17b}$$

Thus, even though, after a sufficiently long time, the approximate period yields a mismatch relative to the true period, it maintains the periodic nature of the exact solution. In Chapter 9, we will learn how to construct this type of solution. ■

One of our goals is to develop expansions in a small parameter ε that are valid for very long times and we have to organize our expansions accordingly. We call such expansions uniform expansions and say that they are uniformly valid in time. This is a difficult concept to keep in mind and it comes up as a central part of our discussion of weakly nonlinear oscillatory systems. In these systems, we perturb a linear oscillator by a small nonlinear term and study the behavior of the system over a long interval of time. We find that this requires a new approach or viewpoint since the tried-and-true perturbation schemes of linear theories prove to be inadequate for describing this long-time behavior. (These systems are discussed in Chapters 8 to 15.)

3.2.2 The Big "*O*" Symbol and the Little "*o*" Symbol

Let us say that we do not know enough about function $f(x)$ to obtain the asymptotic behavior that we identify with the $\sim$ symbol. Or, perhaps, we might be satisfied with a less precise statement regarding the nature of $f(x)$. The symbol "O" can be used in these circumstances. We write

$$f(x) = O[g(x)] \quad \text{as} \quad x \to A \tag{3.18a}$$

where A is a constant, if

$$\frac{|f(x)|}{|g(x)|} \leq K, \tag{3.18b}$$

with K constant as $x \to A$. This says, for instance, if $g(x)$ goes to zero as x approaches A, $f(x)$ cannot go to zero more slowly. For example, if we write

$$f(x) = O(1) \quad \text{as} \quad x \to \infty$$

we know that $f(x)$ is bounded as $x \to \infty$.

We also have the "o" symbol, where we write

$$f(x) = o[g(x)] \qquad \text{as} \quad x \to A \tag{3.19a}$$

if

$$\frac{|f(x)|}{|g(x)|} \to 0 \qquad \text{as} \quad x \to A \tag{3.19b}$$

This says, for instance, that if $g(x)$ goes to zero as x approaches A, $f(x)$ must vanish more rapidly.

Thus, the symbols $\sim$, O, and o characterize different states of our knowledge of the asymptotic behavior of a function:

$\sim$ = we know the most
o = we know less
O = we know the least

Let us look at some examples of the "O" symbol and observe how it gives us only qualitative information about the example under consideration.

Example 2.6 We know that $\sin x$ is bounded for all x. So we can write

$$\sin x = O(1) \qquad \text{for all} \quad x$$

■

Look at the following items (i) through (iv) and observe that the "O" symbol conceals information.

(i) $f(x) = \beta x/(1 + \alpha x) = O(x)$; $x \to \infty$, since $|f(x)|/|x| \le K$ as $x \to \infty$.
(ii) $f(x) = \beta x/(1 + \alpha x) = O(1)$; $x \to \infty$, since $|f(x)|/1 \le K$ as $x \to \infty$.
(iii) $f(x) = \beta x/(1 + \alpha x) = O(1)$; $x \to 0$, since $|f(x)|/1 \le K$ as $x \to 0$.
(iv) $f(x) = \beta x/(1 + \alpha x) = O(x)$; $x \to 0$, since $|f(x)|/|x| \le K$ as $x \to 0$.

Example 2.7 We want to show that

$$I(n) = \int_0^{\alpha} \sin nx \, dx = O\left(\frac{1}{n}\right); \qquad n \to \infty$$

with the limit of integration α fixed. This can be seen by making the change of variables $y = nx$ and noting that the integral of $\sin y$ is $O(1)$. But it is also interesting to observe that the integral can be equal at most to the area of one positive cycle that is bounded by the spacing of successive zeros of $\sin nx$. If we approximate this region by a triangle of height 1 and base equal to π/n, we conclude that $I(n) = O(1/n)$. Similarly, if the integrand were of the form $\sin n^2 x$, we would see that the associated integral was $O(1/n^2)$. ■

In Exercises 2.5 to 2.15, show that the asymptotic behavior is characterized correctly.

EXERCISE 2.5

$$x \ln x = O(1) \qquad \text{as} \quad x \to 0$$ ■

EXERCISE 2.6

$$\cosh x = O(1) \qquad \text{as} \quad x \to 0$$ ■

EXERCISE 2.7

$$\cosh x = O(\exp x) \qquad \text{as} \quad x \to \infty$$ ■

EXERCISE 2.8

$$\sum_0^N n^s = O(N^{s+1}) \qquad \text{as} \quad N \to \infty; \qquad \text{for} \quad s > 0$$ ■

EXERCISE 2.9

$$\Phi(x) = \int_0^x \exp(-t^2)\, dt = O(1) \qquad \text{as} \quad x \to \infty$$ ■

EXERCISE 2.10

$$\sqrt{1 + x} = 1 + O(x) \qquad \text{as} \quad x \to 0$$ ■

EXERCISE 2.11 Look at the integral

$$F(t) = \int_0^t \frac{x \exp(-x^2)}{\ln(2 + x)}\, dx; \qquad t \to \infty$$

First show that $F(t) < 1/(2 \ln 2)$. Then conclude that $F(t) = O(1)$ as $t \to \infty$. ■

EXERCISE 2.12

$$G(t) = \int_0^t \ln(1 + x) \exp(-x^2)\, dx = O(1) \qquad \text{as} \quad t \to \infty$$ ■

EXERCISE 2.13

$$\sin x = x + O(x^2) \qquad \text{as} \quad x \to 0$$ ■

EXERCISE 2.14

$$\sin x = x + O(x^3) \qquad \text{as} \quad x \to 0 \qquad \blacksquare$$

EXERCISE 2.15

$$\ln x = O(x^\varepsilon) \qquad \text{as} \quad x \to \infty; \qquad \varepsilon > 0 \qquad \blacksquare$$

We provided a lot of exercises using the "O" symbol because it is easy to use and occurs more frequently than the "o" symbol.

Now practice using the "o" symbol. Show that each of the correspondences given is correct.

EXERCISE 2.16

$$\frac{\sin x}{x} = o(1) \qquad \text{as} \quad x \to \infty \qquad \blacksquare$$

EXERCISE 2.17

$$\frac{\ln x}{x} = o(1) \qquad \text{as} \quad x \to \infty \qquad \blacksquare$$

EXERCISE 2.18

$$A(x) = \exp x^2 \int_x^\infty \exp(-t^2)\, dt = o(1) \qquad \text{as} \quad x \to \infty$$

Observe that

$$A(x) = \exp x^2 \int_x^\infty \frac{[\exp(-t^2)](2t)}{2t}\, dt$$

$$< \frac{\exp x^2}{2x} \int_x^\infty [\exp(-t^2)](2t)\, dt$$

where we have taken out a factor $1/2x$ that is larger than $1/2t$ inside the interval of integration. We then have

$$A(x) < \frac{\exp x^2}{2x} \exp(-x^2) = \frac{1}{2x} \to 0 \qquad \text{as} \quad x \to \infty \qquad \blacksquare$$

Example 2.8 Often it is useful to use l'Hospital's rule to establish the order of the error term. For example, consider

$$I(x) = \exp(-x) \int_a^x (\exp t) t^{-5}\, dt; \qquad x \to \infty, \quad a > 0$$

We can establish that $I(x) = O(x^{-5})$ by using l'Hospital's rule to evaluate

$$\int_a^x \frac{(\exp t)t^{-5}\,dt}{x^{-5}\exp x}$$

as x tends to infinity. Differentiating, we have

$$\frac{(\exp x)x^{-5}}{(x^{-5} - 5x^{-6})\exp x} = \frac{1}{1 - 5x^{-1}} = O(1)$$

as $x \to \infty$. ■

It takes practice to get used to the symbols and to realize that a lot of information is concealed by the o symbol. We end this section with a word of caution. (This discussion is based on an example in the text by Wilf.) Let us say that we have established that

$$f(x) \sim g(x)$$

as x goes to some value, say x goes to infinity. This *does not* imply that

$$\exp[f(x)] \sim \exp[g(x)]$$

In order for this correspondence to be correct, we need the additional statement that

$$f(x) = g(x) + o(1) \qquad \text{as} \quad x \to \infty$$

In order to see that this could cause some difficulties, note that

$$x^2 + 1 \sim x^2 \qquad \text{as} \quad x \to \infty$$

but $\exp(x^2 + 1)$ is not asymptotically equivalent to $\exp x^2$ as $x \to \infty$.

What we need is that x^2 + "the other piece" is such that the exponential of the other piece goes like 1 as x goes to infinity. So, we see that if

$$f(x) = g(x) + o(1) \tag{3.20a}$$

then

$$\exp[f(x)] = \exp[g(x) + o(1)] = \exp[g(x)] \cdot [1 + o(1)] \tag{3.20b}$$

To summarize: two functions are asymptotically equivalent if their ratio approaches 1 as x tends to its limiting value. But, there may be lower-order terms that this equivalence neglects. Because the exponential function of a sum is the product of the exponentials, we require that the exponential of the suppressed terms look like 1 in the limit in order to maintain the equivalence. This is a difficult but important concept.

Example 2.9 We know that

$$x^2 + x \sim x^2 \qquad \text{as} \quad x \to \infty$$

Yet, $\exp(x^2 + x)$ is not asymptotically equivalent to $\exp x^2$ as $x \to \infty$. ■

Example 2.10 Let us assume that we have already established the asymptotic behavior of $\ln n!$ for large values of n, i.e., Stirling's approximation:

$$\ln n! = \tfrac{1}{2}\ln 2\pi n + n(\ln n) - n + o(1) \qquad \text{as} \quad n \to \infty \tag{3.21a}$$

However, often our primary interest is in the behavior of $n!$ rather than in $\ln n!$. We have just learned that we can exponentiate this result to obtain

$$n! = \sqrt{2\pi n}\, n^n \exp(-n)[1 + o(1)] \qquad \text{as} \quad n \to \infty \tag{3.21b}$$

Furthermore, we will learn, during our study of the Euler–MacLaurin sum expansion in Chapter 4, that it is not difficult to obtain higher-order corrections to $\ln n!$. We can exponentiate these to obtain higher-order terms in the asymptotic expansion of $n!$, whereas other more direct computations of these terms become extremely tedious. ■

3.3 THE ERROR FUNCTION

The error function and its complement occur in a wide variety of problems in science, for example, in the study of probability theory and in the study of the one-dimensional heat-diffusion equation. The error function is defined as

$$\operatorname{Erf}(x) \equiv \frac{2}{\sqrt{\pi}} \int_0^x \exp(-t^2)\, dt \tag{3.22}$$

and we also have its complement:

$$\operatorname{Erfc}(x) \equiv \frac{2}{\sqrt{\pi}} \int_x^\infty \exp(-t^2)\, dt \tag{3.23}$$

so that

$$\operatorname{Erf}(x) + \operatorname{Erfc}(x) = 1 \tag{3.24a}$$

and

$$\operatorname{Erf}(\infty) = \operatorname{Erfc}(0) = 1 \tag{3.24b}$$

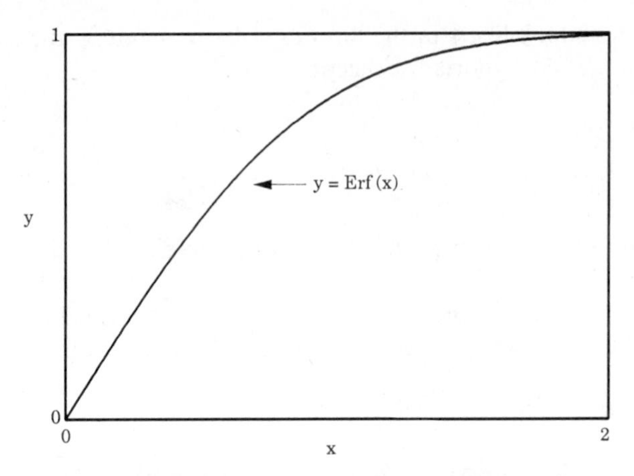

Figure 3.5 Observe how rapidly Erf(x) approaches 1. This is why we generally only find tables of Erf(x) for $x \leq 2$.

TABLE 3.2 Function Erf(x)

x	Erf(x)	x	Erf(x)
0.00	0.00	1.10	0.8802
0.10	0.1125	1.20	0.9103
0.20	0.2227	1.30	0.9340
0.30	0.3286	1.40	0.9523
0.40	0.4284	1.50	0.9661
0.50	0.5205	1.60	0.9763
0.60	0.6039	1.70	0.9838
0.70	0.6778	1.80	0.9891
0.80	0.7421	1.90	0.9928
0.90	0.7969	2.00	0.9953
1.00	0.8427		

The factor $2/\sqrt{\pi}$ is introduced in order to normalize these functions to 1. See Figure 3.5 and Table 3.2.

EXERCISE 3.1 This is an exercise in differentiation. Consider the one-dimensional diffusion equation or the equation of linear heat flow:

$$\frac{\partial^2 T(x,t)}{\partial x^2} = \frac{1}{\kappa}\frac{\partial T(x,t)}{\partial t} \tag{3.25}$$

where $T(x,t)$ is the temperature in degrees Celsius, t is the time, and κ is the diffusivity. (The diffusivity of silver is 1.7 and that of water is 0.0014, in c.g.s. units.) Verify that a solution of the equation is given by

$$T(x,t) = A\,\mathrm{Erf}\left(\frac{x}{2\sqrt{\kappa t}}\right) \tag{3.26}$$

where

$$\operatorname{Erf}\left(\frac{x}{2\sqrt{\kappa t}}\right) = \frac{2}{\sqrt{\pi}} \int_0^{x/2\sqrt{\kappa t}} \exp(-u^2)\, du \tag{3.27}$$

We can alternatively write

$$\operatorname{Erf}\left(\frac{x}{2\sqrt{\kappa t}}\right) = \frac{1}{\sqrt{\pi \kappa t}} \int_0^x \exp\left(\frac{-s^2}{4\kappa t}\right) ds \tag{3.28}$$

by the transformation of variables: $u = s/2\sqrt{\kappa t}$.

Verify that this form of the solution also satisfies Equation (3.25). ■

We want to develop approximations for Erf(x) and Erfc(x) for both large and small values of x. We begin with small values of x.

The Erf function is usually tabulated for values of its argument less than 2 since it is very close to 1 when $x \geq 2$. See Figure 3.5 and Table 3.2.

If we want to know the value of Erf(x) for small values of x, we just expand the exponential under the integral sign and integrate term by term. [Refer to Equation (3.22)]. If x is bounded, the series can be seen to converge uniformly using the ratio test. So we have

$$\begin{aligned} \operatorname{Erf}(x) &= \frac{2}{\sqrt{\pi}} \int_0^x \left[\sum_0^\infty \frac{(-1)^n t^{2n}}{n!} \right] dt \\ &= \frac{2}{\sqrt{\pi}} \sum_0^\infty \frac{(-1)^n x^{2n+1}}{(2n+1)(n!)} \end{aligned} \tag{3.29}$$

Since the series alternates in sign, the error is less in absolute value than the first neglected term. Thus, if we write

$$\operatorname{Erf}(x) = \frac{2}{\sqrt{\pi}} \left[x - \frac{x^3}{3} + \frac{x^5}{10} + O(x^7) \right]$$

the error is less than the next term,

$$\left| \frac{(-)(2/\sqrt{\pi})x^7}{42} \right|$$

We can make the error as small as we wish by taking enough terms in the series. For example, if we wanted to know Erf(any $x < 1$), with an error less than 10^{-4}, we are safe choosing our first neglected term to be $n = 7$, since $1/(15 \cdot 7!) < (1.4) \cdot (10)^{-5}$.

So, the first seven terms in the series are more than adequate to compute the error function to the required accuracy. But if x is not sufficiently small, you will have to take very many terms in order to get a useful result. Thus, we develop a method that is suitable for large values of x.

When we are interested in large values of x, we consider the complementary error function Erfc(x). We begin by noting that it is unreasonable to expand $\exp(-x^2)$ for large x, so we take a different approach. We integrate by parts, obtaining large terms in the denominator.

Our procedure is to multiply both numerator and denominator of the r.h.s. of Equation (3.23) by a factor $2t$ and integrate by parts, as follows:

$$\operatorname{Erfc}(x) = \frac{2}{\sqrt{\pi}} \int_x^\infty \frac{[\exp(-t^2)]2t}{2t}\, dt \tag{3.30}$$

We can then write:

$$\operatorname{Erfc}(x) = \frac{2}{\sqrt{\pi}} \left(\frac{\exp(-x^2)}{2x} - \int_x^\infty \frac{\exp(-t^2)}{2t^2}\, dt \right) \tag{3.31}$$

We can continue to integrate by parts, and each time we gain a factor of $1/x^2$. One factor $1/x$ comes from the creation of the perfect differential $2t[\exp(-t^2)]\, dt$ and its subsequent integration and the other from the differentiation of $1/t^n$. Let us write a few terms, removing the common factors:

$$\begin{aligned} \operatorname{Erfc}(x) = {} & \left[\frac{2}{\sqrt{\pi}} \exp(-x^2)\right] \left(\frac{1}{2x} - \frac{1}{2^2 x^3} + 1 \cdot 3 \frac{1}{(2^3 x^5)} \right) \\ & - \frac{2}{\sqrt{\pi}} \int_x^\infty [\exp(-t^2)] 5!! \frac{1}{2^3 t^6}\, dt \end{aligned} \tag{3.32}$$

(Remember that $5!! = 5 \cdot 3 \cdot 1$.)

This expression is exact. We can estimate the error that we make if we neglect the integral, since

$$\begin{aligned} & \frac{2}{\sqrt{\pi}} \int_x^\infty \frac{[\exp(-t^2)]5!!}{2^3 t^6}\, dt \\ & \quad = \frac{2}{\sqrt{\pi}} \int_x^\infty \frac{[\exp(-t^2)](2t)}{2^4 t^7} 5!!\, dt \\ & \quad < \frac{(2/\sqrt{\pi})(5!!)}{2^4 x^7} \exp(-x^2) \end{aligned} \tag{3.33}$$

which is equal to the next term in the series expansion. Thus, we see that it is possible to bound the error as less in absolute value than the first neglected term. So, if you are given a value of x, you can tell how small the error will be. For example, if $x = 4$:

$$\begin{aligned} & \frac{2}{\sqrt{\pi}} \int_4^\infty \frac{[\exp(-t^2)]5!!}{2^3 t^6}\, dt \\ & \quad < \frac{(2/\sqrt{\pi})[\exp(-16)](15)}{2^4 4^7} \end{aligned} \tag{3.34}$$

We pause to make two comments.

(i) This series expansion diverges. The terms, although initially decreasing in value, eventually increase. Thus, if you bound the error by the integral term, you obtain a poorer approximation for Erfc(x) by taking more terms in the series. This can be seen by applying the ratio test and observing that the ratio of successive terms does not remain less than unity as n tends to infinity. Thus, we usually stop after a few terms in the series. (We illustrate this in Example 3.1.) The series is said to be semiconvergent or asymptotic. You can approximate the function and bound the error, but there is a least term in the series such that if you pass it, the terms grow and your approximation becomes less valid. However, we will see that our asymptotic series is superb as a calculational device to obtain the value of the error function for large values of its argument. We discuss asymptotic series in Section 3.4.

(ii) You cannot decrease the error below a fixed value. This is in contrast with a convergent series. Let us explain this by an extreme example chosen for illustrative purposes.

Example 3.1 Consider

$$\text{Erfc}(x) = \frac{2}{\sqrt{\pi}} \int_x^\infty [\exp(-t^2)]\, dt$$

given by Equation (3.23). Write the two expansions:

$$A_1(x) = \frac{2}{\sqrt{\pi}} \exp(-x^2)\left(\frac{1}{2x} - \frac{1}{2^2x^3}\right) + \frac{2}{\sqrt{\pi}} \int_x^\infty \frac{[\exp(-t^2)]3!!}{2^2t^4}\, dt \tag{3.35a}$$

and

$$A_2(x) = \frac{2}{\sqrt{\pi}} \exp(-x^2)\left(\frac{1}{2x} - \frac{1}{2^2x^3} + \frac{3!!}{2^3x^5}\right) - \frac{2}{\sqrt{\pi}} \int_x^\infty \frac{[\exp(-t^2)]5!!}{2^3t^6}\, dt \tag{3.35b}$$

They are equivalent in that they are both exact for all values of x, namely,

$$\text{Erfc}(x) = A_1(x) = A_2(x) \tag{3.36}$$

These expressions are useful for the purposes of approximating Erfc(x) for large x, but they are still correct for small x. They are both constructed so as to give an integral that is easily bounded. One might think that an improved approximation is always obtained by integrating by parts more times and then bounding the error term by an estimate of the integral. Thus, for example, one expects that the computed terms in $A_2(x)$ give a better approximation to Erfc(x) than the corresponding terms in $A_1(x)$. However, as we show, in this example, this is sometimes not the case. Namely, in asymptotic analysis, you do not always obtain a better approximation by computing more terms in the series. Let us see how this is done. We pick a value of x where the integral in $A_1(x)$ is small; and, thus, if we neglect it, we expect to have a good approximation for the Erfc function. We then show that the corresponding truncation procedure for $A_2(x)$ gives a poorer approximation. Thus, you lose accuracy by computing more terms in the series. Pick $x = 1$. (This is an extreme case since x is not large.) Then we have

$$A_1(1) = \frac{2}{\sqrt{\pi}}[\exp(-1)]\left(\frac{1}{1 \cdot 2} - \frac{1}{2^2 \cdot 1^3}\right) + \text{error}(1) \equiv \tilde{A}_1(1) + \text{error}(1) \tag{3.37a}$$

and

$$A_2(1) = \frac{2}{\sqrt{\pi}}[\exp(-1)]\left(\frac{1}{1 \cdot 2} - \frac{1}{2^2 \cdot 1^3} + \frac{3!!}{2^3 1^5}\right) + \text{error}(2) \equiv \tilde{A}_2(1) + \text{error}(2) \tag{3.37b}$$

Then we have

$$\tilde{A}_1(1) = 0.10; \qquad \tilde{A}_2(1) = 0.26$$

The correct value is

$$\text{Erfc}(1) = 1 - 0.84 = 0.16$$

We should note that we would do poorer still if we continued by taking more terms in the expansion since the next two are $-15/16$ and $105/32$ multiplied by the common factor $(2/\sqrt{\pi})\exp(-1)$. Thus, we see that the presumed correction terms begin to overwhelm the leading terms.

So, in Example 3.1, we do not improve our approximation by taking a second term in the expansion. Also, notice that we do not get a good approximation to Erfc(1). This brings out a general feature of asymptotic expansions. There is a least term and you have to use another method, i.e., not an asymptotic expansion, if you want to get a better approximation in this region of variable x. ■

TABLE 3.3 Terms in the Asymptotic Expansion of $(\sqrt{\pi}/2)(\exp x^2)\,\mathrm{Erfc}(x)$; $x = 2$[a]

Terms	Term ($x = 2$)	$\left\lvert \dfrac{n\text{th term}}{(n-1)\text{st term}} \right\rvert$	Sum
$\dfrac{1}{2x}$	2.50×10^{-1}	...	0.250
$-\dfrac{1}{2^2x^3}$	-3.13×10^{-2}	0.13	0.219
$+\dfrac{3!!}{2^3x^5}$	1.17×10^{-2}	0.37	0.230
$-\dfrac{5!!}{2^4x^7}$	-7.32×10^{-3}	0.63	0.223
$+\dfrac{7!!}{2^5x^9}$	6.41×10^{-3}	0.88	0.230
$-\dfrac{9!!}{2^6x^{11}}$	-7.21×10^{-3}	1.1	0.222
$+\dfrac{11!!}{2^7x^{13}}$	9.91×10^{-3}	1.4	0.232

[a] The correct value of $(\sqrt{\pi}/2)(\exp 4)\,\mathrm{Erfc}(2) \approx 0.2263$.

Let us repeat our analysis of the Erfc function for Erfc(2). We integrate by parts a few times and obtain:

$$\frac{\sqrt{\pi}}{2}(\exp x^2)\,\mathrm{Erfc}(x) = \left(\frac{1}{2x} - \frac{1}{2^2x^3} + \frac{3!!}{2^3x^5} - \frac{5!!}{2^4x^7} + \frac{7!!}{2^5x^9} - \frac{9!!}{2^6x^{11}} + \frac{11!!}{2^7x^{13}}\right) - \exp x^2 \int_x^\infty \frac{[\exp(-t^2)]\,13!!}{2^7t^{14}}\,dt \tag{3.38}$$

We evaluate Equation (3.38) at $x = 2$. Note that we have taken the factor $(\sqrt{\pi}/2)(\exp x^2)$ to the l.h.s. in order to have a clearer view of the series. We see in Table 3.3 that if we integrate by parts seven times, the last term obtained is larger, in absolute value, than the preceding term (viz. $11/8 > 1$).

This pattern continues and is clear if you construct the ratio of the absolute value of the $(n+1)$th term to the nth term:

$$\frac{(2n+1)!!}{2^{n+2}x^{2n+3}} \cdot \frac{2^{n+1}x^{2n+1}}{(2n-1)!!} = \frac{2n+1}{2x^2} \tag{3.39}$$

Thus, when $2n + 1 > 2x^2$, the terms start increasing in size. It is also clear that they continue to increase with increasing n. Remember that the total expression remains exact since the integral term adjusts itself to maintain the equality. But, since a bound on the integral yields a corresponding bound on the difference between the Erfc function and the series, the bound just increases accordingly. Finally, notice that the terms flatten out in the neighborhood of the least term. This is typical of series expansions of this form. The correct answer for the value of $(\sqrt{\pi}/2)(\exp 4)\,\mathrm{Erfc}(2)$ is 0.2263.

Thus, we observe that we cannot, by the method of integration by parts, obtain a result that is correct to more than two significant figures. Furthermore, it is clear from the asymptotic expansion that the most correct value is obtained if we stop with the term $-5!!/2^4x^7$. If we require greater accuracy, we must use a different technique. Recall, in this regard that the Erf and Erfc functions are complementary and that $\mathrm{Erf}(2) = 0.99532\ldots$. Thus, even an accurate integration formula may require many points to obtain Erfc(2) to three or more significant figures.

Let us now return to the general problem of developing an asymptotic expansion for $\mathrm{Erfc}(x)$. We can integrate $\mathrm{Erfc}(x)$ by parts and obtain a series of the form

$$\mathrm{Erfc}(x) = \frac{2}{\sqrt{\pi}}\exp(-x^2)\left(\frac{1}{2x} - \frac{1}{2^2x^3} + \frac{3!!}{2^3x^5} - \cdots + \cdots\right)$$
$$+ \frac{(-1)^n 2(2n-1)!!}{2^n\sqrt{\pi}}\int_x^\infty [\exp(-t^2)]\,t^{-2n}\,dt \tag{3.40}$$

The term

$$\frac{(-1)^n 2(2n-1)!!}{2^n\sqrt{\pi}}\int_x^\infty [\exp(-t^2)]\,t^{-2n}\,dt$$

is easily bounded and, in absolute value, is less than the next term in the series.

Example 3.2 Let us say that we want to know the value of $\mathrm{Erfc}(x)$ for $x > 2$ with an error of less than 10^{-3}. How many terms do we need to take in Equation (3.40) to obtain a suitable approximation for the Erfc function? We use the asymptotic expansion given by Equation (3.40) and include only the leading term:

$$\frac{2}{\sqrt{\pi}}\exp(-x^2)\left(\frac{1}{2x}\right)$$

The error, in absolute value, is then less than the next term:

$$\left|\frac{2}{\sqrt{\pi}}\exp(-x^2)\left(\frac{1}{2^2x^3}\right)\right| < 6\times 10^{-4}; \qquad x \ge 2$$

On the other hand, notice that you have to include many terms in the convergent power-series expansion of Erf(x), given by Equation (3.29), since the term:

$$\frac{2}{\sqrt{\pi}} \frac{t^{21}}{(21)(10!)}; \qquad t = 2$$

corresponding to $n = 10$ is approximately equal to 0.03. ■

Example 3.3 Consider Dawson's integral $D(x)$ that occurs, for example, in problems in heat conduction.

$$D(x) = \exp(-x^2) \int_0^x \exp t^2 \, dt \tag{3.41}$$

It is related to the error function with imaginary argument. We are interested in developing expansions for $D(x)$ for large and small values of x. We begin with large values. We plan to integrate by parts and thus introduce the breakpoint a in order to avoid an artificial singularity at the origin.

$$\begin{aligned} D(x) &= \exp(-x^2) \int_0^x \exp t^2 \, dt \\ &= \exp(-x^2) \left[\int_0^a + \int_a^x \exp t^2 \, dt \right] \end{aligned} \tag{3.42}$$

Quantity a is fixed and finite and it becomes clear that it introduces an exponentially small contribution to the total integral.

The first integral is exponentially small, since

$$\begin{aligned} \exp(-x^2) \int_0^a \exp t^2 \, dt &< \left[\exp(-x^2)\right](\exp a^2) \cdot a \\ &= O\left[\exp(-x^2)\right]; \qquad x \to \infty \end{aligned}$$

We can, for example pick a value of $a = 1$. Then, we have

$$\exp(-x^2) \cdot (\exp 1) \cdot 1$$

This is an e.s.t. and yields a small contribution for large values of x. If we think of computing the integral in steps, we observe that the dominant contributions come from the last steps and this is why the precise value of breakpoint, a, is so unimportant. (It helps to draw the integrand.) Integrate the second integral by parts and obtain the first two terms in the asymptotic expansion:

$$\exp(-x^2) \int_a^x \exp t^2 \, dt = \frac{1}{2x} + \frac{1}{4x^3} + O\left(\frac{1}{x^5}\right); \qquad x \to \infty$$

(We obtain the order of the error term by using l'Hospital's rule. See Example 2.8.)

A fuller analysis shows that, in the limit $x \to \infty$,

$$D(x) \sim \frac{1}{2x} + \sum_{1}^{\infty} \frac{(2n-1)!!}{2^{n+1}x^{2n+1}} \tag{3.43}$$

We can obtain the value of the integral for small values of x by expanding $\exp t^2$ in a power series and integrating term by term. However, an alternate procedure that is often useful is to convert our integral into a differential equation for $D(x)$. To this end, differentiate the integral given in Equation (3.41) and obtain the equation:

$$D'(x) + 2xD(x) = 1; \qquad D(0) = 0 \tag{3.44}$$

We can solve Equation (3.44) by a power-series expansion of the form

$$D(x) = \sum_{0}^{\infty} a_n x^n \tag{3.45}$$

Then, after substituting the series into our equation, we have

$$\sum_{1}^{\infty} n a_n x^{n-1} + 2\sum_{0}^{\infty} a_n x^{n+1} = 1$$

$$\sum_{0}^{\infty} [(n+2)a_{n+2} + 2a_n] x^{n+1} = 0; \qquad a_1 = 1$$

Using the boundary condition, there is no constant term, so $a_0 = 0$. The even coefficients vanish and the odd ones satisfy the recursion relation:

$$(n+2)a_{n+2} + 2a_n = 0$$

Let $n = 2m - 1$ (n is odd):

$$\begin{aligned} a_{2m+1} &= \left(\frac{-2}{2m+1}\right) a_{2m-1} \\ &= \left[\frac{(-2)(-2)}{(2m+1)(2m-1)}\right] a_{2m-3} \end{aligned}$$

Thus, we have

$$D(x) = \sum_{0}^{\infty} \frac{(-1)^m (2^m x^{2m+1})}{(2m+1)!!} \tag{3.46}$$

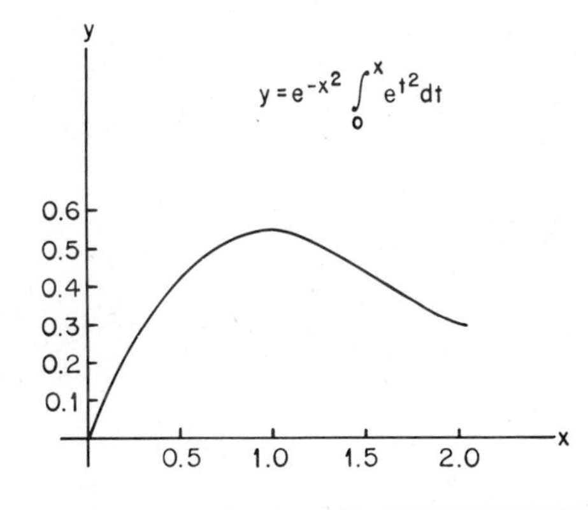

Figure 3.6 Notice how broad is the maximum of the integrand.

Function $D(x)$ has a maximum at $x = 0.92$ and, at this point, has the value 0.54. In Figure 3.6, we show $D(x)$ for $0 \leq x < 2.0$.

Dawson's integral is tabulated in, for example, Abramowitz and Stegun, *Handbook of Mathematical Functions*, pages 297 and 319. ■

3.4 SUMS

We want to understand the structure of sums and we begin by studying the geometric sum since it is easy to grasp and exhibits behavior that is typical of sums in general. In this way, we develop a feel for some of the difficulties and possible methods of attack associated with more complicated problems. Our goals are then as follows:

(a) to understand the geometric sum and use it as a basic tool;
(b) to establish the connection between monotonically increasing (or decreasing) sums and the related integrals plus end-point corrections;
(c) to develop an understanding of the role of a parameter in a sum;
(d) to look for terms that are parameter independent. It is often desirable to separate these terms before evaluating a sum.

3.4.1 The Geometric Sum

The geometric sum is given by

$$F(N, x) = \sum_{0}^{N} x^n = 1 + x + x^2 + \cdots + x^N \tag{3.47}$$

where x can be real or complex and can take on various appearances. For example,

$$x = 2; \qquad F_1(N, 2) = \sum_0^N 2^n = 1 + 2 + 2^2 + 2^3 + \cdots + 2^N \tag{3.48}$$

$$x = \exp(-\alpha); \qquad F_2[N, \exp(-\alpha)] = \sum_0^N \exp(-n\alpha)$$

$$= 1 + \exp(-\alpha) + \exp(-2\alpha) + \cdots + \exp(-N\alpha) \tag{3.49}$$

$$x = \exp i\beta; \qquad F_3[N, \exp i\beta] = \sum_0^N \exp in\beta$$

$$= 1 + \exp i\beta + \exp i2\beta + \cdots + \exp iN\beta \tag{3.50}$$

There are a finite number of terms in the sum, and the first term is independent of x. It has the value 1 for F_1, F_2, and F_3, and, thus, it does not give us any clue regarding the value of the sum. This is an important observation we shall use later when we try to find the approximate value of sums.

There is a trick for evaluating a geometric sum $F(N, x)$. Multiply Equation (3.47) by x and obtain

$$xF(N, x) = x + x^2 + \cdots + x^{N+1} \tag{3.51}$$

If we compare Equations (3.47) and (3.51), we see that $F(N, x)$ has an odd (unpaired) term at the beginning and $xF(N, x)$ has an unmatched term at the end. So, if we subtract the sums, we get

$$F - xF = 1 - x^{N+1}$$

or

$$(1 - x)F = 1 - x^{N+1}$$

Then

$$F = \frac{1 - x^{N+1}}{1 - x} \tag{3.52}$$

Before continuing, let us see what we have. It may appear as though point $x = 1$ gives us difficulty. We know that is not true since $F(N, 1) = N + 1$ because we have $N + 1$ terms, each of which has the value 1. We can see that Equation (3.52) gives this value by writing $x = 1 \pm \mu$; $\mu \to 0$. Then

$$(1 \pm \mu)^{N+1} \approx 1 \pm (N + 1)\mu; \qquad \mu \to 0$$

and

$$F \to \frac{1 - [1 \pm (N + 1)\mu]}{\mp \mu} = N + 1; \qquad \mu \to 0$$

Let us find the values of F_1, F_2, and F_3 in Equations (3.48) to (3.50):

$$F_1(N,2) = \frac{1 - 2^{N+1}}{1 - 2} = 2^{N+1} - 1 \tag{3.53}$$

$$F_2[N, \exp(-\alpha)] = \frac{1 - \exp[-(1 + N)\alpha]}{1 - \exp(-\alpha)} \tag{3.54}$$

$$F_3[N, \exp i\beta] = \frac{1 - \exp[(1 + N)i\beta]}{1 - \exp i\beta} \tag{3.55}$$

Often, we have values of $|x| < 1$ and $N \to \infty$; so

$$F(\infty, x) = F(x) \to \sum_0^\infty x^n = \frac{1}{1 - x}; \qquad |x| < 1 \tag{3.56}$$

If we perform the sum but stop at the term x^N, the remainder is

$$\sum_{n=N+1}^{\infty} x^n$$

or

$$\sum_{n=N+1}^{\infty} x^n = \sum_0^\infty x^n - \sum_0^N x^n = \frac{1}{1-x} - \frac{1 - x^{N+1}}{1 - x}$$
$$= \frac{x^{N+1}}{1 - x}$$

This gives us a measure of the error we make if we stop after the term x^N. For example, if $x = \frac{1}{2}$ and we stop after 10 terms,

$$\sum_0^\infty \tfrac{1}{2}^n \quad \text{and} \quad \sum_0^9 \tfrac{1}{2}^n \quad \text{differ by} \quad \frac{\frac{1}{2}^{10}}{1 - \frac{1}{2}} \approx \frac{1}{500}$$

(Remember $n = 9$ is the tenth term in the sum because the series begins with $n = 0$.) We can also easily find the value of the related sums:

$$G(M, x) = \sum_{n=M}^{\infty} x^n \tag{3.57}$$

$$H(j; N; x) = \sum_{n=j}^{N} x^n \tag{3.58}$$

This gives us enough tools to study the structure of many sums.

Observations. In calculating a sum, you have to be careful to count the number of terms correctly. For example, guess as to the number of terms in the sum in Equation (3.58). The sum can be rewritten as

$$H(x) = x^j \sum_0^{N-j} x^n$$

by factoring the common term x^j. Then the last term is x^{N-j}. We know that $F(N, x)$ has $N + 1$ terms, so $H(j; N; x)$ must have $N - j + 1$ terms, or, equivalently, $(N + 1) - j$ terms. In other words, we are missing the first j terms in $F(N, x)$:

$$x^0 + x^1 + x^2 + \cdots + x^{j-1}$$

which together with the $N + 1 - j$ terms

$$x^j + x^{j+1} + \cdots + x^N$$

account for $N + 1$ terms.

3.4.2 Connection Between a Sum and an Integral: "The Trapezoidal Rule"

We know that sums and integrals are connected since the integral is a limit of a sum. In this section, we want to develop a simple correspondence for smooth monotonically decreasing sums that enables us to compare them with an integral and corrections. It also provides us with insight into some of the techniques of approximation.

In Figure 3.7, we show a curve $y = f(x)$, where, for example, you can think of $f(x)$ as $\exp(-\alpha x)$, $\alpha > 0$. We want to compare

$$\int_0^5 f(x)\,dx$$

with two related sums and establish the plausibility of approximating the integral by a sum and the end-point (e.p.) corrections. To do this, we draw two separate sets of rectangles: one forms an upper bound and the other forms a lower bound. Since f is monotonically decreasing, its value at each of the successive integers 1, 2, 3, 4, and 5 is less than at the preceding integer, so that

$$\int_0^5 f(x)\,dx < [f(0)\cdot 1 + f(1)\cdot 1 + f(2)\cdot 1 + f(3)\cdot 1 + f(4)\cdot 1] \quad (3.59)$$

Note that there are five terms on the right, each of which is multiplied by 1, representing the length, on the axis, of the integer separation. Now we do the

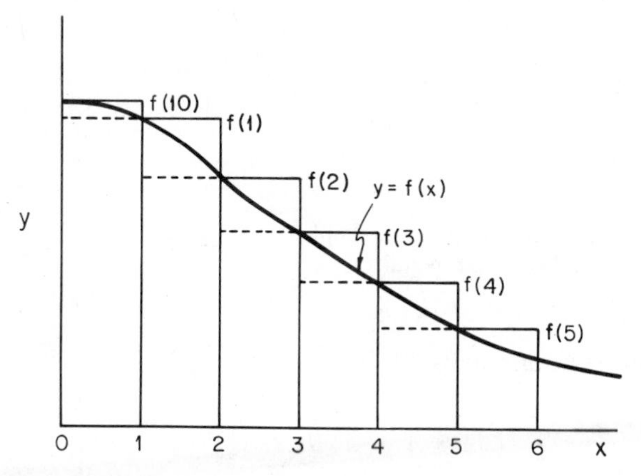

Figure 3.7 Make certain that you understand how the figure leads to the inequality given by Equation (3.60).

same thing, but this time we draw rectangles to the left beginning with $f(1)$ and ending at $f(5)$. Again, there are five terms on the left, each of which has the value of f(integer) multiplied by 1. Each of these rectangles yields an area less than the value of the corresponding integral, so

$$1 \cdot \left[\sum_{1}^{5} f(j) \right] < \int_{0}^{5} f(x)\, dx < \left[\sum_{0}^{4} f(j) \right] \cdot 1 \tag{3.60}$$

We can also bound a monotonically decreasing sum between its two related integrals:

$$\int_{1}^{6} f(x)\, dx < \sum_{1}^{5} f(j) < \int_{0}^{5} f(x)\, dx \tag{3.61}$$

Now think of function $f(x)$ as varying in an approximately linear manner between successive integer values. Then it is plausible to state that

$$\int_{0}^{5} f(x)\, dx = \frac{1}{2}\left(\sum \text{bigger rectangles} + \sum \text{smaller rectangles} \right)$$

Terms $f(0)$ and $f(5)$ each occur once, whereas the other terms occur twice. So

$$\int_{0}^{5} f(x)\, dx \approx \sum_{1}^{4} f(j) + \tfrac{1}{2} \cdot f(0) + \tfrac{1}{2} \cdot f(5)$$

or

$$\int_{0}^{5} f(x)\, dx \approx \sum_{0}^{5} f(j) - \tfrac{1}{2} \cdot f(0) - \tfrac{1}{2} \cdot f(5) \tag{3.62a}$$

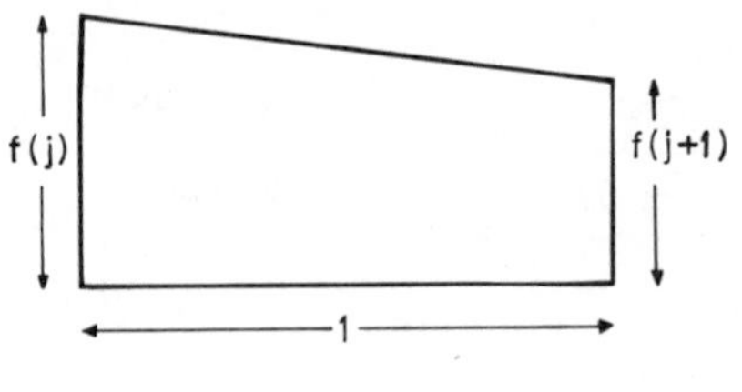

Figure 3.8 The trapezoid provides a simple way to obtain bounds on an integral.

We then also have, alternatively,

$$\sum_{0}^{5} f(j) \approx \int_0^5 f(x)\,dx + \tfrac{1}{2}[f(0) + f(5)] \tag{3.62b}$$

The beginning and ending values, f(initial) and f(final), occur once. The other terms (ordinates) occur twice in the approximation to the integral.

In this analysis, we have established a correspondence between the value of a monotonically decreasing sum and an integral and end-point corrections. Observe that we would get the same result if $f(x)$ was a monotonically increasing function. Just count the terms backwards (5 to 0) and perform the integral from 5 to 0. This is also equivalent to a reflection about the line $x = 2.5$.

In general, we have

$$\sum_{0}^{N} f(j) = \int_0^N f(x)\,dx + \tfrac{1}{2}[f(0) + f(N)] + \text{h.o.t.} \tag{3.63}$$

(The h.o.t. arise from errors in the trapezoidal approximation.)

The relation that we have established between monotonic sums and integrals forms the foundation of the most elementary numerical integration technique, i.e., the trapezoidal rule. In this, we approximate the value of an integral between two successive integers by the area of a trapezoid, as indicated in Figure 3.8.

The usefulness of this equality depends upon the ability to neglect h.o.t. This, in turn, relies on the validity of a linear approximation to the function $f(x)$ between successive integer values.

Now we use this result to gain appreciation for the role of a parameter in determining the dominant behavior of a sum.

Example 4.1 Consider

$$S(N) = \sum_{0}^{N} \exp(-n) \approx \int_0^N \exp(-x)\,dx + \tfrac{1}{2}[\exp(-0) + \exp(-N)]$$

$$= \tfrac{3}{2} - \tfrac{1}{2}\exp(-N) \tag{3.64}$$

The exact result is derivable from the geometric series and is

$$S(N) = \frac{1 - \exp[-(N+1)]}{1 - \exp(-1)} \approx 1.58 - 0.58\exp(-N) \tag{3.65}$$

So Equations (3.64) and (3.65) are in good agreement. Now let us alter $S(N)$ by introducing a parameter $\alpha > 0$ and letting $N \to \infty$.

$$S_{\text{Exact}} = \sum_{0}^{\infty} \exp(-\alpha n) = \frac{1}{1 - \exp(-\alpha)} \tag{3.66}$$

$$\begin{aligned} S_{\text{Approximate}} &= \int_0^{\infty} \exp(-\alpha x)\,dx + \tfrac{1}{2}[\exp(-\alpha \cdot 0) + \exp(-\alpha \cdot \infty)] \\ &= \alpha^{-1} + \tfrac{1}{2} \end{aligned} \tag{3.67}$$

Comparing Equations (3.66) and (3.67), we see that S_A (subscript A means approximate) and S_E (subscript E means exact) differ in functional form. If $\alpha \ll 1$,

$$\begin{aligned} S_E &= \frac{1}{1 - \exp(-\alpha)} \approx \frac{1}{\alpha - \alpha^2/2} = \frac{1 + \dfrac{\alpha}{2}}{\alpha} + O(\alpha) \\ &\approx \alpha^{-1} + \tfrac{1}{2} = S_A(\alpha) \end{aligned} \tag{3.68}$$

This is equal to $S_A(\alpha)$ given in Equation (3.67). Therefore, if α is sufficiently small, the two expressions are equivalent. But the situation is quite different when α is large. For example, if we look at the leading terms in the sum

$$S_E(\alpha) = \sum_{0}^{\infty} \exp(-\alpha n) = 1 + \exp(-\alpha) + \exp(-2\alpha) + \text{h.o.t.}$$

we see that the first term is equal to 1, independent of the value of α. Thus, the sum and the integral approximation given by Equation (3.67) treat the parameter-independent term differently. This can lead to difficulties (or incorrect results), so it is good practice to separate the parameter-independent terms and then use an approximate method to evaluate the remaining terms in the sum. You still have to be careful and perhaps use a more elaborate approximation procedure, as we point out by continuing our study of Equation (3.66). We separate the $n = 0$ term and write

$$S_E(\alpha) = 1 + \sum_{1}^{\infty} \exp(-\alpha n) = 1 + \frac{\exp(-\alpha)}{1 - \exp(-\alpha)} \tag{3.69}$$

We can approximate the sum part of $S_E(\alpha)$ by the trapezoidal rule:

$$\sum_1^\infty \exp(-\alpha n) \approx \int_1^\infty \exp(-\alpha x)\,dx + \tfrac{1}{2}[\exp(-\alpha) + \exp(-\infty)]$$

Thus, our approximate value for the sum $S_E(\alpha)$, given in Equation (3.69), is

$$S_A(\alpha) = 1 + \alpha^{-1}\exp(-\alpha) + \tfrac{1}{2}\exp(-\alpha) \tag{3.70}$$

where $S_A(\alpha)$ indicates that we are using the integral approximation. Expanding the terms on the r.h.s. of Equation (3.69) for small values of α, we also have

$$S_E(\alpha) \approx 1 + \frac{\exp(-\alpha)}{\alpha(1-\alpha/2)} = 1 + \frac{\exp(-\alpha)}{\alpha} + \frac{1}{2}\exp(-\alpha) + \text{h.o.t.} \tag{3.71}$$

leading to agreement with Equation (3.70). Now let us see what happens when α is large. Equation (3.69) has as its leading term $1 + \exp(-\alpha)$, whereas Equation (3.70) becomes

$$S_A(\alpha) \approx 1 + \tfrac{1}{2}\exp(-\alpha)$$

So we see that when α is large, it is a poor approximation to simply replace the sum by an integral. Why? Because the trapezoidal rule matches or tries to match a straight line to the function between successive integer points. When α is large, the exponential is poorly approximated by straight-line segments. You can obtain better agreement by subdividing each interval into many parts. We return to these considerations when we discuss the Euler–MacLaurin sum expansion, in Chapter 4. At that time, we will develop an approximate scheme for evaluating sums that works for both large and small parameter values. We conclude this discussion by drawing attention to Equation (3.61), the basic inequality that relates monotonically decreasing sums and integrals. This relation is often useful for sums that decrease so rapidly that linear interpolation is unsatisfactory. ∎

Example 4.2 Consider the geometric sum

$$\sum_1^\infty \exp(-\alpha n) = \frac{\exp(-\alpha)}{1-\exp(-\alpha)}$$

Observe that we have the inequality

$$\int_1^\infty \exp(-\alpha x)\,dx = \frac{\exp(-\alpha)}{\alpha} < \sum_1^\infty \exp(-\alpha n)$$
$$< \int_0^\infty \exp(-\alpha x)\,dx = \alpha^{-1}$$

We find this inequality to be of value when we discuss the Gaussian sum in Chapter 4. ■

3.5 ASYMPTOTIC SERIES

Consider a series expansion of function $f(x)$:

$$f(x) = \sum_{0}^{N} a_n x^n; \qquad N \to \infty \tag{3.72}$$

If the expansion converges, we know that

$$\left| f(x) - \sum_{0}^{N} a_n x^n \right| \to 0; \qquad N \to \infty, \quad \text{fixed } x. \tag{3.73}$$

Thus, the difference between function $f(x)$ and its series expansion can be made as small as you want by taking a sufficient number of terms in the series. The agreement improves if you take more terms. Asymptotic series are fundamentally different. Given function $g(x)$, we say that the expansion

$$g(x) = \sum_{0}^{N} a_n (x - x_0)^n \tag{3.74}$$

for fixed N and $x \to x_0$ is an asymptotic expansion if for each fixed integer N, we have

$$g(x) = \sum_{0}^{N} a_n (x - x_0)^n + o\left[(x - x_0)^N\right] \tag{3.75}$$

Such a series, in general, does not converge. It is, however, an extremely powerful computational tool in that it provides an excellent way to find the numerical value for a sum or integral. In general, the terms in the series rapidly decrease in size, then level out, and then increase. This was the character of the series we developed for Erfc(x) and for Dawson's integral. Most frequently, we use this type of expansion to develop a series for large values of x. Thus, we write

$$g(x) = \sum_{0}^{N} a_n x^{-n} + o(x^{-N}) \tag{3.76}$$

as $x \to \infty$, for fixed N.

Example 5.1 Consider

$$A(z) = \exp z \int_z^\infty \frac{\exp(-t)}{t}\, dt; \qquad z > 0 \tag{3.77}$$

We are interested in large values of z. We integrate by parts:

$$A(z) = \exp z \left[\frac{(-)\exp(-t)}{t} \Bigg|_z^\infty - \int_z^\infty \frac{\exp(-t)}{t^2}\, dt \right] \tag{3.78}$$

$$A(z) = \frac{1}{z}\left(1 - \frac{1}{z} + \frac{1 \cdot 2}{z^2}\right) - \exp z \int_z^\infty \frac{3!\exp(-t)}{t^4}\, dt$$

We easily bound the integral term

$$\exp z \int_z^\infty \frac{3!\exp(-t)}{t^4}\, dt < \exp z \cdot \exp(-z) \cdot \frac{3!}{z^4}$$

It is clear that if z is sufficiently large, the difference between $A(z)$ and the first few terms in the series is quite small:

$$A(z) = \frac{1}{z}\left(1 - \frac{1}{z} + \frac{1 \cdot 2}{z^2}\right) + o(1/z^3)$$

Observe that we can write the error term as $O(1/z^4)$ or $o(1/z^3)$. So this expansion is a good way to compute $A(z)$ for large values of z. But the series diverges for any fixed value of z, since its terms are of the form $(n-1)!/z^n$.

We observe that the ratio of the nth to $(n-1)$st term is

$$\frac{(n-1)!}{(n-2)!}\,\frac{z^{n-1}}{z^n} = \frac{n-1}{z}$$

Thus, if $n - 1 > z$, the terms increase in magnitude and the series blows up. It is possible, although usually with some difficulty, to find the least term in the series. This would then indicate the optimum point to truncate the series expansion. However, it is a more common practice to usually just compute a few terms and bound the error term. ∎

EXERCISE 5.1 Consider

$$F(x) = \exp(-x) \int_1^x \frac{\exp t}{t}\, dt; \qquad \text{as} \quad x \to \infty$$

Develop a series expansion of the form

$$F(x) = \left(\frac{1}{x} + \frac{1}{x^2} + \frac{2!}{x^3} + \frac{3!}{x^4} \right) + \text{constant} \cdot \exp(-x)$$
$$+ \exp(-x) \int_1^x \frac{4! \exp t}{t^5} \, dt$$

and show that the error is $O(1/x^5)$. Find $F(20)$ correct to two significant figures. (Answer is 0.053.) (Note that it takes some effort to guarantee that the result is correct to two figures. I suggest that you divide the integral into several pieces and obtain a satisfactory bound for each piece separately.) ■

EXERCISE 5.2 Consider the integral

$$I = \int_0^{\sqrt{12}} t^2 \exp\left(\frac{-t^2}{2} \right) dt$$

Find its value to three significant figures. It may be helpful to make a substitution to convert it into the form

$$I = \int_0^6 \sqrt{2} \sqrt{y} \exp(-y) \, dy$$

and then write the integral as

$$I = \sqrt{2} \int_0^\infty \sqrt{y} \exp(-y) \, dy - \sqrt{2} \int_6^\infty \sqrt{y} \exp(-y) \, dy$$

It is then necessary only to integrate by parts and bound the error term. Remember that exp 3 is approximately equal to 20.

Answer: 1.24. ■

EXERCISE 5.3 Recall Example 6.2 of Chapter 0. At that time, we introduced the integral

$$K = \int_0^\infty y \frac{\exp y}{1 + \exp 2y} \, dy \tag{3.79}$$

We convert this to an alternating sum by multiplying through by $\exp -2y$, expanding the integrand, and writing

$$K = \sum_{n=0}^{\infty} (-)^n \int_0^\infty y \exp[-(2n+1)y] \, dy \tag{3.80}$$

We evaluate the integral to obtain the sum whose numerical value is known as Catalan's constant.

$$K = \sum_{n=0}^{\infty} (-)^n (2n+1)^{-2} = 0.9159\ldots \tag{3.81}$$

We have two goals in this exercise:

(a) Observe that the sum K is a rapidly converging alternating sum. Verify its value to three significant figures.
(b) Assume we are interested in finding the value of the integral

$$L(\alpha) = \int_{\alpha}^{\infty} \frac{\ln x}{1+x^2}\, dx \tag{3.82}$$

for $1 \le \alpha < \infty$, correct to two significant figures. From the analysis just given and the results of Example 6.2 of Chapter 0, we know that we can convert the integral in Equation (3.82) into a sum that is related to Equation (3.81). Thus, letting $y = \ln x$ and introducing the expansion that led to Equation (3.80), we obtain

$$L(\alpha) = \sum_{n=0}^{\infty} (-)^n (2n+1)^{-2} \cdot [\alpha^{-(2n+1)}] \cdot [(2n+1)(\ln \alpha) + 1] \tag{3.83}$$

This last result follows from our ability to evaluate the integral

$$F(\beta) = \int_{\beta}^{\infty} x \exp(-x)\, dx = \exp(-\beta) + \beta \exp(-\beta) \tag{3.84a}$$

In our case, we have $\beta = (2n+1)(\ln \alpha)$, with $x = (2n+1)y$. Thus, Equation (3.84a) takes the form

$$F(\beta) = [\alpha^{-(2n+1)}] \cdot [(2n+1)(\ln \alpha) + 1] \tag{3.84b}$$

The sum given by Equation (3.83) converges, and since it is an alternating series, the error due to truncation is less than the first term neglected. If α is close to 1, say $\alpha = 1.05$, then we need 17 terms in the sum to guarantee two significant figures. The sum converges more rapidly for larger values of α. If $\alpha \le 1.05$, we have $L(\alpha) = K = 0.92$. This is seen by considering the integral

$$\begin{aligned} M(\alpha) &= \int_1^{\alpha} \frac{\ln x}{1+x^2}\, dx \le \tfrac{1}{2} \int_1^{\alpha} \ln x\, dx \\ &= \tfrac{1}{2}(\alpha \ln \alpha - \alpha + 1) \end{aligned} \tag{3.85}$$

and showing that it is less than 0.001.

Observe that by implementing these procedures, we would, in principle, be able to construct a table of values of $L(\alpha)$ correct to two significant figures over the entire range of variable α. Confirm that $L(2) = 0.81$, $L(3) = 0.68$, $L(5) = 0.52$, and $L(20) = 0.20$. Notice that $L(20)$ is significant, indicating that a sizable fraction of its total integral for K lies in the region $(20, \infty)$. (Finally, referring to Equations (3.80) and (3.82), we have $y = \ln x$. So when $x = 20$, $y \approx 3$.) ■

BIBLIOGRAPHY

1. De Bruijn, N. G., *Asymptotic Methods in Analysis*, North-Holland, Amsterdam, 1958. Reprinted from the third edition by Dover, New York, 1981. Although this is an advanced text, it has a very readable discussion of the asymptotic symbols and asymptotic series.
2. Wilf, H. S., *Mathematics for the Physical Sciences*, Wiley, New York, 1962. Reprinted by Dover, New York, 1978. There is a very clear discussion of the asymptotic symbols and asymptotic series, including some nice examples. There are solutions given to most of the problems.
3. Lebedev, N. N., *Special Functions and their Applications*, Prentice-Hall, Englewood Cliffs, New Jersey, 1965. Reprinted by Dover, New York, 1972. There is a very clear discussion of the error function and related integrals, including Dawson's integral.
4. Hamming, R. W., *Numerical Methods for Scientists and Engineers*, McGraw-Hill, New York, 1962. Reprinted from the second edition by Dover, New York, 1986. There is a very readable discussion of numerical integration and the trapezoidal rule. The connection between sums and integrals is illustrated with many examples.
5. Bender, C. M., and S. A. Orszag, *Advanced Mathematical Methods for Scientists and Engineers*, McGraw-Hill, New York, 1978. This text has an extensive treatment of asymptotics with many illustrative examples and problems.
6. Olver, F. W. J., *Asymptotics and Special Functions*, Academic Press, New York, 1974. This is an excellent source of material on asymptotics. It has both simple and advanced examples. There is an extensive introductory section on the order symbols.

CHAPTER 4

EVALUATION OF SUMS: THE EULER–MAC LAURIN SUM EXPANSION

We have already learned a simple technique to evaluate sums by comparison with an integral. However, we did not develop either a systematic procedure or a simple way to estimate the errors involved. The Euler–MacLaurin (EM) summation formula fills this need since it establishes a correspondence between a sum and an integral in a form that includes a remainder term that can be easily bounded and it is applicable in a broad spectrum of cases that we will classify and illustrate. It has the following characteristics:

(1) It is applicable to sums that have slowly varying terms and decreasing derivatives. It is easiest to use if the terms are monotonically increasing or decreasing because then the error analysis is particularly simple.
(2) It is easily adaptable to use by a computer or hand calculator.
(3) It can be used to find the numerical value of the finite part of divergent sums. This means, for example, that if we have a sum that diverges logarithmically, we can identify and separate this part and find the remaining finite part.
(4) We can evaluate both finite and infinite sums.

We begin, in Section 4.1, with an orientation to the EM sum formula and include an outline of a proof of the formula and a bound on the error term. We then explain the various parts of the formula, including the error or remainder term, and practice by evaluating a few simple sums. In Section 4.2, we discuss sums that depend on a parameter. We identify the dominant behavior and the order of the neglected terms.

Finally, in Section 4.3, we learn to evaluate the finite part of some divergent sums. At the end of the chapter, we cite literature that contains detailed derivations of the Euler–MacLaurin expansion and discussions of the error analysis. We also give literature on the Poisson sum formula, a complementary method of evaluating sums over infinite and semiinfinite intervals.

4.1 DERIVATION OF THE EULER – MAC LAURIN SUM EXPANSION

There are a few notations in common use, and we use the one in the classic text by Knopp, *Theory and Application of Infinite Series*. First, let us show that generally it is often impractical to just simply evaluate a convergent sum by summing a number of terms by computer and bounding the error by an integral. Consider the sum

$$\zeta(2) = \sum_{n=1}^{\infty} \frac{1}{n^2} \tag{4.1}$$

Assume that we want to know its value with an error less than 10^{-6}. If we try to simply compute a partial sum, we have to include the first million terms since we know that the remainder can be bounded by the integrals

$$\int_{10^6+1}^{\infty} \frac{dx}{x^2} < \sum_{n=10^6+1}^{\infty} \frac{1}{n^2} < \int_{10^6}^{\infty} \frac{dx}{x^2} \tag{4.2}$$

We will see that the EM formula gives us this accuracy by using only a few terms in the sum, evaluating a few derivatives, and applying a simple bound on the remainder. (We do not use the EM formula in its most general form and thus its complete structure is not revealed nor is the asymptotic nature of the remainder term readily apparent. However, these points are discussed in some detail in the books found in the Bibliography.)

Our derivation of the EM formula begins by comparing a sum and its related integral and an easily bounded error term.

$$\sum_{i=j}^{n} f(i) = \int_{j}^{n} f(x)\, dx + \text{remainder} \tag{4.3}$$

The function $f(x)$ is assumed to be continuously differentiable on the entire interval under consideration. We assume that the integral can be evaluated analytically or by accurate numerical integration. Thus, we concentrate on bounding the remainder or error term. The correspondence between a sum and its related integral is developed for functions $f(x)$ that are smoothly varying and have smoothly varying derivatives.

The essence of the formula is contained in the reduction of the error term by evaluating derivatives of $f(x)$. It is often cumbersome to take many derivatives of $f(x)$ and so, almost always, we compute only three. The rationale for this should become clear after a few examples. We begin by introducing a sawtooth function that leads us to the desired formula. (This is the trick.) Consider

$$I = \int_j^{j+1} \left(x - j - \tfrac{1}{2}\right) f'(x)\, dx \tag{4.4}$$

where j is an integer.

The integrand is the product of a straight line that interpolates between $-\frac{1}{2}$ and $+\frac{1}{2}$ over the interval of integration and the derivative of $f(x)$. It follows then, by integration by parts, that this is equal to

$$I = \tfrac{1}{2}[f(j) + f(j+1)] - \int_j^{j+1} f(x)\, dx \tag{4.5}$$

We are going to want to sum over a range of values of x, and so we let $j \to j + 1$. We then have

$$\begin{aligned} J &= \int_{j+1}^{j+2} \left[x - (j+1) - \tfrac{1}{2}\right] f'(x)\, dx \\ &= \tfrac{1}{2}[f(j+1) + f(j+2)] - \int_{j+1}^{j+2} f(x)\, dx \end{aligned} \tag{4.6}$$

We want to join the two integrals, which is usually done through the use of the symbol $[x] \equiv$ the greatest integer less than or equal to x.

So, if

$$\begin{aligned} x &= 3.16; \qquad [x] = 3 \\ x &= 0.14; \qquad [x] = 0 \\ x &= 4; \qquad\;\;\;\; [x] = 4 \end{aligned}$$

Then the sum of integrals $I + J$ can be written as

$$\begin{aligned} I + J &= \int_j^{j+2} \left(x - [x] - \tfrac{1}{2}\right) f'(x)\, dx \\ &= \tfrac{1}{2}f(j) + 2 \cdot \tfrac{1}{2}f(j+1) + \tfrac{1}{2}f(j+2) - \int_j^{j+2} f(x)\, dx \end{aligned} \tag{4.7}$$

We want to sum over many terms, so we write

$$\int_j^{j+n}\left(x-[x]-\tfrac{1}{2}\right)f'(x)\,dx$$

$$=\tfrac{1}{2}f(j)+\sum_{i=1}^{n-1}f(j+i)+\tfrac{1}{2}f(j+n)-\int_j^{j+n}f(x)\,dx \quad (4.8)$$

Notice that each interior point contributes to two terms and, hence, we only have factors of $\frac{1}{2}$ for the initial and final terms in the series. After some rearrangement, we have

$$\sum_{i=0}^{n}f(i+j)=\sum_{i=j}^{n+j}f(i)=\int_j^{n+j}f(x)\,dx+\tfrac{1}{2}[f(j)+f(n+j)]$$

$$+\int_j^{n+j}\left(x-[x]-\tfrac{1}{2}\right)f'(x)\,dx \quad (4.9)$$

It is important to emphasize that the formula is exact. It is possible to generalize it to situations where j is not an integer or the steps are not integers. (See the text by Olver.)

This is the first form of the EM sum formula, but we have to introduce some auxiliary quantities in order to have our remainder term

$$\int_j^{n+j}\left(x-[x]-\tfrac{1}{2}\right)f'(x)\,dx \quad (4.10)$$

in a convenient form. We know that

$$x-[x]-\tfrac{1}{2}; \qquad j\le x\le n+j \quad (4.11)$$

looks like a repeating sawtooth wave with a period equal to 1, interpolating between $-\frac{1}{2}$ and $+\frac{1}{2}$. (Draw it.) It is convenient to represent it by a Fourier series. Thus, we write

$$P_1(x)=-\sum_{n=1}^{\infty}\frac{2\sin 2n\pi x}{2n\pi} \quad (4.12)$$

The series converges to the sawtooth function for all nonintegral values of x. We can integrate it term by term, calling the resulting function

$$P_2(x)=\sum_{n=1}^{\infty}\frac{2\cos 2n\pi x}{(2n\pi)^2}+A \quad (4.13)$$

The constant of integration A is most often taken to be equal to zero. With this choice, the average value of $P_k(x) = 0$ over the unit interval. Since the expansion for $P_2(x)$ on the l.h.s. of Equation (4.13) is obtained from the integration of $P_1(x) = x - \frac{1}{2}$ when x lies between 0 and 1, we also have

$$P_2(x) = \frac{x^2}{2} - \frac{x}{2} + \frac{1}{12} \tag{4.14}$$

Example 1.1 Let us see how we arrive at Equation (4.14). We are given that $P_1(x) = x - \frac{1}{2}$ and that the average value of $P_k(x) = 0$. Then, integrating $P_1(x)$, we have

$$P_2(x) = \int P_1(x)\,dx = \frac{x^2}{2} - \frac{x}{2} + C$$

Constant C is determined by the requirement that the average of $P_2(x)$ over the unit interval is equal to zero. Thus, upon integrating, we have

$$P_3(x) = \int P_2(x)\,dx = \frac{x^3}{6} - \frac{x^2}{4} + Cx$$

Then the requirement of zero average is equivalent to $P_3(1) = P_3(0)$. Thus, we have $C = \frac{1}{12}$. ■

We can continue to integrate our series and continue to choose the constants of integration equal to zero. We then have

$$P_{2k}(x) = (-)^{k+1} \sum_{n=1}^{\infty} \frac{2\cos 2n\pi x}{(2n\pi)^{2k}}; \qquad k = 1, 2, 3, 4, \dots \tag{4.15a}$$

$$P_{2k+1}(x) = (-)^{k+1} \sum_{n=1}^{\infty} \frac{2\sin 2n\pi x}{(2n\pi)^{2k+1}}; \qquad k = 0, 1, 2, 3, 4, \dots \tag{4.15b}$$

$$P'_{k+1}(x) = P_k(x); \qquad k = 1, 2, 3, 4, \dots \tag{4.15c}$$

We can also continue integration of the polynomial forms of $P_k(x)$ and obtain:

$$P_3(x) = \frac{x^3}{6} - \frac{x^2}{4} + \frac{x}{12} \tag{4.16}$$

$$P_4(x) = \frac{x^4}{24} - \frac{x^3}{12} + \frac{x^2}{24} - \frac{1}{720} \tag{4.17}$$

EXERCISE 1.1 Continue the integration of $P_k(x)$ and show that the constant term in $P_6(x) = 1/30{,}240$. ■

(The P functions are related to the Bernoulli polynomials. See any of the texts, e.g., Knopp, Wilf, etc.) We are going to have to know the values of $P_k(x)$ for integer values of the argument. The odd P's vanish and we have

$$
\begin{aligned}
P_{2k}(\text{integer}) &= (-)^{k+1} \sum_{n=1}^{\infty} \frac{2\cos(2n\pi \cdot \text{integer})}{(2n\pi)^{2k}} \\
&= (-)^{k+1} \sum_{n=1}^{\infty} \frac{2}{(2n\pi)^{2k}}
\end{aligned}
\tag{4.18}
$$

We are interested in the numerical value of some of the $P_{2k}(x)$'s. These are most easily obtained from evaluating the constants on integration that were introduced so as to guarantee that the $P_k(x)$'s had an average value equal to zero. Thus, for example, consulting Equations (4.14) (setting $x = 0$) and (4.18), we have

$$
P_2(\text{integer}) = \frac{1}{12} = \sum \frac{2}{(2n\pi)^2} = \frac{1}{2\pi^2} \sum \frac{1}{n^2} \tag{4.19a}
$$

$$
P_4(\text{integer}) = \frac{-1}{720} = -\sum \frac{2}{(2n\pi)^4} = -\frac{1}{8\pi^4} \sum \frac{1}{n^4} \tag{4.19b}
$$

Recall the Riemann zeta function that we introduced in Section 2.5 of Chapter 2. It is defined by

$$
\zeta(k) \equiv \sum_{n=1}^{\infty} n^{-k}; \qquad k > 1 \tag{4.20}
$$

and satisfies the inequality

$$
\zeta(k+1) < \zeta(k) < \zeta(k-1); \qquad k > 2 \tag{4.21}
$$

From Equations (4.19a) and (4.19b) and using the results of Example 1.1 and Exercise 1.1, we see that

$$
\zeta(2) = \frac{\pi^2}{6}; \qquad \zeta(4) = \frac{\pi^4}{90}; \qquad \zeta(6) = \frac{\pi^6}{945} \tag{4.22}
$$

Let us see how we are helped by the introduction of the P functions. We return to our remainder term:

$$
\int_j^{n+j} \left(x - [x] - \tfrac{1}{2}\right) f'(x)\, dx = \int_j^{n+j} P_1(x) f'(x)\, dx \tag{4.23a}
$$

and plan to integrate by parts with the expectation that the next few derivatives of $f(x)$ will both be easy to evaluate and small. (This sometimes requires a trade-off in the calculation and we will see that it is generally best to compute only a few derivatives.) So we write

$$\int_j^{n+j} P_1(x) f'(x)\, dx = f'(x) P_2(x)\Big|_j^{n+j} - \int_j^{n+j} P_2(x) f''(x)\, dx \tag{4.23b}$$

Recall from Equation (4.19a) that $P_2(\text{integer}) = \frac{1}{12}$. Then the term integrated by parts becomes:

$$\tfrac{1}{12}[f'(n+j) - f'(j)] \tag{4.24}$$

After noting that the boundary terms from the next integration by parts vanish since $P_3(\text{integer}) = 0$, we get

$$\int_j^{j+n} P_1(x) f'(x)\, dx = \tfrac{1}{12}[f'(n+j) - f'(j)] + \int_j^{n+j} P_3(x) f'''(x)\, dx \tag{4.25}$$

At this point, we stop and organize ourselves so as to use the EM formula in this form. That is, we have

$$\sum_{i=j}^{n+j} f(i) = \int_j^{n+j} f(x)\, dx + \tfrac{1}{2}[f(j) + f(n+j)]$$
$$+ \tfrac{1}{12}[f'(n+j) - f'(j)] + \int_j^{n+j} P_3(x) f'''(x)\, dx \tag{4.26}$$

The expression given by Equation (4.26) is exact and we hope we have achieved an effective summation method by the introduction of higher derivatives of the function $f(x)$ and the ability to give a simple bound for the P's. We now turn our attention to obtaining the required bound on the remainder term. We introduce

$$\eta \equiv \left| \int_j^{n+j} P_3(x) f'''(x)\, dx \right| \le \max|P_3(x)| \int_j^{n+j} |f'''(x)|\, dx \tag{4.27}$$

If $f'''(x)$ has one sign on the interval of integration, we can write

$$\eta \le \max|P_3(x)| \cdot |f''(n+j) - f''(j)| \tag{4.28}$$

We can bound $P_3(x)$ using the Fourier-series expansion given in Equation (4.15b), after noting that $\sin x$ is bounded by unity and $\zeta(3) = 1.202\ldots$:

$$|P_3(x)| \le \sum_{n=1}^{\infty} \frac{2}{(2n\pi)^3} = \frac{2}{(2\pi)^3}\zeta(3) < 10^{-2} \tag{4.29}$$

(It is possible to obtain a better bound, but we shall not do so. The text by Knopp gives an extensive discussion of the error term.)

So we write our formula as

$$\sum_{i=j}^{n+j} f(i) = \int_j^{n+j} f(x)\,dx + \tfrac{1}{2}[f(j) + f(n+j)] + \tfrac{1}{12}[f'(n+j) - f'(j)] \pm \eta \qquad (4.30)$$

Note that the formula remains valid when the upper limit on the summation is infinity. It is important to emphasize that we can continue to integrate by parts, but it is usually not a preferred procedure. Instead, our tactic is to sum the first few terms by hand so as to obtain a suitable error after evaluating only three derivatives of $f(x)$.

Also note that very often the derivatives of our function $f(x)$ diminish significantly as x increases, so the error is determined primarily by the value at the lower limit of integration.

Let us recapitulate: our formula consists of the following parts:

(a) The sum we want to evaluate.

(b) The integral of the function in the sum. We can either perform the integration analytically or numerically. We can also reverse our problem and express the integral as a sum plus the correction terms.

(c) There are end-point corrections, $\frac{1}{2}[f(j) + f(n+j)]$, that can be expected to be small compared to the integral.

(d) There are end-point derivative corrections, $\frac{1}{12}[f'(n+j) - f'(j)]$ that can also be expected to be small.

(e) We usually assume that it is tedious to take many derivatives and thus we limit ourselves to no higher than the third derivative. Also, it is best that the last derivative taken has one sign over the range of integration as this gives a good bound for the error. This might influence how many terms are summed by hand. These ideas will become clear after we do some examples. We will discuss one problem, in which we calculate Euler's constant γ, where we take higher derivatives so as to obtain a value of the constant correct to eight figures. This will demonstrate the usefulness of the EM expansion for the determination of the value of a sum to many decimal places.

(f) There is a bounded error term. We use a crude error bound since it is easy to apply.

Let us do a simple example that illustrates how we choose the limits of our sum and how we use the EM formula.

Example 1.2 Consider the sum

$$\zeta(2) = \sum_{n=1}^{\infty} \frac{1}{n^2} = \frac{\pi^2}{6} \qquad (4.31)$$

and say that we want an error less than 10^{-6}. We separate the sum into two parts. We sum the first $j-1$ terms directly by addition and then use the EM expansion to evaluate the remaining sum:

$$\sum_{n=j}^{\infty} \frac{1}{n^2}$$

We organize our calculation as follows:

(1) We evaluate the integral:

$$\int_j^{\infty} \frac{dx}{x^2} = \frac{1}{j}$$

(2) We compute the end-point corrections:

$$\frac{1}{2}[f(j) + f(\infty)] = \frac{1}{2j^2}$$

(3) We compute the derivative correction term:

$$\frac{1}{12}[f'(\infty) - f'(j)] = \frac{1}{6j^3}$$

(4) We are now ready to bound the remaining term so as to satisfy our error requirements. This involves choosing an appropriate value for the number of terms that we sum by hand or choosing a value for j. Since $f'''(x)$ has one sign on the interval of integration, our error term is written as

$$\eta < \max|P_3(x)| \cdot |f''(\infty) - f''(j)| < (10^{-2})(2 \cdot 3)j^{-4}$$

and it is clear that if j is equal to 20, we have

$$\eta < 10^{-2} \cdot (6/16)(10^{-4}) < 10^{-6}$$

satisfying the requirements we set.

We are now ready to put all the pieces together. We have

$$\sum_{n=1}^{\infty} \frac{1}{n^2} = \sum_{n=1}^{19} \frac{1}{n^2} + \frac{1}{20} + \frac{1}{2 \cdot 20^2} + \frac{1}{6 \cdot 20^3} + \eta$$

Thus, we sum the first 19 terms and have

$$\sum_{n=1}^{19} \frac{1}{n^2} = 1.593663244$$

and

$$\frac{1}{20} + \frac{1}{800} + \frac{1}{48{,}000} = 0.051270833$$

So the EM expansion gives the result:

$$\sum_{1}^{\infty} \frac{1}{n^2} = 1.644934077 \tag{4.32}$$

The exact result is

$$\frac{\pi^2}{6} = 1.644934067$$

So you notice that it is easy, for such a sum, to get an extremely accurate result with only a little planning and effort. This result is typical for sums of this type. ■

EXERCISE 1.2 Repeat this analysis for the same sum if the error is to be less than 10^{-3}. Observe how few terms are necessary to satisfy the error requirements. ■

EXERCISE 1.3 Evaluate

$$\zeta(3) = \sum_{n=1}^{\infty} \frac{1}{n^3} = 1.20205690$$

correct to six decimals. ■

EXERCISE 1.4 Evaluate the sum

$$\zeta\left(\frac{3}{2}\right) = \sum_{n=1}^{\infty} \frac{1}{(\sqrt{n})^3} = 2.612$$

correct to three decimals. ■

EXERCISE 1.5 Alternating sums are often difficult to evaluate. However, if you have a simple sum, you can evaluate the pieces separately. Thus, given the sum

$$\begin{aligned} S(-) &= \sum_{n=1}^{\infty} (-)^{n+1} \frac{1}{n^2} = 1 - \frac{1}{2^2} + \frac{1}{3^2} - \frac{1}{4^2} + \cdots \\ &= \sum_{n=0}^{\infty} \frac{1}{(2n+1)^2} - \sum_{n=1}^{\infty} \frac{1}{(2n)^2} \end{aligned} \tag{4.33}$$

we already know that

$$\sum_{n=1}^{\infty} \frac{1}{(2n)^2} = \frac{1}{4}\frac{\pi^2}{6} = \frac{\pi^2}{24} \tag{4.34}$$

So we really are finding the value of the sum

$$\sum_{n=0}^{\infty} \frac{1}{(2n+1)^2} = \sum_{n=1}^{\infty} \frac{1}{n^2} - \sum_{n=1}^{\infty} \frac{1}{(2n)^2}$$
$$= \frac{\pi^2}{6} - \frac{\pi^2}{24} = \frac{\pi^2}{8}$$

Therefore,

$$S(-) = \frac{\pi^2}{12} \tag{4.35}$$

Use the EM expansion method to find the value of the sum $S(-)$ correct to three decimals. ■

Example 1.3 The EM sum expansion provides us with a convenient way of evaluating an integral of a smoothly varying function to a given accuracy with a bounded error term. It is precisely because this technique enables us to bound the error term that makes it of interest. However, it is limited to functions whose derivatives can be easily evaluated. It can be used for integrals over a finite or an infinite range. For example, consider

$$I = \int_2^{\infty} \frac{dx}{x^2}$$

and assume that we want to find its value correct to three significant figures. Of course, we know that the exact answer is $I = \frac{1}{2}$. We also know that we can use the EM expansion to write

$$\int_j^{\infty} \frac{dx}{x^2} = \sum_{n=j}^{\infty} \frac{1}{n^2} - \frac{1}{2j^2} - \frac{1}{6j^3} + \eta$$

with $\eta < (10^{-2})(6/j^4)$. The lower limit on the integral is given as $j = 2$ yielding too large an error. To address this difficulty, we introduce a stretching transformation, $y = 7x/2$, so that our integral becomes

$$I = \frac{7}{2}\int_7^{\infty} \frac{dy}{y^2}$$

Then, we can achieve an error less than 2.5×10^{-5}. We then have

$$I = \frac{7}{2}\left[\sum_{n=7}^{\infty} \frac{1}{n^2} - \frac{1}{98} - \frac{1}{2058}\right] + \eta$$

$$= \tfrac{7}{2}(0.142855) = 0.499993 + (\eta < 10^{-4})$$

Observe that the stretching transformation has magnified the error term while keeping it within acceptable limits. ■

EXERCISE 1.6 Use the method of Example 1.3 and the results of Exercise 1.3 to find

$$I = \int_3^{\infty} \frac{dx}{x^3}$$

correct to three significant figures. Since the answer is known to be 0.0556, keep in mind that the error must be such as to guarantee that the result is correct to four decimal places, yielding three significant figures. ■

EXERCISE 1.7 Consider

$$I = \int_1^2 \ln x \, dx = 2 \ln 2 - 1 = 0.3863$$

Use the EM expansion with the stretching transformation $y = 10x$ to find the value of the integral correct to four decimals. ■

Example 1.4 We can handle integrals in which the lower limit is not an integer. For example, consider

$$I = \int_{2.5}^{\infty} \frac{dx}{x^2}$$

and assume we want to find its value to four significant figures. Proceeding as in Example 1.3, we introduce a stretching transformation that gives us the desired accuracy as well as making the lower limit an integer. Thus, let $y = 4x$ and

$$\eta < 4(6 \times 10^{-6}) = 2.4 \times 10^{-5}$$

where the factor "4" comes from the stretching transformation. We then have

$$I = 4\left[\sum_{n=10}^{\infty} \frac{1}{n^2} - \frac{1}{200} - \frac{1}{6000}\right] + \eta$$

$$= 0.4000 + (\eta < 2.4 \times 10^{-5})$$

■

4.2 SUMS THAT DEPEND UPON A PARAMETER

Now let us try the EM expansion method on the geometric sum

$$S(\alpha) = \sum_{n=0}^{\infty} \exp(-\alpha n) \tag{4.36a}$$

We want to see the role of parameter α and the point j at which it is appropriate to initiate the EM sum expansion. For example, we want to show that it is often a good idea to separate the first term, since it is parameter independent, perform the calculation, and put the first term back in, and see if there were any difficulties. So let us write our original problem given by Equation (4.36a) as

$$S(\alpha) - 1 = \sum_{n=1}^{\infty} \exp(-\alpha n) \tag{4.36b}$$

Now we examine the structure of the sum $S(\alpha) - 1$ using the EM expansion:

$$\sum_{n=1}^{\infty} \exp(-\alpha n) = \int_1^{\infty} \exp(-\alpha x)\, dx + \tfrac{1}{2}\exp(-\alpha) + \tfrac{1}{12}\alpha \exp(-\alpha) \pm \eta \tag{4.36c}$$

$$= \left[\alpha^{-1} + \tfrac{1}{2} + \tfrac{1}{12}\alpha\right] \exp(-\alpha) \pm \eta \tag{4.36d}$$

We see by examining the error term

$$\eta < \left|10^{-2}\alpha^2 \exp(-\alpha)\right| \tag{4.37}$$

that for any value of parameter α, we can obtain a good estimate for the sum. Let us illustrate these ideas by considering a few values of parameter α. Consider Equation (4.36b) with $\alpha = \frac{1}{2}$. The sum can be evaluated exactly, yielding

$$\sum_{n=1}^{\infty} \exp\left(\frac{-n}{2}\right) = \frac{\exp\left(-\frac{1}{2}\right)}{1 - \exp\left(-\frac{1}{2}\right)}$$

$$= 1.5415 \tag{4.38a}$$

We now evaluate the sum approximately using the EM expansion as given by Equation (4.36d), noting that the error term is particularly simple to bound:

$$\sum_{n=1}^{\infty} \exp\left(\frac{-n}{2}\right) = \left(2 + \tfrac{1}{2} + \tfrac{1}{24}\right) \exp\left(-\tfrac{1}{2}\right) \pm \eta$$

$$= 1.5416 \pm \left(\eta < 2 \times 10^{-3}\right) \tag{4.38b}$$

Since the sum is slowly varying, we did not have to separate the $n = 0$ term in the EM expansion in order to achieve the required accuracy. However, the practice of separating the parameter-independent term is a sound precautionary measure.

EXERCISE 2.1 Consider

$$\sum_{n=0}^{\infty} \exp(-6n) = 1.0025 \tag{4.39}$$

Use the EM expansion to obtain the value of this sum to the accuracy given in Equation (4.39). Show that it is advisable to pick $j = 2$, rather than $j = 1$, in order to approximate the sum in the region where it is slowly varying. The EM expansion is straightforward:

$$\sum_{n=0}^{\infty} \exp(-6n) = 1 + \exp(-6) + \left(\tfrac{1}{6} + \tfrac{1}{2} + \tfrac{1}{2}\right) \exp(-12)$$

$$= 1.0025$$

■

EXERCISE 2.2 Consider

$$\sum_{n=0}^{\infty} \exp(-10n) = 1.000045 \tag{4.40}$$

Pick an appropriate value of j and use the EM expansion to obtain the value of this sum to the same accuracy. ■

We conclude by noting that we can approximate sums of this type over the entire range of the parameter, if we plan our calculation appropriately. This often means separating the parameter-independent term and evaluating the first few terms by hand to obtain the required bounded-error term.

We now show that everything is not always so simple. There are cases when you have to compute higher derivatives that involve significant algebraic manipulations. In these circumstances, it is often better to use another summation technique such as the Poisson sum formula. (See the Bibliography.)

Example 2.1

$$S(\alpha) = \sum_{n=0}^{\infty} \frac{1}{\alpha^2 + n^2} = \frac{1}{\alpha^2} + \sum_{n=1}^{\infty} \frac{1}{\alpha^2 + n^2} \tag{4.41}$$

If $|\alpha| < 1$, we can expand the sum and express the result in terms of ζ functions:

$$\sum_{n=1}^{\infty} \frac{1}{\alpha^2 + n^2} = \sum_{n=1}^{\infty} \frac{1}{n^2} \frac{1}{1 + (\alpha/n)^2}$$

$$= \sum \frac{1}{n^2}\left[1 - \left(\frac{\alpha}{n}\right)^2 + \left(\frac{\alpha}{n}\right)^4 - \left(\frac{\alpha}{n}\right)^6 + \text{h.o.t.}\right]$$

$$= \sum \left(\frac{1}{n^2} - \frac{\alpha^2}{n^4} + \frac{\alpha^4}{n^6} - \frac{\alpha^6}{n^8} + \text{h.o.t.}\right)$$

$$= \frac{\pi^2}{6} - \frac{\alpha^2 \pi^4}{90} + \frac{\alpha^4 \pi^6}{945} + O(\alpha^6) \tag{4.42a}$$

If $|\alpha| > 1$, we have

$$S(\alpha) = \sum_{n=0}^{\infty} \frac{1}{\alpha^2 + n^2} = \int_0^{\infty} \frac{dx}{\alpha^2 + x^2} + \frac{1}{2\alpha^2} + \text{h.o.t.}$$

$$= \frac{\pi}{2|\alpha|} + \frac{1}{2\alpha^2} + \text{h.o.t.} \tag{4.42b}$$

We see that there is some difficulty if we want to evaluate the sum accurately. The derivatives are annoying to compute and thus the sum is not well suited for the EM expansion method. We have a simple bound on the error term by using the fact that

$$|P_1(x)| \leq \frac{1}{2} \Rightarrow \int_0^{\infty} P_1(x) f'(x)\, dx < \frac{1}{2\alpha^2} \tag{4.43}$$

It is then clear that as $|\alpha|$ goes to infinity, we have the dominant behavior given by

$$S(\alpha) = \sum_{n=0}^{\infty} \frac{1}{\alpha^2 + n^2} \Rightarrow \frac{\pi}{2|\alpha|}; \qquad |\alpha| \to \infty \tag{4.44}$$

It is possible, of course, to compute the next terms in the EM expansion and thus approximate the sum with a bounded-error term when the parameter has any value. We just emphasize that it is cumbersome to use this method. ■

Example 2.2

$$M(\alpha) = \sum_{n=0}^{\alpha} \frac{1}{\alpha^2 + n^2}; \qquad \alpha \to \infty \tag{4.45}$$

This is an interesting sum since the upper limit depends on parameter α. We first evaluate the integral:

$$I(\alpha) = \int_0^{\alpha} \frac{dx}{\alpha^2 + x^2}$$

Let $y = x/\alpha$. Then we have

$$I(\alpha) = \frac{\pi}{4\alpha}$$

So we see that

$$\begin{aligned} M(\alpha) &\approx \frac{\pi}{4\alpha} + \frac{1}{2}\left(\frac{1}{\alpha^2} + \frac{1}{2\alpha^2}\right) \\ &\approx \frac{1}{2}\left[S(\alpha) \approx \frac{\pi}{2\alpha}\right] \end{aligned}$$

where $S(\alpha)$ is given by Equation (4.44).

This leads us to conclude that half of the sum $S(\alpha)$ is contained in the first α terms. Let us see if this makes sense. Try it for $\alpha = 10$:

$$\begin{aligned} M(10) &= \sum_{n=0}^{10} \frac{1}{10^2 + n^2} = 0.086 \\ S(10) &= \sum_{n=0}^{\infty} \frac{1}{10^2 + n^2} = 0.16 \end{aligned}$$

■

Example 2.3 (Complicated) We are interested in studying the sum

$$S(\alpha) = \sum_{n=0}^{\infty} \exp(-\alpha n^2) \tag{4.46}$$

for the entire range of parameter α. This sum occurs in many branches of science since it is the discrete form of the Gaussian distribution. It also is connected to the "theta function." We first observe that when α is very small, we can capture the dominant behavior of the sum by approximating it by an integral.

$$\begin{aligned} S(\alpha) &= \sum_{n=0}^{\infty} \exp(-\alpha n^2) \approx \int_0^{\infty} \exp(-\alpha x^2)\, dx \\ &= \frac{1}{2}\sqrt{\frac{\pi}{\alpha}} \qquad \text{as} \quad \alpha \to 0 \end{aligned} \tag{4.47a}$$

On the other hand, when α is very large, we estimate the sum by the first term, since it is parameter independent, and the remaining terms are exponentially small.

$$S(\alpha) = \sum_{n=0}^{\infty} \exp(-\alpha n^2)$$

$$= 1 + \sum_{n=1}^{\infty} \exp(-\alpha n^2) = 1 + \text{e.s.t.} \qquad \text{as} \quad \alpha \to \infty \qquad (4.47\text{b})$$

We now try to find the other terms in Equations (4.47a) and (4.47b). We always sum the first few terms by hand, put in the appropriate end-point correction, evaluate the term associated with the first derivative, and then evaluate the remainder term. In this last step, we fix the value of j to satisfy our error requirements. So our problem is divided into these parts. The end-point correction and the first error terms are found immediately:

$$\tfrac{1}{2}\exp(-\alpha j^2) + \tfrac{1}{12}(2\alpha j)\exp(-\alpha j^2) \qquad (4.48)$$

The integral is an error function. You recall from Chapter 3, Section 3.3:

$$\text{Erf}(x) = \frac{2}{\sqrt{\pi}} \int_0^x \exp(-t^2)\, dt \qquad (4.49)$$

We studied the behavior for large x and since $\text{Erf}(x) = 1 - \text{Erfc}(x)$, we only have to look at the Erfc function. Recall that we obtained the asymptotic behavior by integrating by parts:

$$\text{Erfc}(x) = \frac{2}{\sqrt{\pi}} \int_x^{\infty} \exp(-t^2) \cdot \frac{2t}{2t}\, dt$$

$$= \frac{2}{\sqrt{\pi}} \left(\frac{\exp(-x^2)}{2x} - \int_x^{\infty} \frac{\exp(-t^2)}{2t^2}\, dt \right) \qquad (4.50)$$

We can construct an asymptotic series that we choose to terminate after the first term by bounding the error term as follows:

$$\int_x^{\infty} \frac{\exp(-t^2)}{2t^2}\, dt < \frac{1}{4x^3} \int_x^{\infty} \exp(-t^2)(2t)\, dt = \frac{1}{4x^3} \exp(-x^2) \qquad (4.51)$$

So we write

$$\text{Erfc}(x) = \frac{2}{\sqrt{\pi}} \exp(-x^2) \left[\frac{1}{2x} + O(x^{-3}) \right]; \qquad x \to \infty \qquad (4.52)$$

If we pick x sufficiently large, we see that $\mathrm{Erfc}(x)$ goes to zero very rapidly. You can gain some insight into its behavior by looking at a table of the Erf function (see Table 3.2 and Figure 3.5). Remember that in our problem, we have $x = j$, an integer.

We now look at the remainder term:

$$\eta = \left| \int_j^{\infty} P_3(x) f'''(x)\, dx \right| \tag{4.53}$$

We have to calculate the first three derivatives and make certain that we choose a value of j that is sufficiently large to guarantee that the integral has one sign over the region of integration. So, first of all, we have

$$f(x) = \exp(-\alpha x^2); \qquad f'(x) = -2\alpha x \exp(-\alpha x^2)$$
$$f''(x) = 2\alpha(2\alpha x^2 - 1)\exp(-\alpha x^2)$$
$$f'''(x) = (12 - 8\alpha x^2)(x\alpha^2)\exp(-\alpha x^2)$$

We see that

$$f'''(j) < 0 \qquad \text{if} \quad x^2 = j^2 > \frac{12}{8\alpha}$$

The choice $j^2 > 2/\alpha$ yields

$$\eta < \left| 10^{-2}(2\alpha)(2\alpha j^2 - 1)\exp(-\alpha j^2) \right| \tag{4.54}$$

Now let us see what happens if α is large, for example, $\alpha = 2$. Assume that we want an error term of less than 10^{-7}. We see that if we pick $j = 3$, we have

$$\eta < \left| 10^{-2}(4)(35)\exp(-18) \right| < 10^{-7} \tag{4.55}$$

Now assume that α is small. For example, $\alpha = \frac{1}{2}$. We can satisfy the same error bounds by choosing $j = 6$, since

$$\eta < 10^{-2}(1)(35)\exp(-18) < 10^{-8} \tag{4.56}$$

So we see that with a little practice, we can always find a value of j matched to the value of α to give us an acceptable error. ■

So to recapitulate: we can approximate our Gaussian sum to any desired accuracy by separating the parameter-independent term, summing the appropriate number of terms by hand, estimating the integral by the asymptotic expansion of the error function, and evaluating the end-point terms and bounding the error.

$$\sum_{n=0}^{\infty} \exp(-\alpha n^2) = \sum_{n=0}^{j-1} \exp(-\alpha n^2) + \int_j^{\infty} \exp(-\alpha x^2)\, dx$$
$$+ \tfrac{1}{2} \exp(-\alpha j^2) + \tfrac{1}{6}(\alpha j) \exp(-\alpha j^2) \pm \eta(j) \quad (4.57a)$$

Recall that

$$\operatorname{Erfc}(x) = \frac{2}{\sqrt{\pi}} \int_x^{\infty} \exp(-t^2)\, dt$$

so

$$\int_j^{\infty} \exp(-\alpha x^2)\, dx = \tfrac{1}{2}\sqrt{\frac{\pi}{\alpha}} \operatorname{Erfc}\left[\sqrt{\alpha}\, j\right]$$
$$= \frac{1}{\sqrt{\alpha}} \exp(-\alpha j^2)\left[\frac{1}{2\sqrt{\alpha}\, j} + O\left(\frac{1}{(\sqrt{\alpha})^3 j^3}\right)\right] \quad (4.57b)$$

(Observe that the error from the integral is negligible compared with the error associated with the remainder term given in Equation (4.54).)

Combining the various terms, we have

$$\sum_{n=0}^{\infty} \exp(-\alpha n^2) = \sum_{n=0}^{j-1} \exp(-\alpha n^2) + \frac{1}{2\alpha j} \exp(-\alpha j^2)$$
$$+ \tfrac{1}{2} \exp(-\alpha j^2) + \frac{\alpha j}{6} \exp(-\alpha j^2)$$
$$+ \left[\exp(-\alpha j^2)\right] O\left(\frac{1}{\alpha^2 j^3}\right) \pm \eta \quad (4.58)$$

with η given by Equation (4.54).

4.3 DIVERGENT SUMS

We now discuss divergent sums. The strategy is to identify the precise nature of the divergence and to separate it from the rest of the sum and then to obtain the numerical value of the remaining sum.

Consider the sum

$$S(N, j) = \sum_{n=1}^{N} \frac{1}{n^j}; \qquad \text{any } j \quad (4.59)$$

We want to evaluate this sum using the EM expansion, knowing that the sum diverges as N goes to infinity, if $j \leq 1$. We begin by identifying the nature of the divergence, i.e., we identify all of those terms in the EM expansion that diverge as N goes to infinity. This is accomplished by first integrating and obtaining the end-point corrections. We see if they are finite or infinite. Let us assume that they are infinite. We then look at the N dependence of the next term and see how it behaves as N goes to infinity. If it goes to zero as N goes to infinity, we know that we can bound our error term and thus we are able to stop worrying and concentrate on finding the value of the constant that is the finite difference between the sum and the divergent terms. If this first-derivative correction term does not have an N dependence of this form, we continue to take derivatives until we obtain the desired result. It becomes clear with a few examples.

Example 3.1

$$S(N,1) = \sum_{n=1}^{N} \frac{1}{n}; \qquad N \to \infty \tag{4.60}$$

We see that the integral diverges while the end-point corrections and derivative terms are finite.

$$\int_1^N \frac{dx}{x} = \ln N \tag{4.61}$$

So we consider the new sum:

$$\begin{aligned} F(N) &= S(N,1) - \ln N = \sum_{n=1}^{N} \frac{1}{n} - \ln N \\ &= \int_1^N \frac{dx}{x} + \frac{1}{2}(1 + N^{-1}) + \frac{1}{12}(1 - N^{-2}) - \ln N \\ &\quad + \text{h.o.t.} \to \text{constant} \qquad \text{as} \quad N \to \infty \end{aligned} \tag{4.62}$$

The sum $F(N)$ converges as N goes to infinity, as is shown in what follows, and we seek its value. This sum plays an important role in analysis. The constant is usually denoted by the symbol γ; it is called the "Euler constant." It comes up in other contexts; for example, we find that it occurs when you evaluate the integral

$$\int_0^\infty (\ln t) e^{-t}\, dt = -\gamma \tag{4.63}$$

(In the Appendix, we show that the integral given by Equation (4.63) does indeed yield the same constant denoted by the symbol γ that we identified in Equation (4.62).) ■

We proceed as indicated and note that we decide how many terms to sum by hand after we know what is the accuracy desired. Assume that we want three significant figures. We then can identify the quantity j that is the lower limit of the integral and therefore generates the corrections. So we can proceed as follows:

$$\underset{N\to\infty}{F(N)} = \sum_{n=1}^{j-1} \frac{1}{n} + \sum_{n=j}^{N} \frac{1}{n} - \ln N; \qquad N \to \infty \tag{4.64a}$$

$$= \sum_{n=1}^{j-1} \frac{1}{n} + \int_j^N \frac{dx}{x} + \frac{1}{2}\left(\frac{1}{j} + \frac{1}{N}\right)$$
$$+ \frac{1}{12}\left(\frac{1}{j^2} - \frac{1}{N^2}\right) - \int_j^N P_3(x) 3! x^{-4}\, dx - \ln N; \; N \to \infty \tag{4.64b}$$

Assume $j = 6$. Then $\eta < 1 \times 10^{-4}$.

$$\underset{(N\to\infty)}{F(N)} = \sum_{n=1}^{5} \frac{1}{n} - \ln 6 + \frac{1}{2 \cdot 6} + \frac{1}{12 \cdot 6^2} \pm \eta$$
$$\approx 0.5772\ldots \equiv \gamma \tag{4.64c}$$

EXERCISE 3.1 Consider the sum

$$S = \sum_{n=0}^{N} \frac{1}{n + \frac{1}{4}}; \qquad N \to \infty$$

(a) Show that the sum diverges as ln N, as N goes to infinity.
(b) Show that the finite part of the sum equals 4.227. ∎

EXERCISE 3.2 Find Euler's constant γ to eight significant figures ($\gamma \approx 0.577215665$). In this calculation, compute higher derivatives in the EM expansion. This is accomplished by returning to the derivation of the EM expansion and integrating the remainder term by parts twice more. Thus, referring to Equation (4.26)

$$\sum_{i=j}^{n+j} f(i) = \int_j^{n+j} f(x)\, dx + \tfrac{1}{2}[f(j) + f(n+j)]$$
$$+ \tfrac{1}{12}[f'(n+j) - f'(j)] + \int_j^{n+j} P_3(x) f'''(x)\, dx \tag{4.26}$$

we integrate the r.h.s. by parts, and obtain

$$\int_j^{n+j} P_3(x) f'''(x)\,dx = P_4(x) f'''(x)\Big|_j^{n+j} - \int_j^{n+j} P_4(x) f^{\text{iv}}(x)\,dx$$

$$= -\frac{1}{720}[f'''(n+j) - f'''(j)] - P_5(x) f^{\text{iv}}(x)\Big|_j^{n+j}$$

$$+ \int_j^{n+j} P_5(x) f^{\text{v}}(x)\,dx \qquad (4.65)$$

Note that $P_5(\text{integer}) = 0$ and the maximum absolute value of $P_5(x)$ satisfies the inequality

$$\max |P_5(x)| < \frac{2}{(2\pi)^5}\zeta(5) < 3 \times 10^{-4} \qquad (4.66)$$

(We know the value of $\zeta(5)$ from tables of the Riemann zeta function.) So we can write the EM formula as

$$\sum_{n=j}^{N} f(n) = \int_j^N f(x)\,dx + \frac{1}{2}[f(j) + f(N)]$$

$$+ \frac{1}{12}[f'(N) - f'(j)] - \frac{1}{720}[f'''(N) - f'''(j)]$$

$$\pm \left[\eta < (3 \times 10^{-4})\left(|f^{\text{iv}}(N) - f^{\text{iv}}(j)|\right)\right] \qquad (4.67)$$

This enables us to satisfy our error requirements by choosing a reasonable value for j. (We assume $f^{\text{v}}(x)$ has one sign on $[j, N]$.) ■

Example 3.2 Consider the alternating sum

$$S(2N+2) = \sum_{n=1}^{2N+2} (-)^{n+1}\frac{1}{n} = 1 - \tfrac{1}{2} + \tfrac{1}{3} - \tfrac{1}{4} + \cdots = \ln 2; \qquad N \to \infty \qquad (4.68)$$

We want to obtain the value of the sum using the EM expansion. We have to be careful because the infinite sum is conditionally convergent. By taking the terms two at a time, the series becomes absolutely convergent. As before, we sum the first few terms by hand and approximate the remaining sum

$$S(2N+2) = \sum_{n=1}^{2j} (-)^{n+1}\frac{1}{n} + \sum_{n=j}^{N}\left(\frac{1}{2n+1} - \frac{1}{2n+2}\right) \qquad (4.69)$$

using the EM method. We begin by calculating the integral

$$\int_j^N \left(\frac{1}{2x+1} - \frac{1}{2x+2} \right) dx = \frac{1}{2} \ln \frac{2N+1}{2N+2} - \frac{1}{2} \ln \frac{2j+1}{2j+2} \tag{4.70}$$

The end-point corrections are easy to compute and we write our error term as

$$\eta < 10^{-2} \left(\frac{2 \cdot 2 \cdot 2}{(2j+1)^3} - \frac{2 \cdot 2 \cdot 2}{(2j+2)^3} \right)$$

If we want $\eta < 10^{-4}$, we can pick $j = 4$ to satisfy our error requirements. We then have:

$$\begin{aligned} S &= \sum_{n=1}^{8} (-)^{n+1} \frac{1}{n} - \frac{1}{2} \ln \frac{9}{10} \\ &\quad + \frac{1}{2} \left(\frac{1}{9} - \frac{1}{10} \right) + \left(\frac{2}{2 \cdot 6} \right) \left(\frac{1}{9^2} - \frac{1}{10^2} \right) \pm \eta \end{aligned} \tag{4.71a}$$

Thus,

$$\begin{aligned} S &= 0.63452 + 0.05268 + \frac{1}{180} + \frac{19}{6 \cdot 8100} \pm \eta \\ &= 0.6931 \end{aligned} \tag{4.71b}$$

which is a satisfactory result. (In fact, one can obtain the same accuracy taking $j = 3$.) ■

It is also interesting to consider the sum

$$S(N, -k) = \sum_{n=0}^{N} n^k; \qquad k \text{ is a positive integer} \tag{4.72}$$

We can evaluate this sum by a variety of ways, but the EM expansion is simple and easy to employ. The integrals, end-point corrections, and derivatives are straightforward to evaluate. We notice that there are only a finite number of nonvanishing derivatives. Therefore, the error can be made identically zero. (Also, we note that there cannot be a constant term, since we are summing from the lower limit equal to 0, and otherwise we would have a finite value for the sum if $N = 0$ terms were taken.) Thus, the EM expansion can be simply used to evaluate the sum of powers of integers. For example:

$$\begin{aligned} S(N, -3) &= \sum_{n=0}^{N} n^3 = \int_0^N x^3 \, dx + \tfrac{1}{2} N^3 + \tfrac{3}{12} N^2 + 0 \\ &= \tfrac{1}{4} N^4 + \tfrac{1}{2} N^3 + \tfrac{1}{4} N^2 \end{aligned} \tag{4.73}$$

We conclude by considering

$$S(N, p) = \sum_{n=1}^{N} \frac{1}{n^p}; \qquad p < 1; \qquad n \to \infty \tag{4.74}$$

It diverges like N^{-p+1} (values for p might be $-\frac{1}{4}, -\frac{3}{2}, +\frac{1}{2}$).

With a knowledge of p, we can easily find how many of the terms in the EM expansion diverge in the limit N goes to infinity. After practicing with a few examples, we often see that a pattern emerges. (It is possible, using the theory of complex variables, to obtain a formula that connects $\zeta(s)$ and $\zeta(1-s)$. See, for example, the text by Wilf.)

Example 3.3 Evaluate the finite part of the sum $S(N, \frac{3}{4})$ to three significant figures. We sum the first $j-1$ terms by hand and then use the EM expansion. Thus,

$$\begin{aligned} S\left(N, \tfrac{3}{4}\right) = \sum_{n=1}^{N} \frac{1}{n^{3/4}} &= 1 + \frac{1}{2^{3/4}} + \frac{1}{3^{3/4}} + \cdots + \frac{1}{(j-1)^{3/4}} \\ &\quad + \int_j^N \frac{1}{x^{3/4}}\, dx + \frac{1}{2}\left(\frac{1}{j^{3/4}} + \frac{1}{N^{3/4}}\right) \\ &\quad + \frac{1}{12} \cdot \frac{3}{4}\left(\frac{1}{j^{7/4}} - \frac{1}{N^{7/4}}\right) \pm \eta \end{aligned}$$

If $j = 16$:

$$\eta < (10^{-2}) \frac{3 \cdot 7}{4 \cdot 4 \cdot 16^{11/4}} < 10^{-5}$$

$$\begin{aligned} S\left(N, \tfrac{3}{4}\right) - 4N^{1/4} &= 4.4957 - 4(16)^{1/4} \\ &\quad + \frac{1/2}{16^{3/4}} + \frac{3/48}{16^{7/4}} \pm \eta \\ &= -3.441 \qquad \text{as} \quad N \to \infty \end{aligned}$$

■

Example 3.4 Evaluate the finite part of the sum $S(N, 1/4)$ to three significant figures.

$$S\left(N, \tfrac{1}{4}\right) = \sum_{n=1}^{N} \frac{1}{n^{1/4}}; \qquad N \to \infty$$

Assume $j = 16$; thus $\eta < 10^{-5}$.

$$\begin{aligned} \left[S\left(N, \tfrac{1}{4}\right) - \tfrac{4}{3}N^{3/4}\right] &= \sum_{n=1}^{15} \frac{1}{n^{1/4}} - \tfrac{4}{3}(16)^{3/4} + \frac{1}{2}\frac{1}{16^{1/4}} + \frac{1}{12 \cdot 4 \cdot 32} \pm \eta \\ &= -0.8133 \end{aligned}$$

■

EXERCISE 3.3 Show that

$$S\left(10^6, \tfrac{1}{3}\right) - \left(\tfrac{3}{2}\right) \cdot 10^4 = \sum_{n=1}^{10^6} \frac{1}{n^{1/3}} - \left(\tfrac{3}{2}\right) \cdot 10^4$$
$$= -0.9684.$$

■

EXERCISE 3.4 Show that

$$S\left(N, -\tfrac{1}{2}\right) = \sum_{n=1}^{N} \sqrt{n}$$
$$= \tfrac{2}{3}\left(\sqrt{N}\right)^3 + \tfrac{1}{2}\sqrt{N} - 0.2079; \qquad N \to \infty$$

■

EXERCISE 3.5 Show that

$$S\left(N, \tfrac{1}{2}\right) = \sum_{n=1}^{N} \frac{1}{\sqrt{n}} = 2\sqrt{N} - 1.460; \qquad N \to \infty$$

■

Example 3.5 How does the sum

$$S(N) = \sum_{n=1}^{N-1} \frac{1}{n^2} \tag{4.75a}$$

differ from the sum

$$S(\infty) = \sum_{n=1}^{\infty} \frac{1}{n^2} = \frac{\pi^2}{6} \tag{4.75b}$$

for large but finite N? We want to exhibit a simple formula that, for example, yields this difference with an error of less than 10^{-5} for all values of $N > 10$. We accomplish this by using the EM formula for the sum

$$F(N) = S(\infty) - S(N) = \sum_{n=N}^{\infty} \frac{1}{n^2} \tag{4.75c}$$

and thereby obtain

$$S(N) = \frac{\pi^2}{6} - \left[\frac{1}{N} + \frac{1}{2 \cdot N^2} + \frac{1}{6 \cdot N^3} \pm \left(\eta < 10^{-5}\right)\right] \tag{4.75d}$$

for $N > 10$.

■

It is also interesting to consider alternating divergent sums of the form

$$S(-, N, -j) = \sum_{n=1}^{N} (-)^n n^j; \qquad j > 0; \qquad N \to \infty \tag{4.76}$$

We are required to stop the Taylor-series expansion of exp $1/(12N)$ after the term $1/(288N^2)$ since the next term in the EM expansion is $o(1/N^2)$. However, notice that we got the $1/(288N^2)$ term free; i.e., without any calculation. It is easy to continue to compute higher-order terms in the EM expansion of ln $N!$ due to the ease of differentiation of $1/x$.

$$S(-, N = 2M, -j) = -\sum_{n=1}^{M} (2n-1)^j + \sum_{n=1}^{M} (2n)^j \tag{4.77}$$

We know that the leading behavior of the sum is given by its integral. Thus, we have

$$\sum_{n=1}^{M} (2n-1)^j \sim \frac{(N-1)^{j+1}}{2(j+1)} \tag{4.78a}$$

and

$$\sum_{n=1}^{M} (2n)^j \sim \frac{N^{j+1}}{2(j+1)} \tag{4.78b}$$

Then the leading term in the sum $S(-, N, -j)$ is

$$N^{j+1}\frac{1-(1-1/N)^{j+1}}{2(j+1)}$$

We write

$$\left(1-\frac{1}{N}\right)^{j+1} \sim 1 - \frac{j+1}{N}$$

and have

$$S(-, N = 2M, -j) \sim \frac{N^j}{2}; \qquad N \to \infty \tag{4.79}$$

Finally, if N is odd, $N = 2M + 1$, we observe that the analysis is the same except that the final result has a change of sign. Thus, we conclude that

$$S(-, N, -j) \sim \pm\frac{N^j}{2}; \qquad N \to \infty \tag{4.80}$$

(It is interesting to refer to Exercise 5.5 of Chapter 2, where we encountered a similar pattern.)

EXERCISE 3.6 Use a hand calculator or simple program to find the value of $S(-, N, -j)$ for $N = 100$ and $j = 1$ and $j = \frac{1}{2}$. Notice how well these sums are approximated by the leading terms in their asymptotic formulas. Finally, observe that the value of the sum changes sign if $N = 101$. ■

We now wish to concentrate on a different class of divergent sums. We begin with the sum related to the development of an asymptotic expansion for the factorial function, namely,

$$N! = N(N-1)(N-2)(N-3)\cdots 3\cdot 2\cdot 1; \qquad N \text{ is an integer}$$

We take the log of both sides and consider the sum:

$$\ln N! = \sum_{j=1}^{N} \ln j \tag{4.81}$$

where N is large. We then exponentiate the resulting asymptotic expansion and thereby obtain Stirling's approximation for $N!$. (The justification for the exponentiation is given in Chapter 3, Section 3.2. See Example 2.8.)

We proceed by first calculating the integral and the end-point corrections. There is a constant term that comes from the lower limit of the integral, the derivative terms, and the remainder term. It can be evaluated numerically, but we will not do so as it is easily obtainable as $\frac{1}{2}\ln 2\pi$ by the Laplace method of evaluating integrals; we discuss that in Chapter 5. The other derivative terms have an N dependence such that they tend to zero as N goes to infinity. We can find them sequentially, but we are satisfied here with finding only the first two correction terms, $1/12N$ and $1/288N^2$. So we have

$$\begin{aligned}\ln N! = \sum_{j=1}^{N} \ln j &= \int_1^N (\ln x)\,dx + \frac{1}{2}\ln N + \frac{1}{12\cdot N} \\ &\quad + \left[\text{a constant} + O\left(\frac{1}{N^3}\right)\right] \\ &= (N\ln N) - N + \frac{1}{2}\ln N + \frac{1}{12\cdot N} \\ &\quad + \tfrac{1}{2}(\ln 2\pi) + O\left(\frac{1}{N^3}\right)\end{aligned} \tag{4.82}$$

(The term $\frac{1}{2}\ln 2\pi$ contains all the end-point, integral, and derivative contributions independent of N.) Note that $O(1/N^3) = o(1/N^2)$. So

$$\begin{aligned} N! &= \sqrt{2\pi N}\, N^N \exp(-N) \exp\frac{1}{12N}\left[1 + o\left(\frac{1}{N^2}\right)\right] \\ &= \sqrt{2\pi N}\, N^N \exp(-N) \\ &\quad \cdot \left[1 + \frac{1}{12 \cdot N} + \frac{1}{288 \cdot N^2} + o\left(\frac{1}{N^2}\right)\right]; \qquad N \to \infty \end{aligned} \tag{4.83}$$

We are required to stop the Taylor-series expansion of $\exp 1/(12N)$ after the term $1/(288N^2)$ since the next term in the EM expansion is $o(1/N^2)$. However, notice that we got the $1/(288N^2)$ term free; i.e., without any calculation. It is easy to continue to compute higher-order terms in the EM expansion of $\ln N!$ due to the ease of differentiation of $1/x$.

EXERCISE 3.7 Consider the sum discussed in the text by Olver, page 285:

$$S(N) = \sum_{j=1}^{N} j \ln j; \qquad N \to \infty$$

Obtain the form of the divergent terms:

$$\tfrac{1}{2}(N^2 \ln N) - \tfrac{1}{4}N^2 + \tfrac{1}{2}(N \ln N) + \tfrac{1}{12}(\ln N)$$

and the value of the finite part, 0.24875. ■

EXERCISE 3.8 Organize and perform a calculation to find the value of the constant term in the expansion of $\ln N!$ correct to three decimal places. We know that its value is $\frac{1}{2}\ln 2\pi$. ■

We conclude our discussion of the EM expansion by noting that we have been discussing sums that, together with their derivatives, have polynomiallike behavior over the interval of summation or integration. We call this property smoothness. If the function has behavior that is not of this class, often a transformation of variables is required. For example, consider the integral

$$\int_0^{10} \sqrt{x}\, dx = \left(\tfrac{2}{3}\right)10^{3/2} \tag{4.84}$$

We are going to use the EM expansion to write the integral as a sum and end-point corrections:

$$\begin{aligned} \int_0^{10} \sqrt{x}\, dx &= \sum_{n=0}^{10} \sqrt{n} \\ &\quad - \left(\frac{1}{2}\sqrt{10} + \frac{1}{12}\frac{1}{2}\left[\frac{1}{\sqrt{10}} - \frac{1}{\sqrt{0}}\right] + \text{h.o.t.}\right) \end{aligned} \tag{4.85}$$

Note that this is unsatisfactory due to the singular nature of the first derivative at $x = 0$. A simple transformation of variables eliminates this difficulty and leads to a resolution. Then we still have to decide how many points are going to be used in the sum. Assume that 100 points is sufficient. We make the transformation of variables:

$$z^2 = 1000x \tag{4.86}$$

then when $x = 10$, $z = 100$. The integral is then written as

$$\begin{aligned} &\frac{1}{1000^{3/2}} \int_0^{100} 2z^2\,dz \\ &\quad = \frac{2}{1000^{3/2}} \left[\sum_{n=0}^{100} n^2 - \frac{1}{2}(100^2) - \frac{1}{12} \cdot 2 \cdot 100 \right] + 0 \end{aligned} \tag{4.87}$$

Of course, this example is a bit artificial because there are only a finite number of derivatives. However, consider the related integral:

$$\int_0^{10} \sqrt{x}\,\ln(2 + x)\,dx \tag{4.88}$$

We employ the same transformation of variables and obtain

$$\begin{aligned} &\frac{1}{1000^{3/2}} \int_0^{100} 2z^2 \ln\left(2 + \frac{z^2}{1000}\right) dz \\ &\quad = \frac{2}{1000^{3/2}} \left[\sum_{n=0}^{100} n^2 \ln\left(2 + \frac{n^2}{1000}\right) \right. \\ &\qquad\qquad \left. - \frac{1}{2}(100^2)\ln 12 - \frac{200}{12}\ln 12 - \frac{2000}{12^2} \right] \\ &\quad + (\text{h.o.t.} = \eta) \end{aligned} \tag{4.89}$$

where the error term can be bounded by the techniques we have used in our previous examples. The message we learn from this example is that the EM expansion assumes that the derivative terms give smaller contributions than the main terms and that higher derivatives are less important than lower derivatives. Thus, it is often necessary to make a transformation of variables in order to cast our integral into proper form. This may not be possible or sufficient. One might then use the technique referred to as weakening the singularity. It is discussed in the text by Krylov.

BIBLIOGRAPHY

1. Wilf, H. S., *Mathematics for the Physical Sciences*, Wiley, New York, 1962. Reprinted by Dover, New York, 1978. This book has a clear derivation of the EM expansion and a lot of very interesting analysis and applications.
2. De Bruijn, N. G., *Asymptotic Methods in Analysis*, North-Holland, Amsterdam, 1958. Reprinted by Dover, New York, 1981. This book has an excellent discussion of the EM method and a clear derivation of the Poisson sum method. The problems are quite difficult.
3. Olver, F. W. J., *Asymptotics and Special Functions*, Academic Press, New York, 1974. We find an exceptionally fine and detailed discussion of the EM and Poisson sum expansions. Many of the details depend upon a knowledge of complex variables, but it is possible to follow the ideas. There are many excellent examples that are worked out in detail.
4. Titchmarsh, E. C., *Theory of the Riemann Zeta Function*, Oxford University Press, London, 1951. This is the classic book on the zeta function.
5. Titchmarsh, E. C., *Theory of the Fourier Integral*, Oxford University Press, London, 1948. You find here a clear discussion of the Poisson sum method.
6. Knopp, K., *Theory and Application of Infinite Series*, 2nd English Ed., Blackie, Glasgow and London, 1958. This book contains an excellent and extensive discussion of the EM expansion and a detailed error analysis. It is not difficult to read.
7. Krylov, V. I., *Approximate Calculation of Integrals*, Macmillan, New York, 1962. This is a comprehensive treatise that discusses the broad subject of numerical methods of integration at a most readable level. The section on weakening singularities is unique to this text.

CHAPTER 5

EVALUATION OF INTEGRALS: THE LAPLACE METHOD

A situation often arises in differential equations where we might want to find the long-time behavior of a system or its behavior at large distances. You might think of these problems as the counterparts to the ones associated with small values of parameters, where we naturally think of power-series solutions. Similarly, we are often interested in obtaining an estimate for the value of an integral that depends upon a large parameter.

In both these situations, we say that we are interested in the asymptotic behavior. This means that we are interested in determining the dominant behavior of the system as some parameter goes to infinity.

This leads us to the study of integrals and differential equations involving a large parameter. We discuss differential equations in Chapter 6, and in this chapter, we discuss a variety of integrals, with primary emphasis on two types.

The first type is represented by

$$F(\lambda) = \int_0^\infty \exp(-\lambda x) f(x)\, dx \tag{5.1}$$

This integral is the Laplace transform of the function $f(x)$, and although, in some instances, it can be evaluated exactly, it is often of interest to know its value for large values of parameter λ. The Laplace transform is used in problems associated with parabolic partial differential equations and differential equations with constant coefficients. However, its applicability is extensive and it is a general working tool of applied mathematics.

The second type of integral is

$$G(\lambda) = \int_0^\infty g(x) \exp(-\lambda x^2)\, dx \tag{5.2}$$

and occurs in a wide variety of physical situations, for example, in the study of the heat equation and diffusion, probability theory, and in statistical mechanics.

We have organized the chapter into four sections, gradually increasing the generality and complexity of the discussion. We begin with an overview of the philosophy behind the Laplace method. This is accomplished by looking at two integrals that depend upon large parameter λ and seeing how they behave as the parameter becomes very large, viz. λ goes to infinity. We then discuss a few more general problems in which the integrals cannot be evaluated in closed form, leading us to seek a satisfactory quantitative estimate of their behavior for large values of an appropriate parameter.

Next, we develop the Laplace method and apply it to a broad class of integrals that are already prepared properly or cast into an immediately usable form.

Then, we turn our attention to problems that involve some preparation before the Laplace method can be employed. This includes problems requiring scaling of the variables and questions associated with movable maxima. Finally, we discuss the computation of higher-order terms and error analysis. We have throughout the analysis restricted our attention to problems that do not involve complex variables in order to keep the discussion on an elementary level and thereby emphasize the basic ideas. However, the Laplace method is generalizeable and has applicability to an extraordinary range of problems. With this in mind, we have, in the Bibliography at the end of the chapter, given texts that have proofs as well as discussions of more advanced topics, including the methods of stationary phase, the method of steepest descent, and the saddle-point method.

5.1 GENERAL IDEAS

5.1.1 Some Observations

We begin by examining two integrals that depend on a parameter λ and see how the integrands behave as the parameter tends toward infinity. We can evaluate each of the integrals in closed form.

$$I(0, \lambda) = \int_0^{\infty} \exp(-\lambda t)\, dt = \frac{1}{\lambda} \tag{5.3}$$

$$J(0, \lambda) = \int_0^{\infty} \exp(-\lambda t^2)\, dt = \frac{1}{2}\sqrt{\frac{\pi}{\lambda}} \tag{5.4}$$

The first integral behaves like $1/\lambda$ and the second integral behaves like $\sqrt{1/\lambda}$.

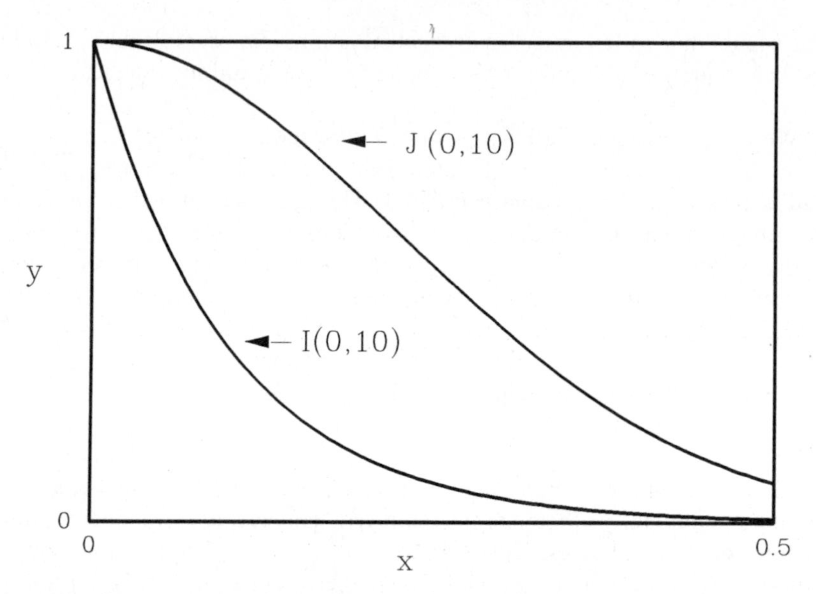

Figure 5.1 Concentrate on the neighborhood of the origin and notice how much more of the integral of $J(0,10)$ is contained there than of $I(0,10)$.

Thus, for example, if λ were equal to 10, we would have

$$I(0,10) = \frac{1}{10}$$

$$J(0,10) = \frac{1}{2}\sqrt{\frac{\pi}{10}} \approx 0.28$$

Figure 5.1 shows a graph of the integrands.
We see that the integrand of $I(0,10)$ falls rapidly from its value of 1 at the origin to $\exp(-10)$, or almost zero, by point $t = 1$. It has the value $\exp(-0.1)$ at $t = 1/100$ and then falls to $\exp(-1)$ at $t = 0.1$. So it yields its primary contribution from values of t in the interval 0 to 0.1.

By contrast, the integrand of $J(0,10)$ has a zero slope at the origin and has a value close to unity for t in the interval 0 to 0.1. In this interval, it has the value $\exp(-0.1) \leq \exp(-10t^2) \leq 1$.

So its slope is flat over a longer region. The function itself then falls to $\exp(-1)$ when $t = 1/\sqrt{10}$ and it reaches the value $\exp(-10)$ when $t = 1$.

Let us see how we can use these observations to estimate the value of

$$C_0(\lambda) = \int_0^\infty (\cos x)\exp(-\lambda x)\,dx = \frac{\lambda}{1+\lambda^2} \tag{5.5a}$$

Example 1.1 We can derive Equation (5.5a) as follows. Consider

$$I(\lambda) = \int_0^{\infty} (\cos x) \exp(-\lambda x)\, dx$$

Integrate by parts twice and obtain

$$I(\lambda) = \lambda - \lambda^2 I(\lambda)$$

or

$$I(\lambda) = \frac{\lambda}{1 + \lambda^2}$$

■

We are interested in large values of λ. Expanding Equation (5.5a), we have

$$I(\lambda) = \frac{1}{\lambda} - \frac{1}{\lambda^3} + O\left(\frac{1}{\lambda^5}\right) \tag{5.5b}$$

When $\lambda = 10$, we have

$$C_0(10) = 0.099$$

Observe that for $\lambda = 10$,

$$\cos x \approx 1 \qquad \text{for} \quad 0 \le x \le \frac{1}{10}$$

and

$$\exp(-1) \le \exp(-10x) \le 1$$

So, in our attempt to find the approximate value of the integral, we treat the cosine factor as always equal to its value at the origin, viz. = 1, and write our integral as:

$$C_0(\text{approximate}; \lambda) = \int_0^{\infty} 1 \cdot \exp(-\lambda x)\, dx = 1/\lambda \tag{5.6a}$$

$$C_0(\text{approximate}; 10) = \frac{1}{10} \tag{5.6b}$$

We see that our approximation gives us a good estimate for the integral $C_0(\lambda)$ and it improves as λ increases.

Comparing Equations (5.5a) and (5.6a), we see that

$$C_0(\text{exact}; \lambda) = C_0(\text{approximate}; \lambda)\frac{\lambda^2}{1 + \lambda^2} \tag{5.7}$$

Now look at the related integral:

$$\begin{aligned} C_1(\lambda) &= \int_0^\infty (\cos x) \exp(-\lambda x^2)\, dx \\ &= \frac{1}{2}\sqrt{\frac{\pi}{\lambda}} \exp \frac{-1}{4\lambda} \end{aligned} \tag{5.8a}$$

Example 1.2 We derive Equation (5.8a) by expanding the factor $\cos x$ as follows:

$$C_1(\lambda) = \int_0^\infty (\cos x) \exp(-\lambda x^2)\, dx = \sum_{n=0}^{\infty} (-)^n \int_0^\infty \frac{1}{(2n)!} x^{2n} \exp(-\lambda x^2)\, dx$$

Let

$$y = \lambda x^2: \qquad dx = \frac{1}{2}\frac{1}{\sqrt{\lambda y}}\, dy$$

$$C_1(\lambda) = \sum_{n=0}^{\infty} (-)^n \frac{1}{(2n)!}\frac{1}{\lambda^n}\frac{1}{2\sqrt{\lambda}}\Gamma\left(n + \frac{1}{2}\right)$$

But

$$\Gamma\left(n + \frac{1}{2}\right) = \frac{(2n)!}{2^n 2^n n!}\sqrt{\pi} = \frac{(2n)!}{4^n n!}\sqrt{\pi}$$

$$\begin{aligned} C_1(\lambda) &= \sum_{n=0}^{\infty} \frac{1}{2}\sqrt{\frac{\pi}{\lambda}} (-)^n \frac{1}{(4\lambda)^n n!} \\ &= \frac{1}{2}\sqrt{\frac{\pi}{\lambda}} \exp \frac{-1}{4\lambda} \end{aligned}$$

■

Let $\lambda = 10$; then $C_1(10) = 0.27$. We attempt an approximation, C_1(approximate), by observing that, for $\lambda = 10$,

$$\exp(-0.1) \approx 0.9 \le \exp(-10x^2) \le 1 \qquad \text{for} \quad 0 \le x \le 0.1$$

and that $\exp(-10x^2) \approx 0$ when $x \ge 1$. (Look at Figure 5.1). So, as our first guess, we replace $\cos x$ by 1 in the integral

$$C_1(\text{approximate}; \lambda) = \int_0^{\infty} 1 \cdot \exp(-\lambda x^2)\, dx$$

$$= \frac{1}{2}\sqrt{\frac{\pi}{\lambda}} \tag{5.8b}$$

$$C_1(\text{approximate}; 10) = 0.28$$

Comparing Equations (5.8a) and (5.8b), we see that

$$C_1(\text{exact}; \lambda) = C_1(\text{approximate}; \lambda) \cdot \exp\frac{-1}{4\lambda}$$

$$= \frac{1}{2}\sqrt{\frac{\pi}{\lambda}}\left[1 - \frac{1}{4\lambda} + O\left(\frac{1}{\lambda^2}\right)\right] \tag{5.9}$$

Our approximation to $C_1(\lambda)$, although fine, is less accurate than our approximation to $C_0(\lambda)$. In $C_0(\lambda)$, the factor $\exp(-\lambda x)$ has a nonzero slope at the origin that becomes steeper as λ increases. See Figure 5.2.

Thus, the multiplier function $\cos x$ really does not have an opportunity to express itself before the exponential factor has fallen to a very small value. By contrast, in our integral $C_1(\lambda)$, the factor $\exp(-\lambda x^2)$ has a zero slope at the origin and maintains a flat region over some distance.

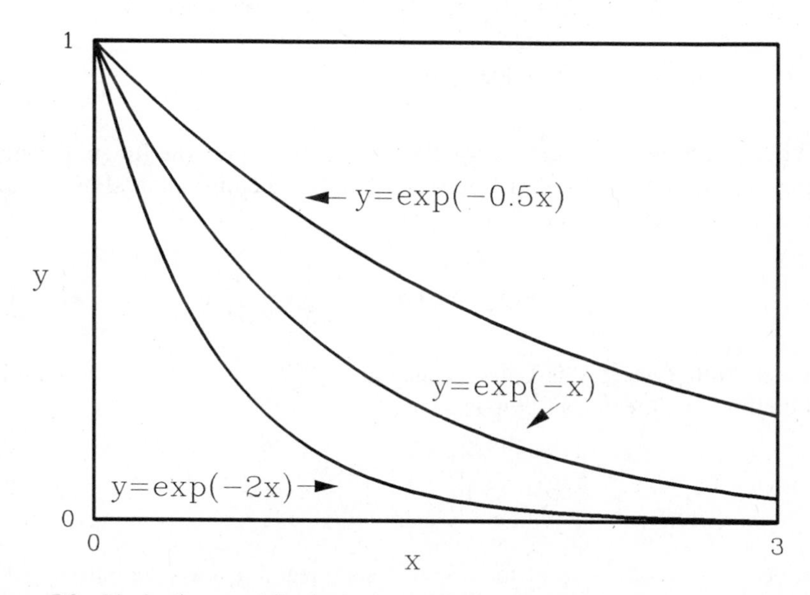

Figure 5.2 Notice how rapidly the integrand falls as the exponential factor grows.

So $\cos x$ has more opportunity to exhibit its deviation from its value at the origin. In Examples 1.1 and 1.2, we have an easy comparison between our approximate and exact results.

If λ is very large, or tends toward infinity, our approximations to both types of integrals become exceedingly good. (We postpone a further examination of the relationship between the approximation and the exact result, until Section 5.4, where we discuss the computation of higher-order terms.)

5.1.2 Introduction of the Breakpoint, ◆

With these observations, let us look at a similar problem:

$$K(\lambda) = \int_0^\infty \ln(2+x)\exp(-\lambda x)\,dx; \qquad \lambda \to \infty \tag{5.10}$$

(The problems are similar because $\cos x$ and $\ln(2 + x)$ are both nonzero at the origin.)

First, we note that we can integrate by parts as many times as we want. (Observe that in doing this in Example 1.1, we obtained the exact answer.) In this way, we can develop an expansion in inverse powers of large parameter λ:

$$\begin{aligned} K(\lambda) &= -\frac{\ln(2+x)}{\lambda}\exp(-\lambda x)\bigg|_0^\infty + \frac{1}{\lambda}\int_0^\infty \frac{1}{2+x}\exp(-\lambda x)\,dx \\ &= \frac{\ln 2}{\lambda} - \left[\frac{1}{\lambda^2}\frac{\exp(-\lambda x)}{2+x}\bigg|_0^\infty\right] - \frac{1}{\lambda^2}\int_0^\infty \frac{\exp(-\lambda x)}{(2+x)^2}\,dx \\ &= \frac{\ln 2}{\lambda} + \frac{1}{2\lambda^2} + \text{h.o.t.} \end{aligned} \tag{5.11}$$

We can use the integration-by-parts method as long as the function multiplying the exponential has finite derivatives over the entire interval of integration. However, for example, the integral

$$\int_0^\infty (\ln x)\exp(-\lambda x)\,dx = -\frac{1}{\lambda}\gamma - \frac{1}{\lambda}\ln\lambda \tag{5.12}$$

although finite (where γ is Euler's constant $= 0.5772\ldots$), cannot be done by this method, as integration by parts yields

$$-\frac{\ln x}{\lambda}\exp(-\lambda x)\bigg|_0^\infty + \frac{1}{\lambda}\int_0^\infty \frac{1}{x}\exp(-\lambda x)\,dx \tag{5.13}$$

where each term blows up at the origin. (The integral given in Equation (5.12) is discussed in the Appendix.)

We will not pursue the integration-by-parts method further since it takes us away from our general theme. We proceed directly to the development of the Laplace method since it applies to a broad spectrum of problems and can be used when the integration-by-parts method fails.

We will go through the analysis, recapitulate the procedure, and do two more examples before going on to the further development of the method. Return to Equation (5.10) and conclude that in the integral $K(\lambda)$, the largest contribution comes from small values of x. So we plan to expand the log term in a Taylor series about the origin, and since it has a finite radius of convergence, we separate the integral into two parts, indicated by $K_1(\lambda)$ and $K_2(\lambda)$:

$$K_1(\lambda) = \int_0^{\blacklozenge} \ln(2 + x) \exp(-\lambda x)\, dx \tag{5.14}$$

$$K_2(\lambda) = \int_{\blacklozenge}^{\infty} \ln(2 + x) \exp(-\lambda x)\, dx \tag{5.15}$$

$$K = K_1 + K_2$$

We choose the breakpoint, which is indicated by the symbol $\blacklozenge$, making certain that it is within the radius of convergence of the Taylor-series expansion of $\ln(2 + x)$. We purposely indicate it in this fashion to draw attention to the fact that it is finite, viz. not infinitesimal, and otherwise not important. It enters into our calculations, together with the parameter λ, as a product term that we require to go to infinity as λ goes to infinity. This becomes clear shortly.

Let us first look at the integral $K_1(\lambda)$. We expand the log term in a Taylor series and keep the first two terms.

$$K_1(\lambda) = \int_0^{\blacklozenge} (\ln 2 + \tfrac{1}{2}x) \exp(-\lambda x)\, dx + K_1(\text{error}) \tag{5.16a}$$

$$= \int_0^{\blacklozenge} (\ln 2) \exp(-\lambda x)\, dx + \tfrac{1}{2}\int_0^{\blacklozenge} x \exp(-\lambda x)\, dx + K_1(\text{error})$$

$$= K_{11} + K_{12} + K_1(\text{error}) \tag{5.16b}$$

We indicate by $K_1(\text{error})$ the contribution to $K_1(\lambda)$ that comes from the neglected terms in the Taylor-series expansion. We give a bound for $K_1(\text{error})$ at the end of this example.

We first consider the integral

$$K_{11}(\lambda) = \int_0^{\blacklozenge} (\ln 2) \exp(-\lambda x)\, dx \tag{5.17a}$$

We know that we can perform the integration exactly, and obtain

$$K_{11}(\lambda) = \frac{1}{\lambda}(\ln 2)[1 - \exp(-\lambda \blacklozenge)] \tag{5.17b}$$

Instead, we choose to make a scale transformation by the substitution $y = \lambda x$ and write

$$K_{11}(\lambda) = \frac{1}{\lambda}\int_0^{\lambda \blacklozenge} (\ln 2) \exp(-y)\, dy \tag{5.17c}$$

This removes the λ dependence in the exponential and puts it in the upper limit of integration. It is now a bit clearer why we indicated the upper limit by a symbol, i.e., $\blacklozenge$, that is finite and within the radius of convergence of the series expansion. Since λ is going to tend toward infinity, we want to think of the upper limit as tending toward infinity. Thus, we would like to write

$$K_{11}(\lambda) \to \frac{1}{\lambda}\int_0^{\infty} (\ln 2) \exp(-y)\, dy = \frac{1}{\lambda} \ln 2; \qquad \lambda \to \infty \tag{5.17d}$$

We make an error in this approximation, which is just $-[(\ln 2)/\lambda] \exp(-\lambda \blacklozenge)$ and is easily evaluated for any λ and $\blacklozenge$. Let us look at $K_{12}(\lambda)$. We write

$$K_{12}(\lambda) = \tfrac{1}{2}\int_0^{\blacklozenge} x \exp(-\lambda x)\, dx \tag{5.18a}$$

and substitute $y = \lambda x$ to obtain

$$K_{12}(\lambda) = \frac{1}{2\lambda^2}\int_0^{\lambda \blacklozenge} y \exp(-y)\, dy \tag{5.18b}$$

We can evaluate the integral exactly and obtain:

$$K_{12}(\lambda) = \frac{1}{2\lambda^2}[1 - (\lambda \blacklozenge) \exp(-\lambda \blacklozenge) - \exp(-\lambda \blacklozenge)] \tag{5.18c}$$

Our transformation of variables once again transferred the λ dependence in the exponential to the limit of integration. Notice that if we let the upper limit $\lambda \blacklozenge \to \infty$, we have the approximation

$$\begin{aligned} K_{12}(\lambda) &\to \frac{1}{2\lambda^2}\int_0^{\infty} y \exp(-y)\, dy \\ &= \frac{1}{2\lambda^2} \qquad \text{as} \quad \lambda \to \infty \end{aligned} \tag{5.18d}$$

where we are neglecting exponentially small terms.

We find that the expression "exponentially small" is pervasive in problems associated with the Laplace method. These terms are neglected since they are insignificant compared with the dominant terms. We abbreviate them as e.s.t. We now turn our attention to the integral

$$K_2(\lambda) = \int_{\blacklozenge}^{\infty} \ln(2 + x) \exp(-\lambda x)\, dx \tag{5.19}$$

It can be easily bounded if we observe the inequality

$$\ln(2 + x) = \ln 2 + \ln\left(1 + \tfrac{1}{2}x\right) \le \ln 2 + \tfrac{1}{2}x$$

or

$$\tfrac{1}{2}x > \ln\left(1 + \tfrac{1}{2}x\right); \qquad x > 0 \tag{5.20}$$

We exponentiate both sides of the inequality given by Equation (5.20) to obtain

$$e^{x/2} > 1 + \tfrac{1}{2}x; \qquad x > 0 \tag{5.21}$$

This last step is justified, since if $f(x) > g(x) > 0$ for all x, then $\exp[f(x)] > \exp[g(x)]$. We see this if we examine the ratio

$$\frac{\exp[f(x)]}{\exp[g(x)]} = \exp[f(x) - g(x)] \tag{5.22}$$

Since it was given in Equation (5.20) that $f(x) > g(x)$, we obtain the desired inequality.

Then, we can easily bound the integral

$$\begin{aligned} K_2(\lambda) &\le \int_{\blacklozenge}^{\infty} \left(\ln 2 + \tfrac{1}{2}x\right) e^{-\lambda x}\, dx \\ &= \frac{1}{\lambda}(\ln 2) e^{-\lambda\blacklozenge} + \frac{1}{2\lambda^2}(\lambda\blacklozenge) e^{-\lambda\blacklozenge} + \frac{1}{2\lambda^2} e^{-\lambda\blacklozenge} \end{aligned} \tag{5.23}$$

So the entire integral $K_2(\lambda)$ gives an exponentially small contribution to the original integral $K(\lambda)$. Finally, we consider the error term that originated with our termination of the Taylor-series expansion of $\ln(2 + x)$ after the second term. We can write

$$\ln(2 + x) = \ln 2 + \frac{x}{2} + O(x^2); \qquad 0 < x < \blacklozenge \tag{5.24}$$

leading us to

$$K_1(\text{error}) = \int_0^{\blacklozenge} O(x^2)e^{-\lambda x}\,dx = O\left(\frac{1}{\lambda^3}\right) \tag{5.25}$$

(Note that since our error term is $O(1/\lambda^3)$, if we are to be consistent, we must neglect all smaller terms.)

So we write our result as

$$\begin{aligned} K(\lambda) &= K_1(\lambda) + K_2(\lambda) + K_1(\text{error}) \\ &= \frac{1}{\lambda}(\ln 2) + \frac{1}{2\lambda^2} + O\left(\frac{1}{\lambda^3}\right) \end{aligned} \tag{5.26}$$

with the agreement that we are always neglecting e.s.t.

To summarize:

(i) We divide the interval of integration into two parts. The first contains the region where we introduce the Taylor-series expansion of the smoothly varying λ independent terms.

We introduce a scale transformation to shift the λ dependence from the exponential to the limit of integration ◆.

We then let the upper limit equal infinity since it consists of the product of the finite quantity ◆ and parameter λ, which we are considering to be very large.

We find that the entire integral associated with the second region yields an exponentially small contribution and is thus neglected.

(ii) The approximation can be improved by taking more terms in the Taylor series.

(iii) We observe that our integral is determined primarily by its behavior near the origin, and that the temporary or breakpoint upper limit of integration ◆ only played a transient role to enable us to organize the calculation. Its precise value is unimportant.

Example 1.3 Now let us practice on a related integral:

$$L(\lambda) = \int_0^{\infty} \ln(2 + x)\exp(-\lambda x^2)\,dx; \qquad \lambda \to \infty \tag{5.27}$$

and compare our result with $K(\lambda)$. We want to show that the integral given by Equation (5.27) is greater than the integral $K(\lambda)$ due to the greater region of flatness of the exponential factor in the integrand. We proceed in the same manner as we did in evaluating the integral $K(\lambda)$. Namely:

(i) We divide the interval of integration by introducing a finite quantity $\blacklozenge$ that is within the radius of convergence of the Taylor-series expansion of the log term.
(ii) We expand $\ln(2 + x)$ and keep the leading two terms.
(iii) We evaluate the associated integrals after the introduction of a scale transformation of the form $y = \sqrt{\lambda}\, x$ that has the effect of transferring the λ dependence from the integrand to the limits of integration. We then observe that we can call the upper limit infinity and in doing so only neglect e.s.t.
(iv) We show that the remaining piece of the original integral, i.e., the part that extends from the breakpoint to infinity, gives only the e.s.t.
(v) Finally, we find the order of the next contribution to $L(\lambda)$. This comes from the next term in the Taylor series.

To begin, we write

$$L(\lambda) = L_1(\lambda) + L_2(\lambda) \tag{5.28}$$

$$L_1(\lambda) = \int_0^{\blacklozenge} \ln(2 + x)\exp(-\lambda x^2)\, dx \tag{5.29}$$

$$L_2(\lambda) = \int_{\blacklozenge}^{\infty} \ln(2 + x)\exp(-\lambda x^2)\, dx \tag{5.30}$$

We know that

$$\ln(2 + x) = \ln 2 + \tfrac{1}{2}x + O(x^2); \qquad 0 < x \leq \blacklozenge \tag{5.31}$$

We then have

$$L_1(\lambda) = L_{11}(\lambda) + L_{12}(\lambda) + L_1(\text{error})$$

$$L_{11}(\lambda) = \int_0^{\blacklozenge} (\ln 2)\exp(-\lambda x^2)\, dx \tag{5.32a}$$

$$L_{12}(\lambda) = \int_0^{\blacklozenge} \tfrac{1}{2}x\exp(-\lambda x^2)\, dx \tag{5.32b}$$

We first consider $L_{11}(\lambda)$ and make the scale transformation, $y = \sqrt{\lambda}\, x$:

$$L_{11}(\lambda) = \int_0^{\sqrt{\lambda}\blacklozenge} \frac{1}{\sqrt{\lambda}}(\ln 2)\exp(-y^2)\, dy$$

or

$$L_{11}(\lambda) = \frac{1}{\sqrt{\lambda}}(\ln 2)\cdot\tfrac{1}{2}\sqrt{\pi} + \text{e.s.t.} \qquad \text{as} \quad \lambda \to \infty \tag{5.33}$$

We can see that we are neglecting an e.s.t. since

$$\int_{\sqrt{\lambda}\blacklozenge}^{\infty} \exp(-y^2)\, dy = \int_{\sqrt{\lambda}\blacklozenge}^{\infty} \frac{1}{2y}(2y)\exp(-y^2)\, dy$$

$$\leq \frac{1}{2\sqrt{\lambda}\blacklozenge} \exp(-\lambda\blacklozenge^2)$$

In a similar manner, we have, for the integral $L_{12}(\lambda)$:

$$L_{12}(\lambda) = \int_0^{\sqrt{\lambda}\blacklozenge} \left(\frac{1}{2}\frac{1}{\lambda}\right) y \exp(-y^2)\, dy \tag{5.34a}$$

$$= \frac{1}{4\lambda}(1 - \exp(-\lambda\blacklozenge^2) \tag{5.34b}$$

or

$$L_{12}(\lambda) = \frac{1}{4\lambda} + \text{e.s.t.} \qquad \text{as} \quad \lambda \to \infty \tag{5.34c}$$

Referring to our discussion of $K(\lambda)$, we see that the entire contribution of $L_2(\lambda)$ yields an e.s.t. Finally, we look at the error term:

$$L_1(\text{error}) = \int_0^{\blacklozenge} O(x^2)\exp(-\lambda x^2)\, dx \tag{5.35a}$$

Again, introducing the scale transformation, $y = \sqrt{\lambda}\, x$, we have

$$L_1(\text{error}) = \int_0^{\sqrt{\lambda}\blacklozenge} \frac{1}{\lambda} O(y^2)\exp(-y^2)\frac{1}{\sqrt{\lambda}}\, dy$$

$$= O\left(\frac{1}{\sqrt{\lambda}^3}\right) \qquad \text{as} \quad \lambda \to \infty \tag{5.35b}$$

So, putting the pieces together, we have

$$L(\lambda) = \frac{1}{2}\sqrt{\pi/\lambda}\,(\ln 2) + \frac{1}{4\lambda} + O\left(\frac{1}{\lambda^{3/2}}\right) \qquad \text{as} \quad \lambda \to \infty \tag{5.36}$$

In conclusion, we emphasize the important observation that the corresponding terms in $L(\lambda)$ are greater than those in $K(\lambda)$ due the flatness of $\exp(-\lambda x^2)$ as compared with $\exp(-\lambda x)$. ■

EXERCISE 1.1 Consider

$$M(\lambda) = \int_0^{\infty} (\cos x) \exp(-\lambda x^4)\, dx; \qquad \lambda \to \infty$$

and show that it is obvious that the leading terms go as $1/\lambda^{1/4}$ and $1/\lambda^{3/4}$, respectively. Then find the coefficients of these terms.

Answer: $\Gamma(5/4)/\lambda^{1/4} - \Gamma(7/4)/6\lambda^{3/4}$. ■

EXERCISE 1.2 Consider the integrals:

$$S(\lambda) = \int_0^{\infty} (\sin x) \exp(-\lambda x)\, dx$$

$$T(\lambda) = \int_0^{\infty} [\ln(1 + x)] \exp(-\lambda x^2)\, dx; \qquad \lambda \to \infty$$

Why does a graph of the integrands in the neighborhood of the origin immediately indicate that the integrals $S(\lambda)$ and $T(\lambda)$ are smaller than $K(\lambda)$ and $L(\lambda)$, respectively?

Find the leading term in each integral as parameter λ tends to infinity.

Answer: $S(\lambda) \sim 1/\lambda^2$ and $T(\lambda) \sim 1/2\lambda$. ■

EXERCISE 1.3 Find the leading two terms in the integral

$$W(\lambda) = \int_0^{\infty} [\ln(2 + x)] \exp(-\lambda\sqrt{x})\, dx; \qquad \lambda \to \infty$$

Comment on the λ dependence of the result.

Answer: $W(\lambda) \sim (2\ln 2)/\lambda^2 + 3!/\lambda^4$. ■

EXERCISE 1.4 The Taylor-series expansion of the Bessel function $J_0(x)$ has an infinite radius of convergence, and its power-series expansion about the origin is given by

$$J_0(x) = \sum_{k=0}^{\infty} (-)^k \frac{(x/2)^{2k}}{k!k!} = 1 - \frac{x^2}{4} + O(x^4)$$

Find the leading two terms in the integrals:

$$P(\lambda) = \int_0^{\infty} J_0(x) \exp(-\lambda x)\, dx; \qquad \lambda \to \infty$$

$$Q(\lambda) = \int_0^{\infty} J_0(x) \exp(-\lambda x^2)\, dx; \qquad \lambda \to \infty$$

Answer: $P(\lambda) \sim 1/\lambda - 1/2\lambda^3$; $\quad Q(\lambda) \sim \frac{1}{2}\sqrt{\pi/\lambda}\,(1 - 1/8\lambda)$. ■

5.1.3 Primary Conclusions

In Section 5.1, we discussed some simple examples of integrals of the form

$$F(\lambda) = \int_0^\infty f(x)\exp(-\lambda x)\,dx \qquad \text{and} \qquad G(\lambda) = \int_0^\infty g(x)\exp(-\lambda x^2)\,dx$$

given by Equations (5.1) and (5.2), respectively. The Taylor-series expansions of $f(x)$ and $g(x)$ about the origin are

$$f(x) = f(0) + xf'(0) + \frac{x^2}{2!}f''(0) + \text{h.o.t.} \tag{5.37}$$

$$g(x) = g(0) + xg'(0) + \frac{x^2}{2!}g''(0) + \text{h.o.t.} \tag{5.38}$$

We concluded the following:

1. The integrals can be well approximated by the structure of the function in the neighborhood of the origin. (There is a lemma called Watson's lemma that gives precise conditions under which the entire asymptotic behavior of an integral is determined by its Taylor-series expansion. (See the Bibliography at the end of the chapter, most particularly, the texts by Bender and Orszag and by Olver for a detailed discussion of Watson's lemma.) Thus, we used the Taylor-series expansion to find the dominant asymptotic behavior and we found the first two nonvanishing contributions to the integrals.
2. Integrals of the form $F(\lambda)$ go like $(1/\lambda)f(0)$. By contrast, those of the form $G(\lambda)$ go like $(1/\sqrt{\lambda})g(0)$, reflecting the presence of a flat region for the exponential in the neighborhood of the origin.
3. The analysis is as follows. First, we consider

$$\begin{aligned} F(\lambda) &= \int_0^\infty f(x)\exp(-\lambda x)\,dx \\ &= \int_0^\blacklozenge [f(0) + xf'(0)]\exp(-\lambda x)\,dx \\ &\quad + \int_0^\blacklozenge O(x^2)\exp(-\lambda x)\,dx + \text{e.s.t.}; \qquad \lambda \to \infty \end{aligned}$$

Let $y = \lambda x$. Then

$$F(\lambda) \sim \int_0^{\lambda\blacklozenge}\left[f(0) + \frac{y}{\lambda}f'(0)\right]\left(\frac{1}{\lambda}\right)\exp(-y)\,dy + O\left(\frac{1}{\lambda^3}\right) \tag{5.39}$$

Neglecting the e.s.t. in Equation (5.39), we have

$$F(\lambda) \sim \frac{1}{\lambda} f(0) + \frac{1}{\lambda^2} f'(0) + O\left(\frac{1}{\lambda^3}\right); \qquad \lambda \to \infty \tag{5.40}$$

In a similar manner, we also have:

$$\begin{aligned} G(\lambda) &= \int_0^\infty g(x) \exp(-\lambda x^2)\, dx \\ &= \int_0^\blacklozenge [g(0) + xg'(0)] \exp(-\lambda x^2)\, dx \\ &\quad + \int_0^\blacklozenge O(x^2) \exp(-\lambda x^2)\, dx + \text{e.s.t.} \end{aligned} \tag{5.41}$$

Let $y = \sqrt{\lambda}\, x$. Then, as with $F(\lambda)$, we rescale our variables and neglect the e.s.t. to obtain

$$G(\lambda) \sim \frac{1}{2}\sqrt{\pi/\lambda}\, g(0) + \frac{1}{2\lambda} g'(0) + O\left(\frac{1}{\lambda^{3/2}}\right); \qquad \lambda \to \infty \tag{5.42}$$

Example 1.4

$$F(\lambda) = \int_0^\infty \frac{\exp(-\lambda x)}{1+x}\, dx; \qquad \lambda \to \infty$$

$$f(0) = 1; \qquad f'(0) = -1$$

$$F(\lambda) \sim \frac{1}{\lambda} - \frac{1}{\lambda^2} + O\left(\frac{1}{\lambda^3}\right); \qquad \lambda \to \infty$$

■

Example 1.5

$$G(\lambda) = \int_0^\infty \frac{\exp(-\lambda x^2)}{1+x^2}\, dx; \qquad \lambda \to \infty$$

$$g(0) = 1; \qquad g'(0) = 0$$

$$G(\lambda) \sim \frac{1}{2}\sqrt{\pi/\lambda} + O\left(\frac{1}{\lambda^{3/2}}\right); \qquad \lambda \to \infty$$

■

Example 1.6 We can extend these results to integrals over a finite region. Consider

$$\begin{aligned} F(\lambda) &= \int_0^{\pi/2} (\sin x) \exp(-\lambda x)\, dx \\ &= \int_0^\blacklozenge \left(x - \frac{x^3}{3!}\right) \exp(-\lambda x)\, dx \\ &\quad + \int_0^\blacklozenge O(x^5) \exp(-\lambda x)\, dx + \int_\blacklozenge^{\pi/2} (\sin x) \exp(-\lambda x)\, dx \end{aligned}$$

We are given that ◆ is small, but nonzero, and we know

$$\left| \int_{\blacklozenge}^{\pi/2} (\sin x) \exp(-\lambda x)\, dx \right| < \int_{\blacklozenge}^{\pi/2} \exp(-\lambda x)\, dx = \text{e.s.t.}$$

so

$$F(\lambda) = \frac{1}{\lambda^2} - \frac{1}{\lambda^4} + O\left(\frac{1}{\lambda^6}\right); \qquad \lambda \to \infty$$

■

Example 1.7

$$F(\lambda) = \int_0^{\pi/4} (\tan x) \exp(-\lambda x)\, dx; \qquad \lambda \to \infty$$

Proceeding as in Example 1.6, we write

$$\tan x = x + \frac{x^3}{3} + O(x^5); \qquad 0 \le x \le \frac{\pi}{4}$$

Then

$$F(\lambda) = \int_0^{\blacklozenge} \left[x + \frac{x^3}{3} + O(x^5) \right] \exp(-\lambda x)\, dx$$
$$+ \int_{\blacklozenge}^{\pi/4} (\tan x) \exp(-\lambda x)\, dx$$

Since $\tan x \le 1$ on $(\blacklozenge, \pi/4)$, neglecting e.s.t., we have

$$F(\lambda) = \frac{1}{\lambda^2} + \frac{2}{\lambda^4} + O\left(\frac{1}{\lambda^6}\right)$$

■

Example 1.8

$$F(\lambda) = \int_0^{\infty} \frac{1}{\lambda + t} \exp(-\lambda t)\, dt; \qquad \lambda \to \infty$$

$$F(\lambda) = \int_0^{\blacklozenge} \frac{1}{\lambda + t} \exp(-\lambda t)\, dt + \int_{\blacklozenge}^{\infty} \frac{1}{\lambda + t} \exp(-\lambda t)\, dt; \qquad 0 < \blacklozenge \ll \lambda$$

We know

$$\int_{\blacklozenge}^{\infty} \frac{1}{\lambda + t} \exp(-\lambda t)\, dt < \frac{1}{\lambda^2} \exp(-\lambda \blacklozenge) = \text{e.s.t.}$$

and

$$\int_0^{\blacklozenge} \frac{1}{\lambda + t} \exp(-\lambda t)\, dt$$
$$= \frac{1}{\lambda} \int_0^{\blacklozenge} \left(1 - \frac{t}{\lambda} + \frac{t^2}{\lambda^2} + \cdots \right) \exp(-\lambda t)\, dt$$

So

$$F(\lambda) = \frac{1}{\lambda^2} - \frac{1}{\lambda^4} + O\left(\frac{1}{\lambda^6}\right); \qquad \lambda \to \infty \qquad \blacksquare$$

At this point we conclude these observations by considering the integrals:

$$I(p, \lambda) \equiv \int_0^{\infty} t^p \exp(-\lambda t)\, dt = \lambda^{-(p+1)} \Gamma(p+1); \qquad \lambda > 0, \quad p > -1 \tag{5.43}$$

$$J(p, \lambda) \equiv \int_0^{\infty} z^p \exp(-\lambda z^2)\, dz$$
$$= \frac{1}{2} \lambda^{-(p+1)/2} \Gamma\left(\frac{p+1}{2}\right); \qquad \lambda > 0, \quad p > -1 \tag{5.44}$$

We can cast the integral $I(p, \lambda)$ in the form of $J(p, \lambda)$ by the change of variables: $z^2 = t$ and $2z\, dz = dt$. Of course, we can proceed in the other direction by the change of variables: $\sqrt{t} = z$ and $1/(2\sqrt{t})\, dt = dz$, yielding,

$$I(p, \lambda) = 2J(2p + 1, \lambda) \tag{5.45}$$
$$J(p, \lambda) = \tfrac{1}{2} I[(p-1)/2, \lambda] \tag{5.46}$$

Example 1.9

$$I(3, \lambda) = \int_0^{\infty} t^3 \exp(-\lambda t)\, dt = 2 \int_0^{\infty} s^7 \exp(-\lambda s^2)\, ds$$
$$= 2J(7, \lambda)$$
$$I(3, \lambda) = \frac{1}{\lambda^4} \Gamma(4)$$

Which form we use depends on the problem we are considering. ■

EXERCISE 1.5 Find the first two terms in the asymptotic expansion of

$$S(\lambda) = \int_{-\infty}^{\infty} [\ln(1 + x + x^2)] \exp(-\lambda x^2)\, dx; \qquad \lambda \to \infty$$

Note that only the even terms contribute to the result.

Answer:

$$S(\lambda) = \frac{1}{4\lambda}\sqrt{\frac{\pi}{\lambda}} + \frac{3}{16\lambda^2}\sqrt{\frac{\pi}{\lambda}} + O\left(\frac{1}{\lambda^{7/2}}\right); \qquad \lambda \to \infty \qquad \blacksquare$$

5.2 THE LAPLACE METHOD

We have already studied integrals of the form

$$F(\lambda) = \int_0^\infty f(x)\exp(-\lambda x)\,dx; \qquad \lambda \to \infty$$

and

$$G(\lambda) = \int_0^\infty g(x)\exp(-\lambda x^2)\,dx; \qquad \lambda \to \infty$$

and found that it was possible to obtain, in a systematic fashion, their behavior as a power series in $1/\lambda$ or $1/\sqrt{\lambda}$. We now want to extend this class of integrals to include

$$\Phi(\lambda) = \int_a^b f(x)\exp[\lambda h(x)]\,dx; \qquad \lambda \to \infty \tag{5.47}$$

We consider functions $h(x)$ that have a single maximum at $x = c$, somewhere in the interval (a, b) including the end points. In addition, this maximum should be the largest value that $h(x)$ attains. See Figure 5.3.

The restriction to integrals that have a single maximum within the range of integration is done solely to simplify some of the arguments. Integrals with

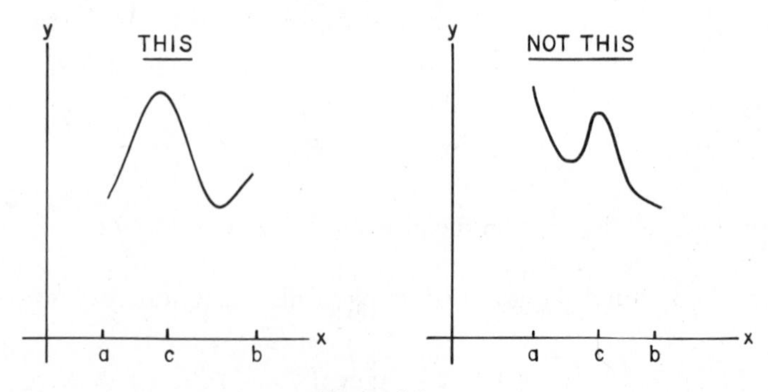

Figure 5.3

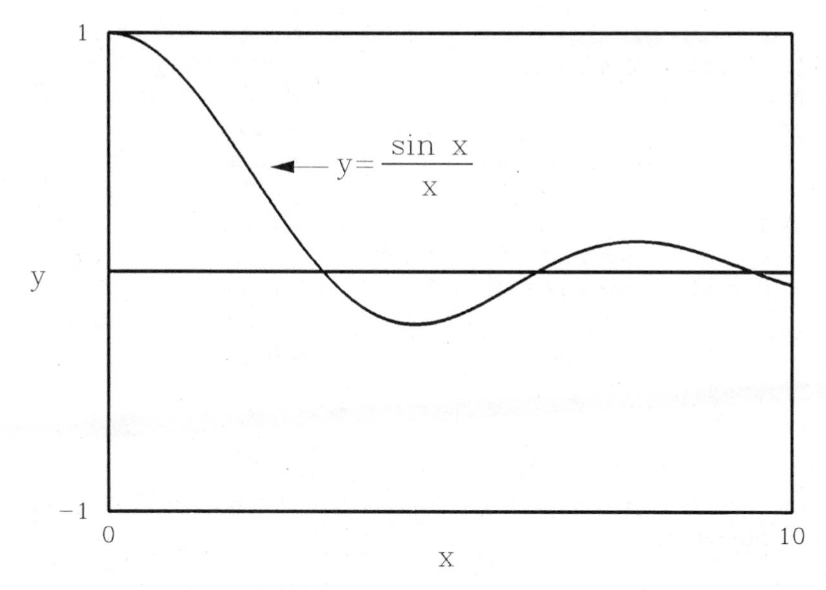

Figure 5.4 The integrand has two maxima; the second yields an exponentially smaller contribution. (See Exercise 2.3.)

several maxima can be treated by a straightforward extension of the methods that we develop. For example, we discuss in Exercise 2.3 the integral

$$\int_0^{10} \exp\left(\lambda \frac{\sin x}{x}\right) dx; \qquad \lambda \to \infty \tag{5.48}$$

The dominant behavior comes from the maximum at $x = 0$. The contribution from the maximum at $x = 7.725$ gives an exponentially smaller contribution. See Figure 5.4.

Also, integrals with integrands whose largest value exceeds its maxima are analyzable by the methods that we have used for the study of forms given by Equation (5.1). Finally, sometimes the sign of parameter λ is given as negative. The only change is to seek the minimum of $h(x)$. The Laplace method enables one to find the asymptotic behavior of the integral through a Taylor-series expansion of functions $f(x)$ and $h(x)$ about the maximum of $h(x)$. In this section, we assume the maximum to be stationary, i.e., fixed in position as λ tends to infinity.

5.2.1 Finding the Dominant Behavior

Let us assume, for the moment, that the maximum of $h(x)$ is interior to the interval (a, b). We then divide the integral into three parts:

$$\Phi(\lambda) = \int_a^{c-\blacklozenge} + \int_{c-\blacklozenge}^{c+\blacklozenge} + \int_{c+\blacklozenge}^{b} f(x) \exp[\lambda h(x)]\, dx \tag{5.49}$$

The breakpoint $\blacklozenge$ is chosen to be small but nonzero and near to the location of the maximum of $h(x)$ at $x = c$. (It plays the same role that it did in the analysis given in Section 5.1.) It is then possible to show that the two exterior integrals give exponentially small contributions relative to the integral that contains the maximum. Our basic line of argument is as follows:

(a) There is no loss of generality in assuming that $h(c) = 0$. (For if $h(c) \neq 0$, we can factor $\exp[\lambda h(c)]$ from the entire integral and replace it inside by the new factor $\exp[h(x) - h(c)]$, which is zero at $x = c$.) We also have $h'(c) = 0$ and $h''(c) < 0$. Furthermore, as $h(x)$ is to have an absolute maximum at $x = c$, we assume that

$$h(x) < -\delta \qquad \text{for} \quad |x - c| > \blacklozenge; \qquad \delta > 0, \quad a \leq x \leq b$$

(b) We assume that all the integrals are finite.

(c) Since $h(x) < 0$ in both the intervals, $(a, c - \blacklozenge)$ and $(c + \blacklozenge, b)$, we can write

$$\begin{aligned}
|\Phi_{\text{outer}}| &\equiv \left| \int_{c+\blacklozenge}^{b} f(x) \exp[\lambda h(x)]\, dx \right| \\
&< |f_{\max}| \int_{c+\blacklozenge}^{b} \exp(-\lambda\delta)\, dx \\
&= |f_{\max}| [b - (c + \blacklozenge)] \exp(-\lambda\delta) = O[\exp(-\lambda\delta)]; \qquad \lambda \to \infty \\
&= \text{e.s.t.}
\end{aligned}$$

Similarly, we conclude,

$$\Phi_{\text{inner}} \equiv \int_{a}^{c-\blacklozenge} f(x) \exp[\lambda h(x)]\, dx = \text{e.s.t.}$$

So we have

$$\Phi(\lambda) \sim \int_{c-\blacklozenge}^{c+\blacklozenge} f(x) \exp[\lambda h(x)]\, dx + \text{e.s.t.}; \qquad \lambda \to \infty \tag{5.50}$$

Thus, we have given a plausibility argument that we can find the leading term in the asymptotic expansion of the integral $\Phi(\lambda)$ by restricting our attention to the interval $(c - \blacklozenge, c + \blacklozenge)$, and in doing so, we are only neglecting e.s.t. Of course, we expect our results to be better, the sharper the maximum of $h(x)$. (It takes some effort to give a mathematically correct proof; see the Bibliography at the end of the chapter.)

Our procedure is to develop $h(x)$ in a Taylor-series expansion about this maximum and stop after the quadratic term. (More complicated situations, such as points of inflection, are discussed later in this section. Also, the computation of higher-order terms is discussed in Section 5.4.)

With these points established, we have

$$h(x) \approx h(c) + (x - c)h'(c) + \frac{(x - c)^2}{2!}h''(c) \tag{5.51a}$$

$$h'(c) = 0; \qquad h''(c) < 0$$

Let

$$\alpha \equiv \left| \frac{h''(c)}{2!} \right| \tag{5.51b}$$

Then we have

$$\Phi(\lambda) \sim \exp[\lambda h(c)] \int_{c-\blacklozenge}^{c+\blacklozenge} f(x) \exp\left[-\lambda\alpha(x - c)^2\right] dx \tag{5.52}$$

We will use Equation (5.52) as the basic form for the situations that we will analyze. Our functions $f(x)$ are assumed to look like products of x^β times polynomials ($\beta > -1$) in the neighborhood of the maximum of $h(x)$. The expansion we use is then somewhat more general than a Taylor series. Possible forms for $f(x)$ are $x^n, \ln(1 + x), \sin x, 1/(1 + x), \sqrt{x}, 1/\sqrt[3]{x}$, whereas functions like $\exp(-1/x)$ are excluded at this time. (We return to consider functions with this behavior in the exercises at the end of this section and in Section 5.3, where we consider problems that involve scaling.)

We generally assume that $h(x)$ has a maximum at $x = 0$, since this can be accomplished by a translation of variable. This means that the exponential factor $\exp[-\lambda\alpha(x - c)^2]$ dominates the integral and the function $f(x)$ is effectively slowly varying in the region near the maximum of $h(x)$. These ideas will become clear after we develop the Laplace method and do a few examples. We consider Equation (5.52) and proceed slightly differently depending on:

(a) whether the maximum of $h(x)$ is interior to the interval or at an end point of integration;
(b) whether $f(x)$ is equal or not equal to zero at the maximum of $h(x)$.

5.2.2 An Interior Maximum

Case I. Assume that $h(x)$ has its maximum at an interior point $x = c$ and that $f(c) \neq 0$. Remember, the breakpoint $\blacklozenge$ is a small but nonzero quantity. This is important, as will become clear in the analysis. We use Equation (5.52) and make the change of variables $s = \sqrt{\lambda\alpha}(x - c)$, leading us to

$$\Phi(\lambda) \sim \frac{\exp[\lambda h(c)]}{\sqrt{\lambda\alpha}} \int_{-\sqrt{\lambda\alpha}\blacklozenge}^{+\sqrt{\lambda\alpha}\blacklozenge} \exp(-s^2) f\left[\left(\frac{s}{\sqrt{\lambda\alpha}}\right) + c\right] ds \tag{5.53}$$

Since λ tends toward infinity, we have

$$\Phi(\lambda) \sim \frac{f(c)\exp[\lambda h(c)]}{\sqrt{\lambda\alpha}} \int_{-\infty}^{\infty} \exp(-s^2)\, ds$$

$$\Phi(\lambda) \approx \frac{\sqrt{2\pi}}{\sqrt{\lambda |h''(c)|}} f(c)\exp[\lambda h(c)]; \qquad \lambda \to \infty \tag{5.54}$$

Thus, we see that with $\blacklozenge$ nonzero, the neighborhood of $x = c$ gives the dominant contribution to the integral. The first correction term comes from the next term in the Taylor-series expansion of $f(x)$.

EXERCISE 2.1 Show that the leading term in

$$A(\lambda) = \int_{-1}^{1} \ln(2 + x)\exp(+\lambda\cos x)\, dx$$

$$\sim \sqrt{\frac{2\pi}{\lambda}}\,[\exp(+\lambda)](\ln 2); \qquad \lambda \to \infty$$ ■

5.2.3 Maximum at an End Point

Case II. Let $h(x)$ have its maximum value at an end point, e.g., $x = a$ and $f(a) \neq 0$. We proceed as before except now we only consider the region $(a, a + \blacklozenge)$ and neglect the contribution from the exterior region, $(a + \blacklozenge, b)$. See Figure 5.5.

With α given by Equation (5.51b) and writing a for c we then have

$$\Phi(\lambda) \sim \int_{a}^{a+\blacklozenge} f(x)\exp[\lambda h(a)]\exp\left[(-\lambda\alpha)\cdot(x-a)^2\right] dx; \qquad \lambda \to \infty \tag{5.55}$$

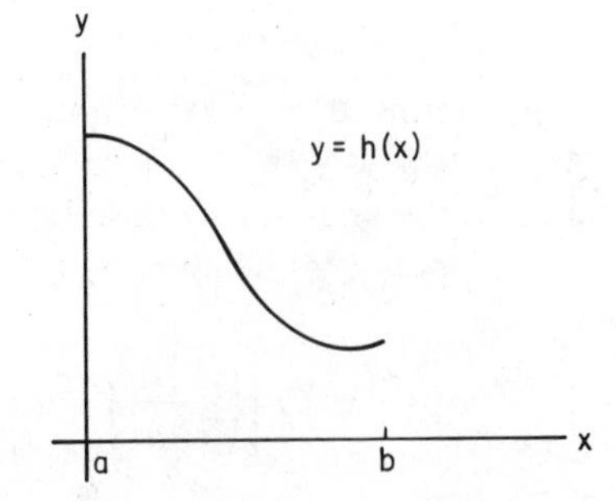

Figure 5.5

Once again, we introduce a change of variables $s = \sqrt{\lambda\alpha}(x - a)$ noting now that end point $s = 0$ does not move as parameter λ tends to infinity. With this in mind, we write

$$\Phi(\lambda) \sim \frac{\exp[\lambda h(a)]}{\sqrt{\lambda\alpha}} \int_0^{\sqrt{\lambda\alpha}\,\blacklozenge} f\left(\frac{s}{\sqrt{\lambda\alpha}} + a\right) \exp(-s^2)\, ds; \qquad \lambda \to \infty$$

Then, always neglecting e.s.t., we have

$$\Phi(\lambda) \sim \frac{1}{\sqrt{\lambda\alpha}} \exp[\lambda h(a)] \cdot f(a) \int_0^{\infty} \exp(-s^2)\, ds$$

$$= \frac{1}{2} \frac{\sqrt{2\pi}}{\sqrt{\lambda |h''(a)|}} \exp[\lambda h(a)] \cdot f(a); \qquad \lambda \to \infty \tag{5.56}$$

This is one-half of the result that we had when the maximum of $h(x)$ was at an interior point. (See Equation (5.54).)

EXERCISE 2.2 Show that

$$B(\lambda) = \int_0^{\pi/2} \frac{1}{\sqrt{1 + x}} \exp(-\lambda \cosh x)\, dx$$

$$\sim \exp(-\lambda) \frac{1}{2} \sqrt{\frac{2\pi}{\lambda}}; \qquad \lambda \to \infty$$

■

EXERCISE 2.3 Consider

$$I(\lambda) = \int_0^{10} \exp\left(\lambda \frac{\sin x}{x}\right) dx; \qquad \lambda \to \infty$$

It has a maximum at $x = 0$ and at $x = 7.725$. Divide the region of integration into two parts so that each part contains a single maximum. Establish that since $(\sin 7.725)/7.725 \approx 0.128$, the contribution from the second maximum is exponentially smaller than the first. Thus, we are safe to estimate the integral by extracting its behavior near the origin. ■

5.2.4 The Next Complications

Case III. We now turn our attention to the situation where $f(x)$ is zero at the maximum of $h(x)$ at an interior point $x = c$. We proceed as before and divide our region of integration into three parts and neglect the contributions from the exterior pieces. We then have, using Equation (5.51b),

$$\Phi(\lambda) \sim \int_{c-\blacklozenge}^{c+\blacklozenge} \exp[\lambda h(c)] f(x) \exp\left[-\lambda\alpha(x-c)^2\right] dx; \qquad \lambda \to \infty \quad (5.57)$$

We expand $f(x)$ in a Taylor series about the point $x = c$ and stop at the first nonvanishing even term. (The odd terms give zero contribution since the exponential factor is an even function of x about $x = c$.) For example, we might have $c = 0$ and

$$f(x) = \ln(1+x) = x - \frac{x^2}{2} + O(x^3) \qquad (5.58a)$$

or

$$f(x) = \frac{x}{1+x} = x - x^2 + O(x^3) \qquad (5.58b)$$

If $f(x)$ is itself a polynomial in x, the largest contribution comes from the lowest even-power term, e.g.,

$$f(x) = x + x^2 + x^3 + x^4 \qquad (5.59)$$

The x^2 term gives the dominant contribution. To see this, compare

$$\int_{-\infty}^{\infty} x^2 \exp(-\lambda x^2)\, dx = \frac{1}{\lambda^{3/2}}\Gamma\left(\frac{3}{2}\right) \qquad (5.60a)$$

with

$$\int_{-\infty}^{\infty} x^4 \exp(-\lambda x^2)\, dx = \frac{1}{\lambda^{5/2}}\Gamma\left(\frac{5}{2}\right) \qquad (5.60b)$$

Let us assume that

$$f(x) = f(c) + (x-c)f'(c) + \frac{(x-c)^2}{2!}f''(c) + \text{h.o.t.}$$

and $f''(c) \neq 0$; $f(c) = 0$. Then, with α given by Equation (5.51b), we have

$$\Phi(\lambda) \sim [\exp \lambda h(c)] f''(c) \int_{c-\blacklozenge}^{c+\blacklozenge} \frac{(x-c)^2}{2!} \exp\left[-\lambda\alpha(x-c)^2\right] dx \quad (5.61)$$

Proceeding as before, we scale our variables $s = \sqrt{\lambda\alpha}\,(x-c)$, and neglecting e.s.t., we have

$$\Phi(\lambda) \sim \frac{\exp[\lambda h(c)]}{\sqrt{\lambda\alpha}} \frac{f''(c)}{2!} \int_{-\sqrt{\lambda\alpha}\,\blacklozenge}^{\sqrt{\lambda\alpha}\,\blacklozenge} \frac{s^2}{\lambda\alpha} \exp(-s^2)\, ds$$

$$\sim \frac{\exp[\lambda h(c)]}{2(\lambda\alpha)^{3/2}} f''(c) \int_{-\infty}^{\infty} s^2 \exp(-s^2)\, ds$$

$$\Phi(\lambda) \sim \frac{1}{2} \exp[\lambda h(c)]\, f''(c) \left[\frac{2}{\lambda|h''(c)|}\right]^{3/2} \Gamma\left(\frac{3}{2}\right) \qquad \lambda \to \infty \quad (5.62)$$

EXERCISE 2.4 Show that

$$c(\lambda) = \int_{-1}^{1} \ln(1 + t^2) \exp(-\lambda \sinh^2 t)\, dt$$

$$\sim \frac{1}{\lambda^{3/2}} \Gamma(3/2)$$

■

Case IV. We now consider situations where the maximum of $h(x)$ is at end point $x = a$, and $f(a) = 0$. We include odd terms in the Taylor-series expansion of $f(x)$ about $x = a$ and then have

$$f(x) = f(a) + (x - a) f'(a) + \text{h.o.t.}; \qquad f(a) = 0, \quad f'(a) \neq 0 \quad (5.63)$$

Proceeding as we did before, we have

$$\Phi(\lambda) \sim f'(a) \exp[\lambda h(a)] \int_{a}^{a+\blacklozenge} (x - a) \exp\left[-\lambda\alpha(x - a)^2\right] dx \quad (5.64)$$

$$\Phi(\lambda) \sim \frac{1}{\lambda|h''(a)|} f'(a) \exp[\lambda h(a)]; \qquad \lambda \to \infty \quad (5.65)$$

Also note that it is possible to extend the class of functions $f(x)$ to include ones that behave like $(x - a)^{\beta}$, $\beta > -1$, in the neighborhood of the maximum of $h(x)$, at $x = a$. This follows from the finiteness of $\Gamma[(p + 1)/2]$ for $p > -1$. See Equation (5.44). We must keep $f(x)$ inside the integral except when β is a nonnegative integer.

EXERCISE 2.5 Show that

$$A(\lambda) = \int_{0}^{1} \sqrt{x} \exp(-\lambda \tan^2 x)\, dx \sim \frac{1}{2} \Gamma\left(\frac{3}{4}\right) \frac{1}{\lambda^{3/4}}; \qquad \lambda \to \infty$$

■

Case V. Consider a situation where $h(x)$ has a point of inflection at the lower limit of integration, $x = a$. Then $h(x) = h(a) + [(x-a)^3/3!]h'''(a) +$ h.o.t.; $h'(a) = h''(a) = 0$; $h'''(a) < 0$. Proceeding as before, we confine our attention to the interval $(a, a + \blacklozenge)$. We introduce $\alpha \equiv |h'''(a)|/3!$ and the change of variables: $s = \sqrt[3]{\lambda\alpha}(x-a)$. Then the analog of Equation (5.52) takes the form

$$\Phi(\lambda) \sim \exp[\lambda h(a)]\frac{1}{\sqrt[3]{\lambda\alpha}}\int_0^{\sqrt[3]{\lambda\alpha}\,\blacklozenge} f\left(a + \frac{s}{\sqrt[3]{\lambda\alpha}}\right)\exp(-s^3)\,ds$$

Assuming $f(a) \neq 0$, we obtain

$$\Phi(\lambda) \sim \frac{\exp[\lambda h(a)]}{\sqrt[3]{\lambda\alpha}}\Gamma\left(\frac{4}{3}\right)\cdot f(a); \qquad \lambda \to \infty \tag{5.66}$$

5.3 PROBLEMS THAT NEED SOME PREPARATION

We now turn our attention to problems that have to be organized into the proper form for the Laplace method to be applicable. The problems we treat have movable maxima and, thus, require scaling of the variables. This follows since the Laplace method requires that the location of the maximum of the integrand remain fixed in position as the large parameter tends to infinity. Why? With a fixed maximum, the integrand has the potential to become more and more sharply peaked as λ tends to infinity. Then it looks more and more like a delta function and our approximation is quite satisfactory. By contrast, if the maximum moves as λ increases, the dominant contribution tends to elude our grasp. (See Exercise 3.1.)

We illustrate this type of problem by developing an asymptotic expression for $n!$; viz. Stirling's approximation for $n!$. We want to know the behavior of $n!$ defined by the integral

$$n! \equiv \int_0^\infty t^n \exp(-t)\,dt \tag{5.67}$$

for large values of n. Our integral is not in the standard form since we do not have an integrand that is the product of a polynomiallike term multiplied by an exponential function that has a single sharp maximum. In addition, t^n grows rapidly as n increases and $\exp(-t)$ falls, so the position of the maximum shifts as n tends toward infinity. To see this clearly, we write our integral as

$$n! = \int_0^\infty \exp[n(\ln t) - t]\,dt = \int_0^\infty \exp[g(t)]\,dt \tag{5.68}$$

We locate the maximum of $g(t)$:

$$g'(t) = \frac{n}{t} - 1 = 0, \qquad \text{when} \quad t = n \tag{5.69}$$

Thus, the maximum moves as n tends to infinity.

EXERCISE 3.1 Consider $n!$ as given by Equation (5.67) for $n = 2$, 5, and 10. In each case, locate the maximum of the integrand. Then obtain a crude measure of the region that yields the dominant contribution by finding where the integrand falls to one-half of its maximum value. Draw a graph so as to conclude that there is a significant broadening of the peak as n increases. It is with this result in mind that we see how essential it is to rescale our variables so as to force the peak to sharpen as n increases and thereby obtain a good estimate of the integral through the Laplace approximation.

We rescale the variable t by the change of variables $t = ny$, $dt = n\,dy$, observing that we now have a fixed maximum since our integral becomes

$$\begin{aligned} n! &= \int_0^\infty \exp[n(\ln n) + n \ln y - ny]\, n\,dy \\ &= n \exp(n \ln n) \int_0^\infty \exp[n(\ln y - y)]\, dy \\ &= n^{n+1} \int_0^\infty \exp[nh(y)]\, dy \end{aligned} \tag{5.70}$$

Then

$$h'(y) = \frac{1}{y} - 1 = 0 \qquad \text{when} \quad y = 1$$

$$h''(y) = \frac{-1}{y^2} = -1 \qquad \text{when} \quad y = 1$$

We expand $h(y)$ about the point $y = 1$, neglecting higher than quadratic terms in the exponential

$$n! \sim n^{n+1} \int_0^\infty \exp\left\{ n\left[\ln 1 - 1 + 0 - \frac{1}{2!}(y-1)^2 \right] \right\} dy \tag{5.71a}$$

We now make the change of variable, $w = y - 1$, and get

$$n! \sim n^{n+1} \exp(-n) \int_{-1}^\infty \exp\left(-\tfrac{1}{2} n w^2\right) dw \tag{5.71b}$$

The lower limit is now -1, rather than 0 and the maximum, at $w = 0$, is therefore interior to the interval of integration. This integral then falls under the category of Case I. Thus, using Equation (5.54), we have

$$n! \sim n^{n+1} \exp(-n) \sqrt{\frac{2\pi}{n}}; \qquad n \to \infty$$

or

$$n! \sim \sqrt{2\pi n}\, n^n \exp(-n); \qquad n \to \infty \tag{5.72}$$

Finally, we note that we did not use the fact that n was an integer, so we also have

$$\Gamma(\lambda + 1) = \sqrt{2\pi\lambda}\, \lambda^\lambda \exp(-\lambda); \qquad \lambda \to \infty \qquad \blacksquare \tag{5.73}$$

Example 3.1 We are interested in the asymptotic behavior of

$$I(\lambda) = \int_0^\infty \exp\left(-\lambda x - \frac{1}{x}\right) dx; \qquad \lambda \to \infty \tag{5.74}$$

Our integral is not in standard form since the multiplier function $\exp[-(1/x)]$ does not behave like a polynomial near $x = 0$. Noting this, we write it as

$$I(\lambda) = \int_0^\infty \exp[g(x)]\, dx \tag{5.75}$$

We locate the maximum,

$$g'(x) = -\lambda + \frac{1}{x^2} = 0, \qquad \text{when} \quad x = \frac{1}{\sqrt{\lambda}}$$

and notice that it is dependent on variable λ.

We make it stationary by rescaling through the change of variables, $x = (1/\sqrt{\lambda}) \cdot t$, since then t takes on the value 1 at the maximum. (Note that we performed a similar scaling in our development of an asymptotic approximation for $n!$.) Our integral becomes

$$\begin{aligned} I(\lambda) &= \frac{1}{\sqrt{\lambda}} \int_0^\infty \exp\left[-\sqrt{\lambda} \cdot \left(t + \frac{1}{t}\right)\right] dt \\ &= \frac{1}{\sqrt{\lambda}} \int_0^\infty \exp[-\sqrt{\lambda}\, h(t)]\, dt \end{aligned} \tag{5.76}$$

We now locate the minimum of $h(t)$:

$$h'(t) = 1 - \frac{1}{t^2} = 0, \qquad \text{when} \quad t = 1$$

$$h''(t) = \frac{2}{t^3} = 2, \qquad \text{when} \quad t = 1$$

We then expand $h(t)$ about $t = 1$ and proceed with the Laplace method and write

$$I(\lambda) = \frac{1}{\sqrt{\lambda}} \int_0^\infty \exp\left(-\sqrt{\lambda}\left[1 + 1 + 0 + \frac{2}{2!}(t-1)^2\right]\right) dt \tag{5.77}$$

We introduce the breakpoint ◆ and the change of variables $y = \lambda^{1/4} \cdot (t - 1)$ to obtain

$$\begin{aligned} I(\lambda) &\sim \frac{1}{\lambda^{3/4}} \exp(-2\sqrt{\lambda}) \int_{-\lambda^{1/4}◆}^{\lambda^{1/4}◆} \exp(-y^2)\, dy \\ &\sim \frac{1}{\lambda^{3/4}} \exp(-2\sqrt{\lambda}) \int_{-\infty}^{\infty} \exp(-y^2)\, dy \\ &\sim \frac{1}{\lambda^{3/4}} \sqrt{\pi} \exp(-2\sqrt{\lambda}) \end{aligned}$$

■

EXERCISE 3.2 Show that the leading term in the asymptotic expansion of

$$J(\lambda) = \int_0^\infty \exp\left(-\lambda t - \frac{8}{t^2}\right) dt; \qquad \lambda \to \infty$$

is

$$J(\lambda) \sim \sqrt{\frac{\pi}{3}} \frac{1}{\lambda^{2/3}} 2^{7/6} \exp\left[-3(2\lambda^2)^{1/3}\right]; \qquad \lambda \to \infty$$

■

(Note that problems with movable maxima require more than the usual care. Refer to the texts by Olver and Bender and Orszag for a discussion of more advanced problems and situations.)

5.4 HIGHER-ORDER TERMS

We will now explain how to find higher-order terms in the asymptotic expansion of integrals of the form

$$\Phi(\lambda) = \int_a^b f(x) \exp[\lambda h(x)]\, dx; \qquad \lambda \to \infty \tag{5.78}$$

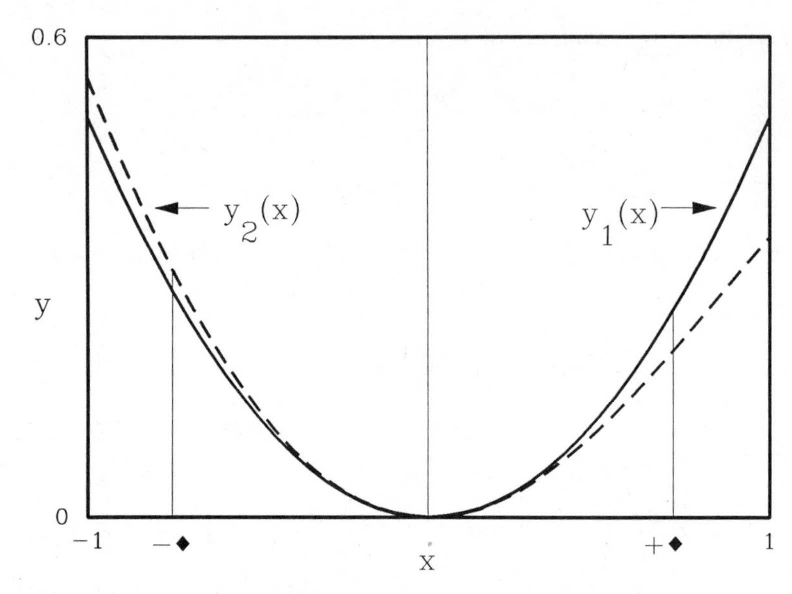

Figure 5.6 Observe how the dominant contribution comes from the quadratic term. It is then our responsibility to organize our calculation to reflect this. $y_1 = \frac{1}{2}x^2$; $y_2 = \frac{1}{2}x^2 - \alpha x^3 - \beta x^4$; $\alpha = 0.1$; and $\beta = 0.05$.

with $h(x)$ having an absolute maximum at $x = c$ and where $f(x)$ looks like $(x - c)^\beta$, $\beta > -1$, in the vicinity of the maximum of $h(x)$. It is algebraically cumbersome to develop the method in general, so we take the approach of explaining it by means of a few examples. The books in the Bibliography have more detailed discussions.

Example 4.1 We discuss a standard problem in statistical mechanics in which one is asked to find the contribution of a given potential $V(x)$ to the heat capacity. In our analysis, we follow the discussion of this problem as it is given in *Statistical Mechanics* by Ryogo Kubo, North-Holland, Amsterdam, 1965, p. 135. Consider a one-dimensional system described by an almost quadratic potential of the form

$$V(x) = \tfrac{1}{2}x^2 - \alpha x^3 - \beta x^4 \tag{5.79}$$

See Figure 5.6.

The system is in a heat bath maintained at a temperature T and we want to evaluate the integral

$$Z(T) = \int_{-\infty}^{\infty} \exp\left(-\frac{V(x)}{kT}\right) dx \tag{5.80}$$

where k is Boltzmann's constant, in the region of very low temperatures, viz. $T \to 0$. We begin by making the observation that the quadratic term is dominant and yields its major contribution in the region $|x| < \sqrt{2kT}$. The cubic and quartic terms in $V(x)$ are corrections to the quadratic term $\frac{1}{2}x^2$. This leads us to require that the coefficients of the cubic term α and the quartic term β respectively satisfy the inequalities

$$\frac{1}{2}\frac{1}{\sqrt{2kT}} \gg \alpha, \qquad \frac{1}{4kT} \gg \beta \tag{5.81}$$

The reason for choosing this form of the inequalities will become clear shortly. Note that since we are interested in the range of temperatures near $T = 0$, the inequalities are quite liberal. We divide our integral into three parts:

$$Z(T) = \int_{-\infty}^{-\blacklozenge} + \int_{-\blacklozenge}^{\blacklozenge} + \int_{\blacklozenge}^{\infty} \tag{5.82}$$

The cutoff $\blacklozenge$ is a fixed nonzero quantity, perhaps as shown in Figure 5.6.

As T tends toward 0, $\exp(-x^2/2kT)$ will have fallen to a value near zero at points $x = \pm\blacklozenge$. We already know, from our previous analysis, that the integrals

$$\int_{-\infty}^{-\blacklozenge} \quad \text{and} \quad \int_{\blacklozenge}^{\infty}$$

lead to e.s.t., and so we neglect them. Let us examine the three factors in the integral

$$\int_{-\blacklozenge}^{\blacklozenge} \exp\left(-\frac{1}{kT}\frac{1}{2}x^2\right) \exp\frac{\alpha x^3}{kT} \exp\frac{\beta x^4}{kT}\, dx \tag{5.83}$$

We know that the term $x^2/2kT$ is significant when $|x| < \sqrt{2kT}$, and that over this entire region, the cubic and quartic terms, αx^3 and βx^4, respectively, are required to lead to small corrections as T tends toward 0. This leads to the inequalities

$$\frac{x^2}{2kT} \gg \frac{\alpha x^3}{kT}$$

or

$$\frac{1}{2x} \gg \alpha \quad \text{or} \quad \frac{1}{2}\frac{1}{\sqrt{2kT}} \gg \alpha \tag{5.84a}$$

and

$$\frac{x^2}{2kT} \gg \frac{\beta x^4}{kT}$$

or

$$\frac{1}{2x^2} \gg \beta \quad \text{or} \quad \frac{1}{4kT} \gg \beta \tag{5.84b}$$

With these inequalities established, we expand both the cubic and quartic terms in a Taylor-series expansion about $x = 0$, the position of the maximum of the quadratic term. It should now be clear why we imposed the original inequalities on α and β.

$$Z(T) \sim \int_{-\blacklozenge}^{\blacklozenge} \left(\exp \frac{-x^2}{2kT}\right)\left(1 + \frac{\alpha x^3}{kT} + \frac{\alpha^2 x^6}{2!(kT)^2}\right) \cdot \left(1 + \frac{\beta x^4}{kT}\right) dx + \text{h.o.t.} \tag{5.85}$$

This method of expanding the correction terms is the central point in the computation of higher-order terms in the asymptotic expansion of Laplace-type integrals. Specifically, the correction terms, or, equivalently, the higher-order terms in the Taylor-series expansion of the function that has a maximum, e.g., the potential $-V(x)$, are themselves expanded in a Taylor series about the maximum. This procedure guarantees that these terms remain correction terms as we extend the region of integration from $-\infty$ to $+\infty$. We make the scale transformation $y = x/\sqrt{2kT}$, and since we are interested in the region where T tends toward zero, we write

$$Z(T) \sim \int_{-\infty}^{\infty} \exp(-y^2)\left[1 + 2\alpha y^3\sqrt{2kT} + \frac{4\alpha^2 y^6}{2!} 2kT\right] \cdot \left[1 + 2\beta y^4(2kT)\right]\sqrt{2kT}\, dy \tag{5.86}$$

neglecting higher-order terms in the Taylor-series expansion and the e.s.t. arising from the intervals

$$\left(-\infty, -\blacklozenge/\sqrt{2kT}\right) \quad \text{and} \quad \left(\blacklozenge/\sqrt{2kT}, \infty\right).$$

The integrals are gamma functions, and we find that

$$Z(T) \sim \sqrt{2\pi kT}\left[1 + \left(3\beta + \frac{15}{2}\alpha^2\right)kT\right] + O(T^{5/2}); \qquad T \to 0 \tag{5.87}$$

Finally, we make the observation that in order to compute the correction term proportional to T, it is necessary to include the deviations from the quadratic term arising from both the cubic and quartic terms. This is a general result and can be traced to the fact that the maximum of the function $-V(x)$ was interior to the interval of integration, resulting in an integral from $-\infty$ to ∞. Then, as the lowest order correction is associated with an odd term, it vanishes upon integration. If, the maximum had been at the boundary, we would have had a contribution from this order term, i.e., $O(T)$. ■

Example 4.2 We seek higher-order terms in Stirling's approximation to

$$n! = \int_0^\infty t^n \exp(-t)\, dt = \int_0^\infty \exp[n(\ln t) - t]\, dt \tag{5.88}$$

We show that

$$n! \sim \sqrt{2\pi n}\, n^n \exp(-n)\left[1 + \frac{1}{12n} + O(n^{-2})\right] \tag{5.89}$$

Briefly reviewing our previous discussion of this problem, we note that the integrand has its maximum value at $t = n$, requiring us to introduce the scaling transformation $t = ny$ in order that the maximum be stationary. We then write our integral as

$$n! = n^{n+1} \int_0^\infty \exp[n(\ln y - y)]\, dy \tag{5.90}$$

and noting that the maximum of $h(y)$ is at $y = 1$, we divide our region of integration into three parts:

$$\int_0^{1-\blacklozenge} + \int_{1-\blacklozenge}^{1+\blacklozenge} + \int_{1+\blacklozenge}^{\infty}$$

and neglect the contributions from the regions exterior to the maximum of $h(y)$. The breakpoint $\blacklozenge$ is chosen to be very close to $y = 1$. The reason for this will be clearer after we compute the higher derivatives of $h(y)$. We evaluate all of the derivatives at $y = 1$:

$$\begin{aligned}
h(y) &= \ln y - y && = -1 \\
h'(y) &= \frac{1}{y} - 1 && = 0 \\
h''(y) &= \frac{-1}{y^2} && = -1 \\
h^n(y) &= (-)^{n-1}\frac{(n-1)!}{y^n} && = (-)^{n-1}(n-1)! \quad \text{for} \quad n \geq 2
\end{aligned}$$

Thus, we have

$$h(y) = -1 - \tfrac{1}{2}(y-1)^2 + \tfrac{1}{3}(y-1)^3 - \tfrac{1}{4}(y-1)^4 + O\left[(y-1)^5\right] \tag{5.91}$$

With these points established, we go back to Equation (5.90) for $n!$ and include the higher-order terms in the Taylor series:

$$n! \sim n^{n+1} \exp(-n) \int_{1-\blacklozenge}^{1+\blacklozenge} \exp\left[-\frac{n}{2}(y-1)^2\right] \cdot \exp\left[\frac{n}{3}(y-1)^3\right] \exp\left[-\frac{n}{4}(y-1)^4\right] dy \tag{5.92}$$

We introduce the change of variables $w = y - 1$:

$$n! \sim n^{n+1} \exp(-n) \int_{-\blacklozenge}^{\blacklozenge} \exp\frac{-nw^2}{2} \exp\frac{+nw^3}{3} \exp\frac{-nw^4}{4}\, dw \tag{5.93}$$

Since the coefficients of the exponential terms alternate in sign, at first glance one might think that the term $\exp(+nw^3/3)$ might overwhelm the term $\exp(-nw^2/2)$ and lead to a result that a correction term is larger than the primary term. However, remember that the maximum is stationary at $y = 1$, and the breakpoint $\blacklozenge$ is chosen to be sufficiently small. Indeed, we pick it to be such that $\exp(+nw^3/3)$ and $\exp(-nw^4/4)$ are very close to 1 for $|w| \le \blacklozenge$. For example, choose $\blacklozenge \ll 1$. Then the correction terms take on their largest value at the end points, where we have the inequalities

$$\left|-\frac{n\blacklozenge^2}{2}\right| > \left|\frac{n\blacklozenge^3}{3}\right| > \left|-\frac{n\blacklozenge^4}{4}\right|$$

Since the choice of the breakpoint is at our disposal, we can be confident that the correction terms are indeed small over the entire range of integration and that it is appropriate to develop them in a Taylor-series expansion. (There is a detailed discussion of this procedure and its justification in the text by Bender and Orszag.) With these observations, we write

$$\exp\frac{nw^3}{3} = 1 + \frac{nw^3}{3} + \frac{n^2w^6}{9 \cdot 2!} + O(n^3)$$

$$\exp\frac{-nw^4}{4} = 1 - \frac{nw^4}{4} + O(n^2)$$

So we have

$$n! \sim n^{n+1} \exp(-n) \int_{-\blacklozenge}^{\blacklozenge} \left(\exp\frac{-nw^2}{2}\right)\left(1 + \frac{nw^3}{3} + \frac{n^2w^6}{18}\right) \cdot \left(1 - \frac{nw^4}{4}\right) dw + \text{h.o.t.} \tag{5.94}$$

Introduce the change of variables $s = \sqrt{n/2}\,w$, leading to

$$n! \sim n^{n+1} \exp(-n) \sqrt{\frac{2}{n}} \int_{-\sqrt{n/2}\,\blacklozenge}^{+\sqrt{n/2}\,\blacklozenge} \exp(-s^2)$$

$$\cdot \left[\left(1 + \sqrt{\frac{2}{n}} \frac{2s^3}{3} + \frac{4s^6}{9n} \right) \left(1 - \frac{s^4}{n} \right) + O\left(\frac{1}{n^2} \right) \right] ds \tag{5.95}$$

We observe that both the third and fourth derivative terms give corrections of $O(1/n)$. We cannot take more terms in the expansions of $\exp(nw^3/3)$ and $\exp(-nw^4/4)$ since they would be terms of $O(1/n^2)$. If we included them, it would be necessary to compute both the fifth and sixth derivatives of $h(y)$ since they would contribute terms of equal order. With these points established, we write our Laplace approximation as

$$n! \sim n^{n+1} \exp(-n) \sqrt{\frac{2}{n}} \int_{-\infty}^{\infty} \exp(-s^2)$$

$$\cdot \left[1 + \sqrt{\frac{2}{n}} \frac{2s^3}{3} + \frac{1}{n} \left(\frac{4}{9} s^6 - s^4 \right) \right] ds \tag{5.96}$$

Since the integral goes from $-\infty$ to $+\infty$, the s^3 integral vanishes. In addition, we know that

$$\int_{-\infty}^{\infty} \exp(-s^2)\, ds = \sqrt{\pi}\,; \qquad \int_{-\infty}^{\infty} \exp(-s^2) s^6\, ds = \Gamma\left(\frac{7}{2} \right);$$

$$\int_{-\infty}^{\infty} \exp(-s^2) s^4\, ds = \Gamma\left(\frac{5}{2} \right)$$

yielding

$$n! \sim n^{n+1} \exp(-n) \sqrt{\frac{2}{n}} \cdot \left[\sqrt{\pi} + \frac{1}{n} \left(\frac{4 \cdot 5}{9 \cdot 2} - 1 \right) \Gamma\left(\frac{5}{2} \right) \right] \tag{5.97}$$

and our final result is

$$n! \sim \sqrt{2\pi n}\, n^n \exp(-n) \left[1 + \frac{1}{12n} + O\left(\frac{1}{n^2} \right) \right] \qquad \blacksquare$$

EXERCISE 4.1 Compute the next correction term $= 1/288n^2$. As a side remark, recall that the Euler–MacLaurin expansion yields these terms with a minimum of effort. However, it is only able to find the numerical value of the constant $\sqrt{2\pi}$. ■

Example 4.3 Consider the integral

$$I_0(\lambda) = \frac{1}{\pi}\int_0^{\pi} \exp(\lambda \cos \alpha)\, d\alpha; \qquad \lambda \to \infty \tag{5.98}$$

where $I_0(\lambda)$ is the modified Bessel function of zero order. We want to find the leading two terms in the asymptotic expansion. We know that the maximum of $\cos\alpha$ is at $\alpha = 0$. We divide the region of integration into two parts: $(0, \blacklozenge)$ and $(\blacklozenge, \pi)$, where $\blacklozenge$ has a small but nonzero value. The Taylor-series expansion of $\cos\alpha$ is

$$\cos\alpha = 1 - \frac{\alpha^2}{2!} + \frac{\alpha^4}{4!} + \text{h.o.t.}$$

We stop at the quartic term and note that $\blacklozenge$ is chosen so that the quadratic term is significantly bigger than the quartic one. This obliges us to expand the quartic term in a Taylor series, and to write our integral as

$$\begin{aligned} I_0(\lambda) &\sim \frac{1}{\pi}\int_0^{\blacklozenge} \exp\lambda \exp\frac{-\lambda\alpha^2}{2!} \exp\frac{+\lambda\alpha^4}{4!}\, d\alpha \\ &= \frac{1}{\pi}\exp\lambda \int_0^{\blacklozenge}\left(\exp\frac{-\lambda\alpha^2}{2}\right)\left(1 + \frac{\lambda\alpha^4}{24}\right) d\alpha + \text{h.o.t.} \end{aligned} \tag{5.99}$$

It is important to note that we are not permitted to continue the expansion by computing the term $(1/2!)(\lambda\alpha^4/24)^2$ since we stopped our Taylor-series expansion of $\cos\alpha$ at the fourth derivative, thereby neglecting terms of order α^6 arising from the sixth derivative. We introduce the change of variables $z = \sqrt{\lambda/2}\,\alpha$, and our integral becomes

$$I_0(\lambda) = \frac{1}{\pi}(\exp\lambda)\sqrt{\frac{2}{\lambda}} \cdot \int_0^{\sqrt{\lambda/2}\,\blacklozenge} \exp(-z^2)\left(1 + \frac{1}{24}\frac{4}{\lambda}z^4\right) dz \tag{5.100}$$

The integrals are gamma functions that yield

$$I_0(\lambda) \sim \frac{1}{\sqrt{2\pi\lambda}}(\exp\lambda)\left[1 + \frac{1}{8\lambda} + O\left(\frac{1}{\lambda^2}\right)\right]; \qquad \lambda \to \infty \quad \blacksquare \tag{5.101}$$

EXERCISE 4.2 Given

$$I(\lambda) = \int_0^{\pi/2} \exp(-\lambda \sin^2 t)\, dt$$

show that

$$I(\lambda) = \frac{1}{2}\sqrt{\frac{\pi}{\lambda}}\left[1 + \frac{1}{4\lambda} + O\left(\frac{1}{\lambda^2}\right)\right]; \qquad \lambda \to \infty \qquad \blacksquare$$

EXERCISE 4.3 Given

$$I(\lambda) = \int_0^{\pi^2/4} \exp(\lambda \cos \sqrt{t})\, dt$$

show that

$$I(\lambda) = \frac{2 \exp \lambda}{\lambda}\left[1 + \frac{1}{3\lambda} + \frac{4}{15\lambda^2} + O\left(\frac{1}{\lambda^3}\right)\right]; \qquad \lambda \to \infty$$

■

EXERCISE 4.4 Given

$$I(\lambda) = \int_0^{\infty} t^{\lambda} \exp(-t)(\ln t)\, dt$$

show that

$$I(\lambda) \sim \sqrt{2\pi}\, \lambda^{\lambda+1/2} \exp(-\lambda)(\ln \lambda); \qquad \lambda \to \infty$$

■

EXERCISE 4.5 Given

$$I(\lambda) = \int_1^{\lambda} \left(1 + \frac{1}{t}\right)^t dt$$

show that

$$I(\lambda) \sim e\lambda - \frac{e}{2} \ln \lambda + O(1); \qquad \lambda \to \infty$$

Note: This integral is not of the form that we have previously discussed. However, it can be readily evaluated if one writes the integrand as $\exp[t \ln(1 + 1/t)]$ and expand it in a Taylor series. ■

5.5 CONCLUSIONS

We have a systematic and powerful method to develop the asymptotic expansion of integrals of the form

$$\Phi(\lambda) = \int_a^b f(x) \exp[\lambda h(x)]\, dx$$

where $h(x)$ has an absolute maximum interior to or at the end point of the interval of integration.

The leading term in the asymptotic expansion is found by the following:

(a) Dividing the integral into several parts and eliminating the terms that will yield exponentially small contributions.
(b) Transforming the dominant part into a Gaussian integral, which is easily evaluated.
(c) The factor $\sqrt{2\pi}$ almost always appears and is characteristic of the Laplace method. (It is absent if there is an inflection point rather than a maximum and it is replaced by an appropriate gamma function.)
(d) The higher-order corrections are found through a Taylor-series expansion of $h(x)$ about its maximum. These correction terms, in the form of exponential terms, are then redeveloped as a Taylor-series expansion in inverse powers of the large parameter. The introduction of the multiplier function $f(x)$ does not substantially change the analysis. Of course, it is necessary to expand this function in a Taylor series as one computes the higher-order corrections.

BIBLIOGRAPHY

Except for the book by Nayfeh, all of the texts give a detailed mathematical proof of the Laplace approximation. Each is at a different level of difficulty and has a slightly different approach. The book by Nayfeh has a wealth of problems and worked out examples.

1. Nayfeh, A. H., *Introduction to Perturbation Techniques*, Wiley, New York, 1981.
2. De Bruijn, N. G., *Asymptotic Analysis*, North-Holland, Amsterdam, 1958. The Third Edition has been reprinted by Dover, New York, 1981.
3. Wilf, H. S., *Mathematics for the Physical Sciences*, Wiley, New York, 1962. Reprinted by Dover, New York, 1978.
4. Bender, C. M., and S. A. Orszag, *Advanced Mathematical Methods for Scientists and Engineers*, McGraw-Hill, New York, 1978.
5. Olver, F. W. J., *Asymptotics and Special Functions*, Academic Press, New York, 1974.

CHAPTER 6

DIFFERENTIAL EQUATIONS

We encounter second-order linear differential equations in almost every aspect of our studies in science and engineering. They arise, for example, in electromagnetic theory through the equations of Laplace and Poisson; in quantum mechanics through the Schrödinger equation; in fluids and heat, acoustics, etc. Therefore, it comes as no surprise to realize that these equations have been studied extensively and are an essential component of courses in differential equations. With this in mind, we concentrate on some general ideas associated with asymptotic solutions and to those topics that arise in our studies of nonlinear oscillatory systems.

Our approach is to consider a second-order linear differential equation of a particular class and to introduce some transformations so as to put it in a form suitable for our analysis. This is done by eliminating the middle term and then, if necessary, introducing a change in the time scale. We then prove an oscillation theorem that gives us a way of comparing various equations and enables us to reach qualitative and quantitative conclusions regarding their solutions. This is followed by a discussion of Gronwall's lemma and its role in establishing the boundedness of solutions. The discussion includes an introduction to the powerful WKB (Wentzel–Kramers–Brillouin) approximation method with application to eigenvalue problems.

We conclude the chapter with an introduction to a perturbation technique that is based on converting a differential equation into an integral equation. The solution is then sought in terms of an expansion in terms of a small parameter. We review this technique in considerable detail in Chapter 8 as preparation for methods that we use in Chapters 9 through 15 for the study of weakly nonlinear oscillatory systems. The books in the Bibliography either bear directly on these topics or provide collateral reading.

6.1 ELIMINATION OF THE MIDDLE TERM

6.1.1 Transformation of Variables

It is often convenient to compare various equations without their middle terms. Thus, we would like to introduce a transformation of variables that converts an equation of the form

$$\frac{d^2z}{dt^2} + p(t)\frac{dz}{dt} + q(t)z = f(t) \tag{6.1}$$

where the coefficients $p(t)$ and $q(t)$ have no singularities on the interval under consideration, to one of the form

$$\frac{d^2u}{dt^2} + Q(t)u = g(t)$$

Often $Q(t)$ behaves like a constant for large times, $t \to \infty$, and, therefore, asymptotically the solutions are expected to look like sines and cosines. We accomplish the desired transformation of Equation (6.1) by introducing the change of variables:

$$u = z \exp\left[\tfrac{1}{2}\int_{t_0}^{t} p(t')\,dt'\right] \tag{6.2}$$

The lower limit of the integral, t_0, is an arbitrary constant whose effect is to multiply the solution by some other arbitrary constant. This transformation has no zeros or singularities and thus u and z have the same zeros. Equation (6.1) becomes:

$$\left\{\frac{d^2u}{dt^2} + \left[q(t) - \frac{1}{2}\frac{dp}{dt} - \frac{p^2}{4}\right]u\right\} \cdot \exp\left[-\frac{1}{2}\int_{t_0}^{t} p(t')\,dt'\right] = f(t)$$

Let

$$Q \equiv q(t) - \frac{1}{2}\frac{dp}{dt} - \frac{p^2}{4}$$

$$g(t) \equiv f(t)\exp\left[+\tfrac{1}{2}\int_{t_0}^{t} p(t')\,dt'\right]$$

Then we have

$$\frac{d^2u}{dt^2} + Q(t)u = g(t) \tag{6.3}$$

EXERCISE 1.1 Show that the transformation given by Equation (6.2) converts Equation (6.1) into Equation (6.3). ■

EXERCISE 1.2 Consider the damped harmonic oscillator

$$\frac{d^2x}{dt^2} + 2\varepsilon \frac{dx}{dt} + x = 0; \qquad 0 < \varepsilon \ll 1$$

Eliminate the middle term by the transformation

$$v(t) = (\exp \varepsilon t)x(t)$$

and show that $v(t)$ satisfies the simple harmonic oscillator equation. Then obtain the solution $x(t)$ and observe that it is given as the product of a decaying exponential and an oscillatory term. In other words, the transformation from $x(t)$ to $v(t)$ separated these two aspects of the motion. ■

Example 1.1 Consider Bessel's equation of order n:

$$x^2\frac{d^2y}{dx^2} + x\frac{dy}{dx} + (x^2 - n^2)y = 0 \tag{6.4}$$

whose solution is written as $J_n(x)$. Eliminate the middle term through the transformation:

$$v = y \cdot \exp\left(\tfrac{1}{2}\int^x \frac{dx'}{x'}\right) = y \cdot \sqrt{x}$$

and obtain the equation

$$\frac{d^2v}{dx^2} + \left(1 + \frac{\frac{1}{4} - n^2}{x^2}\right)v = 0 \tag{6.5}$$

We observe the following:

(a) If $n = \pm\frac{1}{2}$, the solution is a linear combination of sines and cosines. Thus,

$$J_{\pm\frac{1}{2}}(x) = \frac{1}{\sqrt{x}}(A \sin x + B \cos x)$$

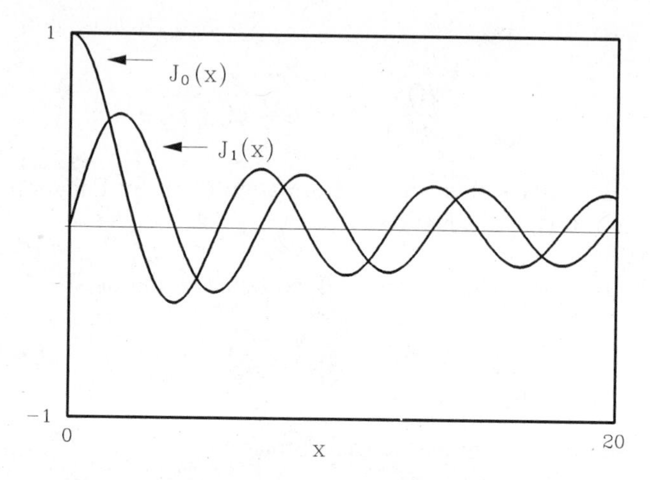

Figure 6.1 Observe $J_0(x)$ and $J_1(x)$ have the same asymptotic behavior. Also note that the zeros of $J_0(x)$ and $J_1(x)$ interlace.

(b) If $(\frac{1}{4} - n^2)/x^2 \ll 1$, Equation (6.5) is, approximately,

$$\frac{d^2v}{dx^2} + v = 0 \tag{6.6}$$

and thus we expect

$$y \sim \frac{1}{\sqrt{x}}(A \sin x + B \cos x); \qquad x \to \infty \tag{6.7}$$

This gives us a feel for the asymptotic behavior of Bessel functions. See Figure 6.1. ■

Example 1.2 Consider Legendre's equation of order n whose solution is written as $P_n(x)$:

$$(1 - x^2)\frac{d^2y}{dx^2} - 2x\frac{dy}{dx} + n(n+1)y = 0; \qquad y \equiv P_n(x), \quad |x| \le 1 \tag{6.8a}$$

Let $x = \cos\theta$. Then Equation (6.8a) becomes

$$\frac{d^2y}{d\theta^2} + \text{ctn}\,\theta\frac{dy}{d\theta} + n(n+1)y = 0; \qquad 0 \le \theta \le \pi \tag{6.8b}$$

We want to study the solution in the region where θ is far from both 0 and π. (The reason for this will be clearer when we look at Equation (6.9).) We eliminate the middle term through the substitution:

$$v = \exp\left(\tfrac{1}{2}\int^{\theta} \operatorname{ctn}\theta'\, d\theta'\right) y = \sqrt{\sin\theta}\, y$$

We then have

$$\frac{d^2v}{d\theta^2} + \left[\left(n + \frac{1}{2}\right)^2 + \frac{1}{4\sin^2\theta}\right] v = 0 \tag{6.9}$$

Thus, for sufficiently large n and θ sufficiently far from 0 and π:

$$v \sim A_n \cos\left[\left(n + \tfrac{1}{2}\right)\theta + \phi_n\right]$$

and

$$y \sim \frac{1}{\sqrt{\sin\theta}} A_n \cos\left[\left(n + \tfrac{1}{2}\right)\theta + \phi_n\right] \tag{6.10}$$

Coefficients A_n and ϕ_n explicitly depend upon n. It is possible to show through a more extensive analysis that

$$A_n \to \sqrt{\frac{2}{\pi n}}; \qquad \phi_n \to -\frac{\pi}{4}; \qquad n \to \infty$$

(For example, see the text by Bender and Orszag, page 231.) ■

EXERCISE 1.3 Given Hermite's equation

$$H_n''(x) - 2xH_n'(x) + 2nH_n(x) = 0$$

where $' = d/dx$.

Eliminate the middle term and obtain the equation

$$v''(x) + (2n + 1 - x^2)v(x) = 0$$

and the asymptotic behavior

$$H_n(x) \sim A_n \sin\left(\sqrt{2n+1}\, x + \phi_n\right) \exp\frac{x^2}{2}; \qquad 2n + 1 \gg x^2$$ ■

6.1.2 An Oscillation Theorem

It is often convenient to compare the structure of a given equation with an equation whose solution is known. In this regard, there is a powerful theorem due to Sturm that gives a condition for the existence of oscillatory solutions.

(A solution is said to be oscillatory if it has an infinite number of distinct zeros. For example, sines and cosines are oscillatory.)

Theorem 1.1 Consider the equation

$$\frac{d^2y}{dx^2} + \Phi(x)y = 0 \tag{6.11}$$

where we are given that all of its solutions are oscillatory. We also have the equation

$$\frac{d^2v}{dx^2} + \Omega(x)v = 0 \tag{6.12}$$

where $\Omega(x) \geq \Phi(x)$ for all x. Then all of the solutions of Equation (6.12) are oscillatory. ■

Proof: (We follow the line of argument given in the text by Bellman on Stability theory, page 119.) We want to show that $v(x)$ must change sign between every two successive zeros of $y(x)$. This enables us to conclude that $v(x)$ is oscillatory. To accomplish our goal, we construct an exact differential by multiplying Equation (6.11) by v and Equation (6.12) by y and subtracting to obtain

$$yv'' - vy'' + (\Omega - \Phi)vy = 0; \qquad ' = d/dx \tag{6.13}$$

We now integrate Equation (6.13) between two successive zeros of $y(x)$, x_1 and x_2:

$$\int_{x_1}^{x_2} (yv'' - vy'')\,dx + \int_{x_1}^{x_2} (\Omega - \Phi)vy\,dx \equiv I + J = 0 \tag{6.14}$$

In the given interval, $y(x)$ has one sign that we take as positive. We can perform the integral I noting that $y'(x_2) < 0$ and $y'(x_1) > 0$ to obtain

$$I = (yv' - vy')|_{x_1}^{x_2} = v(x_1)y'(x_1) + v(x_2)|y'(x_2)| \tag{6.15}$$

We observe that if $v(x)$ does not change sign on (x_1, x_2), then I has that sign. See Figure 6.2.

We now turn our attention to the second integral, J, in Equation (6.14) and recall that we are given that $\Omega \geq \Phi$. If $v(x)$ does not change sign on (x_1, x_2), then integral J also has the sign of $v(x)$. This leads to a contradiction. Thus, we conclude that $v(x)$ must change sign in the interval (x_1, x_2). ■

This is a most powerful theorem in that it enables us to make a qualitative statement about the solution of an equation without knowing its detailed behavior. The theorem can be extended considerably, as is done, for example, in the text by Bellman.

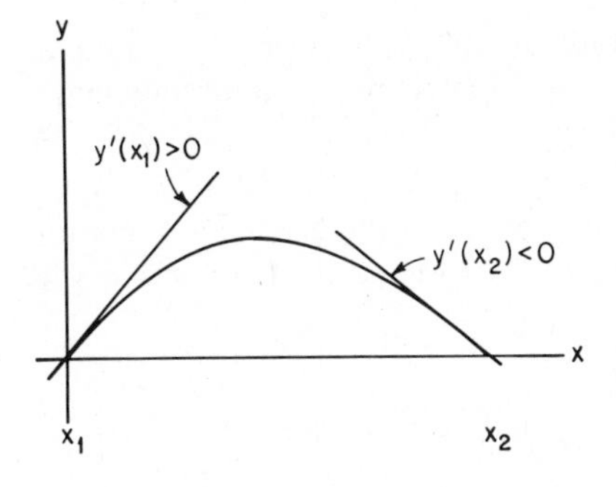

Figure 6.2 It is clear that the slope at x_1 is positive and at x_2 is negative. Also one sees that the curve obtained by a reflection about the x axis leads to another way to establish the theorem.

Example 1.3 Consider an Euler equation of the form

$$x^2 \frac{d^2y}{dx^2} + \frac{k^2}{4} y = 0$$

or

$$\frac{d^2y}{dx^2} + \frac{k^2}{4x^2} y = 0 \tag{6.16}$$

This is called an Euler equation and can be solved by the substitution $x = \exp t$. (An Euler equation is one in which each derivative term of order n is multiplied by the nth power of the independent variable. A typical term is $x^n \, d^n y/dx^n$.) Making the substitution, we obtain

$$\frac{d^2y}{dt^2} - \frac{dy}{dt} + \frac{k^2}{4} y = 0 \tag{6.17}$$

We can solve Equation (6.17) by the trial solution $y = \exp mt$. We then have the algebraic equation

$$m^2 - m + \frac{k^2}{4} = 0 \tag{6.18}$$

that can be solved for the eigenvalues m. Note that they are complex if $k^2 > 1$, and when this criterion is satisfied, Equation (6.17) and, consequently, Equation (6.16) have oscillatory solutions. Using Theorem 1.1 and making a comparison between Equation (6.16) and an equation of the form

$$\frac{d^2y}{dx^2} + f(x) y = 0 \tag{6.19}$$

we conclude that the latter has oscillatory solutions if $f(x) > 1/4x^2$. ■

EXERCISE 1.4 Consider Bessel's equation in the form given by Equation (6.5). Referring to Theorem 1.1, conclude that for all real n and sufficiently large x, $J_n(x)$ has oscillatory solutions. ■

EXERCISE 1.5 Consider Bessel's equation in the form given by Equation (6.5). Use Theorem 1.1 to show that the zeros of $J_0(x)$ and $J_1(x)$ interlace. ■

6.2 INHOMOGENEOUS EQUATIONS

6.2.1 The Second Solution

We are considering equations of the form given by Equation (6.1), where we assume that we have eliminated the middle term and that our equation has become

$$\frac{d^2u}{dt^2} + Q(t)u = g(t); \qquad u(0) = c_1, \quad \dot{u}(0) = c_2 \tag{6.20}$$

(Equation (6.20) is the same as Equation (6.3) except that now we include the initial conditions.) Let us assume that we have found a solution of the homogeneous equation, u_1, perhaps by a power-series expansion. We obtain a second solution, u_2, by writing

$$u_2 = u_1 r \tag{6.21}$$

(It is an independent solution if r is not a constant.)We then have

$$\dot{u}_2 = \dot{u}_1 r + u_1 \dot{r} \tag{6.22a}$$

$$\ddot{u}_2 = \ddot{u}_1 r + 2\dot{u}_1 \dot{r} + u_1 \ddot{r} \tag{6.22b}$$

Using the fact that u_1 and u_2 are solutions of the homogeneous equation, i.e., Equation (6.20) with $g(t) = 0$, we have

$$2\dot{u}_1 \dot{r} + u_1 \ddot{r} = 0 \tag{6.23a}$$

or

$$\frac{\ddot{r}}{\dot{r}} = \frac{-2\dot{u}_1}{u_1} \tag{6.23b}$$

and, hence,

$$\dot{r} = \frac{a_1}{u_1^2} \qquad \text{or} \qquad u_2 = u_1 \left[\int^t \frac{a_1}{u_1^2}\, dt' + a_2 \right] \tag{6.24}$$

where constants a_1 and a_2 are found from the initial conditions satisfied by u_2.

6.2.2 Method of Variation of Parameters

Now that we have two independent solutions of the homogeneous equation, we can use them to find the solution to the inhomogeneous differential equation, Equation (6.20), by the method of variation of parameters. One lets

$$u = w_1 u_1 + w_2 u_2 \tag{6.25}$$

where u_1 and u_2 are the solutions of the homogeneous equation. The functions w_1 and w_2 are determined so that u satisfies Equation (6.20). Differentiating Equation (6.25), we have

$$\dot{u} = \dot{w}_1 u_1 + \dot{w}_2 u_2 + w_1 \dot{u}_1 + w_2 \dot{u}_2 \tag{6.26}$$

Since we have introduced a new equation and we have only to satisfy two initial conditions, we are free to impose a condition on w_1 and w_2. We choose

$$\dot{w}_1 u_1 + \dot{w}_2 u_2 = 0 \tag{6.27}$$

Differentiating Equation (6.26) and using Equation (6.27), we have

$$\ddot{u} = \dot{w}_1 \dot{u}_1 + \dot{w}_2 \dot{u}_2 + w_1 \ddot{u}_1 + w_2 \ddot{u}_2 \tag{6.28}$$

So after substitution of Equations (6.25) and (6.28) into Equation (6.20), we have

$$\dot{w}_1 \dot{u}_1 + \dot{w}_2 \dot{u}_2 = g(t) \tag{6.29}$$

Observe that Equations (6.27) and (6.29) are two equations for two unknowns, $\dot{w}_1$ and $\dot{w}_2$. We solve them and obtain

$$\dot{w}_1 = \frac{-u_2 g}{W} \tag{6.30a}$$

$$\dot{w}_2 = \frac{u_1 g}{W} \tag{6.30b}$$

$$W \equiv \begin{vmatrix} u_1 & u_2 \\ \dot{u}_1 & \dot{u}_2 \end{vmatrix} \tag{6.31}$$

Quantity W is called the Wronskian. It is a constant determined by the initial conditions imposed on u_1 and u_2. That W is a constant is clear if we differentiate Equation (6.31) and use Equation (6.20) with $g = 0$.

$$\frac{dW}{dt} = \begin{vmatrix} \dot{u}_1 & \dot{u}_2 \\ \dot{u}_1 & \dot{u}_2 \end{vmatrix} + \begin{vmatrix} u_1 & u_2 \\ \ddot{u}_1 & \ddot{u}_2 \end{vmatrix} = 0 + Q(t) \cdot 0 = 0 \tag{6.32}$$

Integrating Equations (6.30a) and (6.30b), we have

$$w_1 = -\int_0^t \frac{u_2(s)g(s)}{W}\,ds + c_1 \tag{6.33a}$$

$$w_2 = +\int_0^t \frac{u_1(s)g(s)}{W}\,ds + c_2 \tag{6.33b}$$

where c_1 and c_2 are constants.

We choose our two solutions to the homogeneous equation, u_1 and u_2, to satisfy the special initial conditions:

$$\begin{aligned} u_1(0) &= 1; \qquad u_2(0) = 0 \\ \dot{u}_1(0) &= 0; \qquad \dot{u}_2(0) = 1 \end{aligned} \tag{6.34}$$

(Note: We restrict our attention to problems for which it is possible to impose these special initial conditions. Other situations are discussed in more detailed treatments such as the ones found in the texts by Olver and by Bender and Orszag.) This makes constants c_1 and c_2 the original choices for the initial conditions for u given by Equation (6.20). With this, $W = 1$. We now substitute w_1 and w_2 in Equation (6.25) and obtain

$$u = w_1 u_1 + w_2 u_2 = c_1 u_1 + c_2 u_2 - u_1 \int_0^t u_2(s)g(s)\,ds + u_2 \int_0^t u_1(s)g(s)\,ds \tag{6.35a}$$

or

$$u(t) = c_1 u_1(t) + c_2 u_2(t) + \int_0^t k(t,s)g(s)\,ds \tag{6.35b}$$

We have introduced the kernel

$$k(t,s) \equiv u_2(t)u_1(s) - u_1(t)u_2(s) = \begin{vmatrix} u_1(s) & u_2(s) \\ u_1(t) & u_2(t) \end{vmatrix} \tag{6.36a}$$

It satisfies the conditions

$$k(s,s) = 0; \qquad \left.\frac{dk}{dt}\right|_{t=s} = \begin{vmatrix} u_1(s) & u_2(s) \\ \dot{u}_1(s) & \dot{u}_2(s) \end{vmatrix} = W = 1 \tag{6.36b}$$

(Most likely the concept of the kernel is unfamiliar to readers. It plays a central role in integral equations. In the treatment in this text, it appears only in the form given by Equation (6.36a).)

EXERCISE 2.1 Consider the equation for a forced simple harmonic oscillator:

$$\frac{d^2u}{dt^2} + u = f(t) \tag{6.37a}$$

$$u(0) = c_1; \qquad \dot{u}(0) = c_2 \tag{6.37b}$$

For simplicity, we have chosen the natural frequency of the oscillator equal to 1. Obtain the solution of Equation (6.37a) by choosing solutions of the force-free problem as $u_1 = \cos t$ and $u_2 = \sin t$. This choice satisfies the initial conditions given by Equation (6.34). Show that the kernel is

$$k(t, s) = \begin{vmatrix} \cos s & \sin s \\ \cos t & \sin t \end{vmatrix} = \sin(t - s) \tag{6.38}$$

and the solution is

$$u(t) = c_1 \cos t + c_2 \sin t + \int_0^t \sin(t - s) f(s)\, ds \qquad \blacksquare \tag{6.39}$$

EXERCISE 2.2 Often we are asked to consider a forced harmonic oscillator whose natural frequency ω_0 is $\neq 1$.

$$\frac{d^2u}{dt_1^2} + \omega_0^2 u = f(t_1); \qquad \omega_0^2 \neq 1 \tag{6.40a}$$

$$u(0) = c_1; \qquad \dot{u}(0) = c_2 \tag{6.40b}$$

The change of time scale $t = \omega_0 t_1$ leads to the equation

$$\frac{d^2u}{dt^2} + u = \frac{f(t/\omega_0)}{\omega_0^2}$$

Referring to Equations (6.37a) and (6.39), show that

$$u(t_1) = c_1 \cos \omega_0 t_1 + \frac{c_2}{\omega_0} \sin \omega_0 t_1 + \frac{1}{\omega_0} \int_0^{t_1} \sin[\omega_0(t_1 - s)] f(s)\, ds$$

$$\blacksquare \tag{6.41}$$

6.3 THE LIOUVILLE - GREEN TRANSFORMATION

In this section, we continue our analysis of homogeneous equations of the form

$$\frac{d^2u}{dt^2} + p(t)\frac{du}{dt} + q(t)u = 0 \tag{6.42}$$

which, after elimination of the middle term, become

$$\frac{d^2v}{dt^2} + Q(t)v = 0 \tag{6.43}$$

$$v \equiv u \exp\left[\tfrac{1}{2}\int^t p(t')\,dt'\right]$$

$$Q(t) \equiv q(t) - \frac{1}{2}\frac{dp}{dt} - \frac{1}{4}p^2$$

We concentrate our attention on determining the asymptotic, or long-time, behavior, $t \to \infty$, of the solutions of Equation (6.43). To attain our goals, we introduce a transformation of the time variable that brings our equation into a form that is more amenable to analysis.

We are also interested in a related equation in which a large parameter α enters in the equation as a multiplier of the term $Q(t)$:

$$\frac{d^2\Phi(x)}{dx^2} + \alpha^2 Q(x)\Phi(x) = 0; \qquad Q(x) > 0; \quad \alpha \to \infty; \quad x \text{ finite}$$

We study this type of equation in Section 6.5. In our analysis, we use variables x or t interchangeably to emphasize that these equations occur in problems with spatial or temporal variables.

We consider two different forms of the term denoted as $Q(t)$ in Equation (6.43). We begin with the situation where $Q(t) > 0$, for $t > t_0$, and $Q(t)$ has other behavior to be prescribed. Our approach is to generally follow the discussion in the books by Bellman; see the Bibliography at the end of the chapter. However, all of the books use the same basic methods. The techniques are so universal that they have been discovered and rediscovered often enough to be known by a host of initials, e.g., WKJB (Wentzel–Kramers–Brillouin–Jeffreys) and permutations thereof.

We begin by introducing a change of variables in Equation (6.43) that is called a Liouville or Liouville–Green transformation. We determine, after obtaining Equation (6.56), under what conditions the Liouville–Green transformation brings Equation (6.43) into the form of a perturbed harmonic oscillator. The transformation introduces a new time or spatial scale:

$$s = \int_{t_0}^{t} \sqrt{Q(t')}\,dt' \tag{6.44}$$

(We already introduced a simplified version of this transformation in Exercise 2.2, where we had $Q(t) = \omega_0^2$, a constant. In that case, the rescaled time variable was simply proportional to t.)

Our derivatives are now taken with respect to s:

$$\frac{dv}{dt} = \frac{dv}{ds}\sqrt{Q(t)} \tag{6.45a}$$

$$\frac{d^2v}{dt^2} = \frac{d^2v}{ds^2}Q(t) + \frac{1}{2}\frac{dQ}{dt}\frac{1}{\sqrt{Q}}\frac{dv}{ds} \tag{6.45b}$$

and Equation (6.43) becomes

$$\frac{d^2v}{ds^2} + \frac{1}{2}\frac{dQ}{dt}\frac{1}{Q^{3/2}}\frac{dv}{ds} + v = 0 \tag{6.46}$$

It seems odd to introduce a middle term into an equation that has none. However, we have introduced it in terms of the variable s, which represents a new time or spatial dependence. The resulting equation has the coefficient 1 multiplying the displacement v. This encourages us to look for oscillatory solutions since the equation, except for the middle term, is the same as the harmonic oscillator equation with natural frequency equal to 1. (Keep in mind that Theorem 1.1 gave us a simple criterion to determine whether our equation has oscillatory solutions by comparing it to the oscillator equation.) With these thoughts expressed, we eliminate the new middle term in Equation (6.46) and see what conclusions we can draw.

Example 3.1 Before we develop our general results, things might be made clearer by discussing an illustrative example. Consider the equation

$$\frac{d^2v}{dt^2} + t^2v = 0; \qquad Q(t) = t^2 \tag{6.47}$$

We are interested in the long-time behavior, $t \to \infty$. (Note that the term $Q(t)$, goes to infinity for large t.) We will see that Equation (6.47) has bounded oscillatory solutions that are most easily revealed through a transformation of variables. Thus, we write

$$s = \int^t t'\,dt' = \frac{t^2}{2} \tag{6.48}$$

and using Equation (6.46), we obtain

$$\frac{d^2v}{ds^2} + \frac{1}{2}(2t)\frac{1}{t^3}\frac{dv}{ds} + v = 0 \tag{6.49a}$$

or

$$\frac{d^2v}{ds^2} + \frac{1}{2s}\frac{dv}{ds} + v = 0 \tag{6.49b}$$

Referring to Equations (6.42) and (6.43), we know that the middle term in Equation (6.49b) can be eliminated by the transformation of variables:

$$y = \exp\left(\tfrac{1}{2}\int^{s}\frac{1}{2s'}\,ds'\right)v(s) = s^{1/4}v(s)$$

We obtain

$$\frac{d^2y}{ds^2} + \left(1 + \frac{3}{16s^2}\right)y = 0 \tag{6.50}$$

Observe that Equation (6.50) is of the form of Equation (6.19). Thus, by Theorem 1.1, we conclude that $y(s)$ is oscillatory. If we confine our interest to large values of s, or seek solutions as s tends toward infinity, we expect y to behave like a combination of sines and cosines:

$$y \sim A_1 \sin s + B_1 \cos s; \qquad s \to \infty \tag{6.51}$$

or

$$v(s) \sim (A_1 \sin s + B_1 \cos s)\frac{1}{s^{1/4}} \tag{6.52a}$$

Finally, we return to our variable t and write:

$$v(t) \sim \frac{A \sin(t^2/2) + B \cos(t^2/2)}{\sqrt{t}}; \qquad t \to \infty \tag{6.52b}$$

(Note that we have absorbed constants into A and B.) So to recapitulate, we had Equation (6.47) without a middle term, but it was given in terms of a variable t that did not reveal the underlying structure of the solution. The transformation to the new "time" s led immediately to a solution. These ideas are developed in more detail later in the chapter. ■

We return to Equation (6.46), and eliminate the middle term by the change of variable:

$$y = \exp\left[\frac{1}{2}\int^{s}\left(\frac{dQ}{dt}\,\frac{1}{2Q^{3/2}}\right)ds'\right]v(s) \tag{6.53}$$

The integral is with respect to the variable s' rather than the variable t. So we write

$$ds' = \frac{ds'}{dt'}\,dt' = \sqrt{Q(t')}\,dt' \tag{6.54}$$

and then observe that our transformation is really of the form

$$y = \exp\left(\frac{1}{4}\int^t \frac{dQ}{dt'}\frac{1}{Q}\,dt'\right)v(s)$$
$$= Q(t)^{1/4}v(s) \tag{6.55}$$

Equation (6.46) then takes the form

$$\frac{d^2y}{ds^2} + \left[1 - \frac{1}{2}\frac{dj(s)}{ds} - \frac{1}{4}j^2(s)\right]y = 0 \tag{6.56}$$

where

$$j(s) \equiv \frac{1}{2}\frac{dQ}{dt}\frac{1}{Q^{3/2}} \tag{6.57}$$

We would like to determine under what conditions we obtain bounded solutions. We comment at this point that we introduce, in Section 6.4, Bellman's inequality, or Gronwall's lemma, that enables us to establish bounds on the solution to Equation (6.56) with rather lenient conditions imposed on $j(s)$. It is our expectation that if $j(s) \to 0$, as $s \to \infty$, Equation (6.56) can be written as

$$\frac{d^2y}{ds^2} + y = 0; \qquad s \to \infty \tag{6.58}$$

and then

$$y \sim A_1 \sin s + B_1 \cos s \tag{6.59}$$

and

$$v(t) \sim \frac{A\cos\left[\int^t \sqrt{Q(t')}\,dt'\right] + B\sin\left[\int^t \sqrt{Q(t')}\,dt'\right]}{Q(t)^{1/4}} \tag{6.60}$$

It will take some development to establish Equation (6.60). This is done in the arguments leading up to Equation (6.82). At this point, we wish to show the plausibility of Equation (6.60) by developing an asymptotic expansion in the form

$$y(s) = A(\sin s)\left[1 + \frac{a_1}{s} + \frac{a_2}{s^2} + O\left(\frac{1}{s^3}\right)\right]$$
$$+ B(\cos s)\left[1 + \frac{b_1}{s} + \frac{b_2}{s^2} + O\left(\frac{1}{s^3}\right)\right] \tag{6.61}$$

We substitute $y(s)$ as given by Equation (6.61) into Equation (6.56) and solve for coefficients a_i and b_i. This is called a Stokes' expansion. (See the book by Jeffreys and Jeffreys, page 521.) We explain the technique of the Stokes' expansion through the analysis of an illustrative example.

Example 3.2 Consider Bessel's equation of zero order:

$$\frac{d^2y}{dx^2} + \frac{1}{x}\frac{dy}{dx} + y = 0 \tag{6.62}$$

Let $v = \sqrt{x}\,y$. Then we have

$$\frac{d^2v}{dx^2} + \left(1 + \frac{1}{4x^2}\right)v = 0 \tag{6.63}$$

We know by Theorem 1.1 that $v(x)$ is oscillatory. Observe that Equation (6.63) is of the same form as Equation (6.56). We seek a solution in the form given by Equation (6.61), retaining only terms $O(1/x^2)$.

$$v(x) \sim A(\sin x)\left(1 + \frac{a_1}{x} + \frac{a_2}{x^2}\right) + B(\cos x)\left(1 + \frac{b_1}{x} + \frac{b_2}{x^2}\right) \tag{6.64}$$

We substitute Equation (6.64) into Equation (6.63) and neglect all terms $O(1/x^3)$. We have

$$\begin{aligned}\frac{dv}{dx} &= A(\cos x)\left(1 + \frac{a_1}{x} + \frac{a_2}{x^2}\right) + A(\sin x)\left(-\frac{a_1}{x^2}\right)\\ &\quad - B(\sin x)\left(1 + \frac{b_1}{x} + \frac{b_2}{x^2}\right) + B(\cos x)\left(-\frac{b_1}{x^2}\right) + O\left(\frac{1}{x^3}\right)\end{aligned} \tag{6.65a}$$

$$\begin{aligned}\frac{d^2v}{dx^2} &= -A(\sin x)\left(1 + \frac{a_1}{x} + \frac{a_2}{x^2}\right) + 2A(\cos x)\left(-\frac{a_1}{x^2}\right)\\ &\quad - B(\cos x)\left(1 + \frac{b_1}{x} + \frac{b_2}{x^2}\right) - 2B(\sin x)\left(-\frac{b_1}{x^2}\right) + O\left(\frac{1}{x^3}\right)\end{aligned} \tag{6.65b}$$

We now substitute Equations (6.64) and (6.65b) into Equation (6.63). Constants a_1 and b_1 are found from the equation

$$-\frac{1}{4x^2}(A\sin x + B\cos x) = 2A(\cos x)\left(\frac{-a_1}{x^2}\right) - 2B(\sin x)\left(\frac{-b_1}{x^2}\right)$$

This yields

$$a_1\frac{A}{B} = -b_1\frac{B}{A} = \frac{1}{8} \tag{6.66}$$

Equation (6.64) is then written as

$$v(x) \sim A\left(\sin x - \frac{\cos x}{8x}\right) + B\left(\cos x + \frac{\sin x}{8x}\right) + O\left(\frac{1}{x^2}\right); \qquad x \to \infty \tag{6.67}$$

This procedure can be continued, with increasing labor, to a higher order. ■

EXERCISE 3.1 Consider Equation (6.64) and continue the expansion procedure. Verify that the next term in Equation (6.67) has coefficients

$$(-A\sin x)\frac{9}{2(8x)^2} \qquad \text{and} \qquad (-B\cos x)\frac{9}{2(8x)^2}$$ ■

6.4 BELLMAN'S INEQUALITY — GRONWALL'S LEMMA

We have established the central results that enable us to describe the asymptotic behavior of differential equations of the form

$$\frac{d^2v}{dt^2} + Q(t)v = 0 \tag{6.68}$$

as variable t or x tends toward infinity. We still have to establish the conditions under which the solutions are bounded. Also, we have to be assured that the neglected terms are smaller than those that we include.

We begin by giving a proof of an inequality due to Bellman, originally published in the *Duke Mathematical Journal*, Vol. 10, 1943, p. 643. It is also contained in a slightly different form in the texts of Bellman that are given in the Bibliography at the end of the chapter. It is also called Gronwall's lemma as a slight variant was published much earlier by Gronwall in *Annals of Mathematics*, Vol. 20, 1918, pp. 292–296. We use the form given by Bellman in part because we are following the development he gives in his texts. It is clear that many people discovered it and developed generalizations. (The results given in this section are obtained in the text by De Bruijn, pp. 189–198, in a manner similar to that used by Bellman.)

Theorem 4.1 If $y(x)$, M, and $k \geq 0$, and we have the inequality

$$y(x) \leq M + k\int_{x_0}^{x} |f(x')|y(x')\,dx' \tag{6.69}$$

then

$$y(x) \leq M\exp\left(k\int_{x_0}^{x}|f(x')|\,dx'\right) \tag{6.70}$$ ■

Proof: We introduce

$$v(x) = \int_{x_0}^{x} y(x')|f(x')|\,dx'; \qquad v(x_0) = 0 \tag{6.71}$$

and differentiate $v(x)$ and use Equation (6.69) to obtain

$$\frac{dv(x)}{dx} = y(x)|f(x)| \le |f(x)|[M + kv(x)] \tag{6.72a}$$

$$\frac{dv(x)/dx}{M + kv(x)} \le |f(x)| \qquad \text{or} \qquad \frac{k\,dv(x)/dx}{M + kv(x)} \le k|f(x)| \tag{6.72b}$$

Equation (6.72b) is integrated to yield

$$\ln \frac{M + kv(x)}{M} \le k \int_{x_0}^{x} |f(x')|\,dx' \tag{6.73a}$$

Exponentiating, we have

$$M + kv(x) \le M \exp\left[k \int_{x_0}^{x} |f(x')|\,dx'\right] \tag{6.73b}$$

Using Equation (6.69) and the definition of $v(x)$ given by Equation (6.71), Equation (6.70) is established. Thus, $y(x)$ is bounded if Equation (6.69) is satisfied and

$$\int_{x_0}^{\infty} |f(x')|\,dx' < \infty \tag{6.74}$$

[Note the absolute value sign in Equation (6.74). It is a necessary condition for boundedness as is pointed out in a simple example given on page 113 of the text on stability theory by Bellman.] ■

We use this result to establish conditions under which solutions to equations of the form

$$\frac{d^2y}{dx^2} + [1 + f(x)]\,y(x) = 0 \tag{6.75a}$$

$$y(0) = c_1; \qquad \dot{y}(0) = c_2 \tag{6.75b}$$

are bounded. We write Equation (6.75a) as

$$\frac{d^2y}{dx^2} + y(x) = -f(x)\,y(x) \tag{6.75c}$$

and then convert it to an integral equation. At this point, we recall the analysis given in Section 6.2, and observe that Equation (6.75c) is of the same form as Equation (6.37a). Its solution can then be written as

$$y(x) = c_1 \cos x + c_2 \sin x - \int_0^x f(x') y(x') \sin(x - x')\, dx' \tag{6.76}$$

Take the absolute values of each side of Equation (6.76) and note that for any a and b,

$$|a - b| \leq |a| + |b|$$

and since $\cos x$ and $\sin x$ are bounded, we have the inequality

$$|y(x)| \leq c + \int_0^x |f(x')| \, |y(x')| \, dx' \tag{6.77}$$

with $c = \text{constant} > 0$. Then if $f(x)$ satisfies the condition given by Equation (6.74), the solution $y(x)$ is bounded.

Example 4.1 Gronwall's inequality enables us to establish the uniqueness of the solution to a Volterra integral equation of the second kind. (We discuss integrals of this class in Section 6.6.) That is to say, given an integral of the form

$$y(x) = g(x) + \alpha \int_0^x k(x, s) y(s)\, ds \tag{6.78}$$

where $g(x)$ and $k(x, s)$ are continuous and bounded on $(0, x)$, the solution is unique. For if $z(x)$ also is a solution, satisfying the same initial conditions, we have, after subtracting and taking absolute values,

$$\begin{aligned} |y(x) - z(x)| &= \left|\alpha \int_0^x k(x, s)[y(s) - z(s)]\, ds\right| \\ &\leq |\alpha| \cdot |k(x, s)|_{\max} \int_0^x |y(s) - z(s)|\, ds \\ &\leq |\alpha| \cdot |K| \int_0^x |y(s) - z(s)|\, ds \end{aligned} \tag{6.79}$$

where $|K| = |k(x, s)|_{\max}$.

Equation (6.79) is of the form of Equation (6.69), with $M = 0$ and $f(x) = 1$. Thus, the bound given by Equation (6.70) yields

$$|y(x) - z(x)| \leq 0$$

enabling us to conclude that $y(x) = z(x)$. ■

Example 4.2 Gronwall's lemma enables us to determine the long-time behavior of the solution of Equation (6.75a) or, equivalently, Equation (6.75c). We write the integrals in Equation (6.76) as

$$\int_0^x f(x')y(x')(\sin x')\,dx' = \int_0^\infty - \int_x^\infty = a(0) - a(x)$$

and

$$\int_0^x f(x')y(x')(\cos x')\,dx' = b(0) - b(x)$$

Note that $a(x)$ and $b(x)$ go to zero as x tends to infinity. Then we can write Equation (6.76) as

$$\begin{aligned} y(x) = {} & c_1 \cos x + c_2 \sin x - b(0)\sin x + b(x)\sin x \\ & + a(0)\cos x - a(x)\cos x \end{aligned} \tag{6.80}$$

Since we already have established that $y(x)$ is bounded, we have the result

$$y(x) \sim [c_1 + a(0)]\cos x + [c_2 - b(0)]\sin x + o(1); \qquad x \to \infty \qquad \blacksquare$$

We are now in a position to establish the asymptotic form of the solution to Equation (6.68). Namely, given

$$\frac{d^2v}{dt^2} + Q(t)v = 0; \qquad \sqrt{Q(t)} > 0, \quad t \to \infty \tag{6.68}$$

and

$$s = \int_{t_0}^t \sqrt{Q(t')}\,dt'; \qquad y(s) = v(s)[Q(t)]^{1/4}$$

together with the boundedness conditions imposed on $j(s)$, defined by Equation (6.57),

$$\int_{t_0}^\infty \left| \frac{d}{ds'}\left(\frac{dQ}{dt}\frac{1}{Q^{3/2}}\right)\right| ds' < \infty; \qquad \int_{t_0}^\infty \left|\frac{dQ}{dt}\frac{1}{Q^{3/2}}\right| ds' < \infty$$

we have

$$y(s) \sim d_1 \sin s + d_2 \cos s + o(1) \tag{6.81}$$

or

$$\begin{aligned} v(t) \sim \frac{1}{Q(t)^{1/4}} \Bigg\{ & d_1 \cos\left[\int_{t_0}^t \sqrt{Q(t')}\,dt'\right] \\ & + d_2 \sin\left[\int_{t_0}^t \sqrt{Q(t')}\,dt'\right]\Bigg\} + o\left[\frac{1}{Q(t)^{1/4}}\right] \end{aligned} \tag{6.82}$$

EXERCISE 4.1 Consider Airy's equation:

$$\frac{d^2z}{dx^2} + xz = 0 \tag{6.83}$$

We are interested in the asymptotic behavior as $x \to \infty$. Show that the solution is

$$z(x) \sim \left(c_1 \cos\frac{2}{3}x^{3/2} + c_2 \sin\frac{2}{3}x^{3/2}\right)\frac{1}{x^{1/4}} + o\left(\frac{1}{x^{1/4}}\right) \quad \blacksquare \tag{6.84}$$

6.5 THE WKB APPROXIMATION

We have studied equations of the form

$$\frac{d^2v}{dt^2} + Q(t) \cdot v(t) = 0; \qquad Q(t) > 0 \tag{6.68}$$

and found the leading term in their asymptotic solution as the independent variable tends to infinity. We now turn our attention to equations of the form

$$\frac{d^2\Phi(x)}{dx^2} + \alpha^2 \cdot Q(x) \cdot \Phi(x) = 0; \qquad \alpha \to \infty, \;\; Q(x) > 0, \;\; 0 \le x \le 1 \tag{6.85}$$

The independent variable x is now restricted to a finite region, and we are interested in situations where $Q(x)$ has no zeros or singularities in the interval under consideration and parameter α tends to infinity. This class of problems occurs in quantum mechanics in the classical limit, and in optics in the geometrical region or approximation, as well as in boundary-layer problems.

Example 5.1 Consider the one-dimensional Schrödinger equation:

$$\frac{-\hbar^2}{2m}\frac{d^2\psi}{dx^2} + V(x) \cdot \psi = E\psi \tag{6.86a}$$

with $[E - V(x)] > 0$; $0 \le x \le 1$.

(Parameter $\hbar$ is Planck's constant divided by 2π.) Equation (6.86a) can be reformulated by dividing by the small parameter in front of the second derivative and obtaining an equation of the form of Equation (6.85). We write Equation (6.86a) as

$$\frac{d^2\psi}{dx^2} + \alpha^2 \cdot Q(x) \cdot \psi = 0$$

$$\alpha^2 = \frac{2m}{\hbar^2}; \qquad Q(x) = [E - V(x)] > 0 \tag{6.86b}$$

and often seek a solution in the so-called classical limit, $\hbar$ tends to zero or α tends to infinity. Equations of this form are called singular since the small parameter $\hbar$ multiplies the highest derivative. ■

We have found an asymptotic solution to Equation (6.68) by techniques developed in Section 6.4 and we introduce analogous transformations in Equation (6.85). Thus, we introduce the rescaled space variable

$$s = \alpha \int_{x_0}^{x} \sqrt{Q(x')}\, dx' \tag{6.87}$$

Differentiating, we have

$$\frac{ds}{dx} = \alpha \sqrt{Q(x)}$$

We then introduce

$$y(s) = [Q(x)]^{1/4} \cdot z(s) \tag{6.88}$$

and obtain

$$\frac{d^2y}{ds^2} + \left[1 - \frac{1}{2\alpha}\frac{d}{ds}\left(\frac{1}{2}\frac{dQ}{dx}\frac{1}{Q^{3/2}}\right) - \frac{1}{4\alpha^2}\left(\frac{1}{2}\frac{dQ}{dx}\frac{1}{Q^{3/2}}\right)^2\right] y(s) = 0 \tag{6.89}$$

Equation (6.89) is of the form

$$\frac{d^2y}{ds^2} + y(s) + O\left(\frac{1}{\alpha}\right) \cdot y(s) = 0 \tag{6.90}$$

and thus we expect the leading term of the solution to be

$$y(s) \sim A \cos s + B \sin s + O\left(\frac{1}{\alpha}\right); \qquad \alpha \to \infty \tag{6.91}$$

Expressing this result in terms of our original variables, we have

$$z(x) \sim \frac{1}{[Q(x)]^{1/4}}\left\{A \cos\left[\alpha \int_0^x \sqrt{Q(x')}\, dx'\right] + B \sin\left[\alpha \int_0^x \sqrt{Q(x')}\, dx'\right]\right\} + O\left(\frac{1}{\alpha}\right); \qquad \alpha \to \infty \tag{6.92}$$

The correction terms are obtainable through an expansion in inverse powers of α. The WKB expansion, as expressed by Equation (6.92), is a valid asymptotic expansion in inverse powers of the large parameter α, provided that

$$\int_0^{x=1}\left|\frac{d}{ds}\left(\frac{dQ}{dx}\frac{1}{Q^{3/2}}\right)\right| ds < \infty; \qquad \int_0^{x=1}\left|\frac{dQ}{dx}\frac{1}{Q^{3/2}}\right|^2 ds < \infty \tag{6.93}$$

There is a rich literature on the WKB approximation and the Bibliography presents a representative selection. The most interesting subject we have omitted is the so-called "turning-point" problem. This situation arises in the region where the function $Q(x)$ either vanishes or is approximately zero, the so-called boundary layer or transition region. In such cases, there is an interchange of scales since the large parameter α is no longer the sole scale-determining factor. (See the Bibliography.)

We turn our attention to Equation (6.85) in the form of an eigenvalue problem:

$$\frac{d^2\Phi}{dx^2} + \alpha^2 \cdot Q(x) \cdot \Phi = 0; \qquad \Phi(0) = \Phi(1) = 0;$$
$$0 \le x \le 1; \quad \alpha \to \infty; \quad Q(x) > 0 \tag{6.94}$$

It is possible to show that there are an infinite number of distinct positive solutions or eigenvalues for α. We seek an approximation for the large eigenvalues using the WKB method. Our solution for the associated eigenfunctions $\Phi(x)$ is given by Equation (6.92), where we require $\sqrt{Q(x)}$ to be not too small as compared to $1/\alpha$. (It is essential to avoid the region near the turning points.) We impose the boundary conditions on the solution given by Equation (6.92). Thus, since $\Phi(0) = 0$, we have

$$A = 0; \quad \Phi(x) \sim B\frac{1}{Q(x)^{1/4}}\sin\left[\alpha\int_0^x \sqrt{Q(x')}\, dx'\right]; \qquad \alpha \to \infty \tag{6.95}$$

The boundary condition at $x = 1$ requires

$$\Phi(1) = 0 \Rightarrow \sin\left[\alpha\int_0^1 \sqrt{Q(x')}\, dx'\right] = 0 \tag{6.96}$$

The condition given by Equation (6.96) is satisfied if the parameter α takes on values such that

$$\alpha\int_0^1 \sqrt{Q(x')}\, dx' = n\pi; \quad n = 1, 2, \ldots \tag{6.97}$$

Notice that even though our approximation is valid only in the regime of large values of parameter α, we impose the boundary condition on all values of α.

We expect our approximation to improve as n increases. Our solution is written

$$\Phi_n(x) = \frac{B}{Q(x)^{1/4}} \sin\left[\frac{n\pi}{\int_0^1 \sqrt{Q(x')}\,dx'} \cdot \int_0^x \sqrt{Q(x')}\,dx'\right]; \qquad n = 1, 2, \ldots \tag{6.98}$$

where we have introduced the subscript n to order the solutions.

EXERCISE 5.1 Consider

$$\frac{d^2\Phi}{dx^2} + \alpha^2 \cdot Q(x) \cdot \Phi = 0; \qquad \Phi(0) = \Phi(1) = 0;$$

$$0 \le x \le 1; \quad \alpha \to \infty; \quad Q(x) = (1 + x)^2$$

Show that the solution is given by

$$\Phi_n(x) = B\frac{1}{\sqrt{1 + x}} \sin\left[\frac{2}{3}(n\pi) \cdot x\left(1 + \frac{x}{2}\right)\right]$$

■

EXERCISE 5.2 Consider

$$\frac{d^2\Phi}{dx^2} + \alpha^2 \cdot Q(x) \cdot \Phi = 0; \qquad \Phi(1) = \Phi(2) = 0;$$

$$1 \le x \le 2, \quad \alpha \to \infty, \quad Q(x) = x^2$$

Show that the eigenvalues, α_n, are equal to $(2/3)n\pi$. How would the eigenvalues change if we retained the form of $Q(x)$ but changed the interval of definition to (2, 3) with the boundary conditions $\Phi(2) = \Phi(3) = 0$?

Answer: $\alpha_n = 2n\pi/5$. ■

We can also consider equations of the same form as Equation (6.94), but with the boundary conditions

$$\Phi'(0) = \Phi(1) = 0 \tag{6.99}$$

Our general solution is unchanged, but the boundary conditions now require

$$B = 0 \qquad \text{and} \qquad \alpha \int_0^1 \sqrt{Q(x')}\,dx' = (2n + 1)\frac{\pi}{2} \tag{6.100}$$

(Refer to Equation (6.92) and observe that the derivatives of the trigonometric functions are $O(\alpha)$, whereas the derivative of $1/Q(x)^{1/4}$ is $O(1)$ and thus can be neglected.)

Thus, we have the eigenvalues

$$\alpha_n = \frac{(2n+1)(\pi/2)}{\int_0^1 \sqrt{Q(x')}\,dx'}; \qquad n = 0, 1, 2, 3 \tag{6.101}$$

EXERCISE 5.3 Consider

$$\frac{d^2\Phi}{dx^2} + \alpha^2 \cdot Q(x) \cdot \Phi = 0;$$

$$\Phi'(0) = \Phi(1) = 0, \quad 0 \le x \le 1, \quad \alpha \to \infty, \quad Q(x) = (1+x)^2$$

What are the eigenvalues, α_n? What is the form of the eigenfunctions? Compare your results with those obtained in Exercise 5.1. ■

EXERCISE 5.4 Consider

$$\frac{d^2\Phi}{dx^2} + \alpha^2 \cdot Q(x) \cdot \Phi = 0;$$

$$\Phi'(1) = \Phi(2) = 0, \quad 1 \le x \le 2, \quad \alpha \to \infty, \quad Q(x) = x^2$$

Find the eigenvalues and compare your results with those obtained in Exercise 5.2.

Answer: $\alpha_n = (2n+1)\pi/3$. ■

The WKB approximation usually gives a fine approximation for even the low-lying eigenvalues. However, in the event that $Q(x) \approx 0$ anywhere on its interval of definition, one must use alternative procedures. Techniques for handling this situation are discussed under the topic of "turning-points" in the Bibliography. See, for example, the books by Nayfeh, Olver, and Bender and Orszag.

6.6 PERTURBATION THEORY

In this section, we introduce the basic elements of the traditional perturbation method used to find the solution of inhomogeneous linear differential equations. We outline it briefly in general terms and illustrate its basic elements by means of a simple example. This enables us to point out its shortcomings as preparation for the development of a more efficient perturbation scheme given in Chapters 9 to 14. First, we need a few results:

(a) We have to know how to differentiate an integral of the form

$$\int_0^t f(s, t)x(s)\, ds \tag{6.102a}$$

Differentiating, we have

$$\frac{d}{dt}\int_0^t f(s, t)x(s)\, ds = f(t, t)x(t) + \int_0^t \frac{d}{dt} f(s, t)x(s)\, ds \tag{6.102b}$$

where $f(s, t)$ is well behaved (e.g., it satisfies the first mean-value theorem). This result is simple to obtain by using the delta process to find derivatives. We just evaluate

$$\begin{aligned}\int_0^{t+\delta t} f(s, t+\delta t)x(s)\, ds &- \int_0^t f(s, t)x(s)\, ds \\ &= \int_0^t [f(s, t+\delta t) - f(s, t)]\, x(s)\, ds \\ &\quad + \int_t^{t+\delta t} f(s, t+\delta t)x(s)\, ds\end{aligned} \tag{6.103}$$

Keeping in mind that δt tends to zero, we have the result given by Equation (6.102b).

(b) Consider the first-order linear differential equation:

$$\frac{dx(t)}{dt} = \varepsilon f(t)x(t); \qquad x(0) = A \tag{6.104}$$

where $f(t)$ is a known function, and ε is a small parameter. It is always possible to directly integrate this equation and obtain a solution in the form

$$x(t) = A \exp\left[\varepsilon \int_0^t f(s)\, ds\right] \tag{6.105}$$

We can check, by differentiation, that this is correct. Then, given the function $f(t)$, we can always obtain the solution.

Example 6.1 If $f(t) = -2t$, we have the equation

$$\frac{dx(t)}{dt} = -2\varepsilon t\, x(t); \qquad x(0) = A \tag{6.106}$$

with the solution

$$x(t) = A\exp(-\varepsilon t^2) \qquad \blacksquare \qquad (6.107)$$

We can also integrate Equation (6.104) by transferring the time-derivative operator to the r.h.s. to obtain our solution in the form of an integral equation:

$$x(t) = A + \varepsilon\int_0^t f(s)x(s)\,ds \qquad (6.108)$$

It is simple to check, by differentiation, that this is a solution.

It is possible to show that every linear differential equation of order n, with continuous coefficients, can be transformed into an integral equation of the form

$$x(t) = f(t) + \varepsilon\int_0^t k(t, s)x(s)\,ds \qquad (6.109)$$

There are four parts to the integral equation given by Equation (6.109): a known function $f(t)$, which is often called the inhomogeneous term; the kernel $k(t, s)$, which plays the role of the differential operator and characterizes the equation; the unknown function $x(t)$; and, finally, the expansion parameter ε, which is often used to denote the order of the iteration in the approximation scheme. An equation of the form given by Equation (6.109) is called a "Volterra integral equation of the second kind." It is a Volterra equation because the limit of the integration is the variable t rather than a constant. It is of the second kind because $f(t)$ is not identically equal to zero. (The solution is unique. We showed this in Example 4.1 of Section 6.4. where we used Gronwall's lemma to establish that the solution to a Volterra integral equation of the second kind has a unique solution.)

The transformation of a differential equation to an integral equation is traditionally discussed at the beginning of a course in integral equations. We use only the most basic elements of the theory and we derive and explain all of the pertinent material. An extensive treatment of integral equations, together with the relevant proofs, can be found, for example, in the books by Tricomi and by Lovitt.

Our procedure involves a shift of interest from solving a differential equation to obtaining a solution of the associated integral equation given by Equation (6.109). Often, the latter form admits a simpler perturbation expansion. In part, this is because integration is a smoothing process, whereas differentiation is not.

EXERCISE 6.1 Consider the differential equation

$$\frac{d^2x}{dt^2} + x = f(t); \qquad x(0) = A, \quad \dot{x}(0) = 0 \qquad (6.110)$$

It is equivalent to the integral equation

$$x(t) = A\cos t + \int_0^t k(t, s)f(s)\, ds \tag{6.111}$$

where the kernel $k(t, s) = \sin(t - s)$. Check by differentiating Equation (6.111) twice; one obtains Equation (6.110). (Recall that we considered this problem in Exercise 2.1, leading to Equation (6.39).) ■

Let us take Equation (6.109) as the starting point of a problem in which the parameter ε is small. We know that when $\varepsilon = 0$, $x(t) = f(t)$ and the next term is a correction. Thus, we would like to develop a systematic perturbation expansion that would give us the solution in the form:

$x(t) = f(t) +$ [something we can evaluate and thus include]

+ [something we neglect]

It would be understood that the neglected terms should be small over the entire time interval for which we want the expansion to be valid. If we construct an expansion of this form, we say that it is uniformly valid in time. (This concept, together with its implications, is fully developed in Chapters 8 to 15.) Let us see how this works by developing a perturbation expansion for $x(t)$ given by Equation (6.108). We do this by substituting, in the integral, the lowest approximation to $x(t)$. This means that we write our solution in the form

$$x(t) = A + \varepsilon\int_0^t f(s)\left[A + \varepsilon\int_0^s f(u)x(u)\, du\right] ds \tag{6.112a}$$

Our formal solution, given by Equation (6.112a), is still exact. We rewrite Equation (6.112a) as

$$x(t) = A + \varepsilon\int_0^t f(s)A\, ds + \varepsilon^2\int_0^t f(s)\int_0^s f(u)x(u)\, du\, ds \tag{6.112b}$$

First, we verify by differentiation that this is a solution. It satisfies the initial condition $x(0) = A$. We differentiate and obtain

$$\begin{aligned}\frac{dx}{dt} &= \varepsilon f(t)A + \varepsilon^2 f(t)\int_0^t f(u)x(u)\, du\\ &= \varepsilon f(t)\left[A + \varepsilon\int_0^t f(u)x(u)\, du\right]\\ &= \varepsilon f(t)x(t)\end{aligned}$$

This is precisely Equation (6.104).

We emphasize that the solution of Equation (6.104) as given by Equation (6.112b) is exact and is separated into three terms: the first represents the initial condition and is independent of the parameter ε; the second is proportional to ε; and the last term is proportional to ε^2. Now, if ε is sufficiently small, we would think that we can obtain an approximation to the solution by truncating our expansion after two terms. Doing this, we have

$$x(t) \approx A + \varepsilon \int_0^t A f(s)\, ds + \text{h.o.t.} \tag{6.113}$$

We know $f(s)$, so we can evaluate the integral and obtain an approximation to $x(t)$. We also note, from Equation (6.112b), that the higher-order terms in the expansion can be generated by continuing the substitution procedure. Each integral can be evaluated since it contains the known function $f(s)$. In this way, we expect to generate a better and better approximation to $x(t)$.

Example 6.2 Consider

$$u(t) = 1 + \varepsilon^2 \int_0^t (s - t) u(s)\, ds; \qquad (\text{note: } u(0) = 1) \tag{6.114}$$

We want to find the first two terms in a perturbation expansion that is to be valid for short times. We introduce the substitution process and write

$$\begin{aligned} u(t) &= 1 + \varepsilon^2 \int_0^t (s - t)\left[1 + \varepsilon^2 \int_0^s (x - s) u(x)\, dx\right] ds \\ &= 1 + \varepsilon^2 \int_0^t (s - t)\, ds + \text{neglected terms} \end{aligned}$$

Thus, we say that

$$u(t) = 1 - \frac{\varepsilon^2 t^2}{2!}$$

for times t such that the quantity $|\varepsilon t| \ll 1$. (The next term in the expansion is found to be $\varepsilon^4 t^4/4!$ and is then much less than the included terms. At this point, it is assumed that

$$u = 1 - \frac{\varepsilon^2 t^2}{2!} + \frac{\varepsilon^4 t^4}{4!}$$

will be recognized as the first few terms in the Taylor-series expansion of the exact solution, $\cos \varepsilon t$.) ■

EXERCISE 6.2 Verify that $\cos \varepsilon t$ is the exact solution of Equation (6.114) by substituting $u(s) = \cos \varepsilon s$ in the integral and evaluating the terms. ■

EXERCISE 6.3 Consider the integral equation

$$u(t) = \varepsilon t + \varepsilon^2 \int_0^t (s - t) u(s)\, ds \tag{6.115}$$

Show that the first two terms in the perturbation expansion, valid for times t, such that $|\varepsilon t| \ll 1$, are

$$\varepsilon t - \frac{\varepsilon^3 t^3}{3!}$$

Then, as a separate procedure, differentiate Equation (6.115) to obtain the simple harmonic oscillator equation of frequency ε with solution $u(t) = \sin \varepsilon t$. Finally, substitute $u(s) = \sin \varepsilon s$ in Equation (6.115) and show that $u(t)$ is indeed $\sin \varepsilon t$. ■

We continue our development by considering the integral equation

$$x(t) = 1 + \varepsilon \int_0^t x(s)\, ds; \qquad 0 < \varepsilon \ll 1 \tag{6.116}$$

It originated from the differential equation

$$\frac{dx(t)}{dt} = \varepsilon x(t); \qquad x(0) = 1 \tag{6.117}$$

whose solution is

$$x(t) = \exp \varepsilon t \tag{6.118}$$

It is clear that if $|\varepsilon t| \ll 1$, we can approximate Equation (6.118) by the first two terms in the Taylor series

$$x(t) \approx 1 + \varepsilon t \tag{6.119}$$

We attempt to solve our integral equation, Equation (6.116), by the iteration scheme

$$\begin{aligned} x(t) &= 1 + \varepsilon \int_0^t \left[1 + \varepsilon \int_0^s x(u)\, du \right] ds \\ &= 1 + \varepsilon t + \varepsilon^2 \int_0^t \int_0^s x(u)\, du\, ds \end{aligned} \tag{6.120}$$

noting that Equation (6.120) is exact. (This can be easily verified by showing that $x(t)$ satisfies Equation (6.117).)

If ε is small, it seems correct to terminate the expansion after the first integral and say that the error is proportional to ε^2. Thus, we would approximate Equation (6.120) by

$$x(t) = 1 + \varepsilon t + \text{error proportional to } \varepsilon^2 \tag{6.121}$$

It is important to note that the error term is correct but misleading, as is seen from a comparison with the exact solution given by Equation (6.118). The neglected term is proportional to the square of the product of the smallness parameter ε and the length of the time integration t. In fact, we know that the general term in the Taylor series is of the form $(\varepsilon t)^n/n!$. Thus, ε and t always occur together, and we are only able to truncate our approximation if the time interval under consideration, t, is such that $|\varepsilon t| \ll 1$. Our expansion, given by Equation (6.121), is said *not* to be uniformly valid in time. (We will see, in our discussions of nonlinear systems, that it leads to an incorrect picture of the solution.) The natural question to ask is: With all of these difficulties, why do we use this expansion technique at all? There are at least four reasons.

(1) It is very simple to apply and it can be shown that it always converges to the correct solution. (For a proof, see either the book by Lovitt or by Tricomi.)
(2) It works very satisfactorily if one keeps in mind that there is the coupling between the expansion parameter ε and time t.
(3) Often, one can obtain, from the first few terms in the expansion, a hint regarding the nature of the full solution.
(4) Most better perturbation techniques are significantly more difficult to use.

With these thoughts in mind, we develop in Chapters 9 to 15 a perturbation expansion in powers of a small parameter ε that retains the essence of this method of solution while avoiding its shortcomings.

BIBLIOGRAPHY

1. Bellman, R., *Stability Theory of Differential Equations*, McGraw-Hill, New York, 1953. *Perturbation Techniques in Mathematics, Engineering and Physics*, Holt, Rinehart and Winston, New York, 1966. These books are easy to read and present the elements of the subject without too many details. Our discussion is based on Bellman's approach.
2. Bender, C., and S. A. Orszag, *Advanced Mathematical Methods for Scientists and Engineers*, McGraw-Hill, New York, 1978. The discussion is thorough and comprehensive, with many examples. There is also a connection given between the WKB method and boundary-layer theory and the method of multiple time scales.

3. De Bruijn, N. G., *Asymptotic Methods in Analysis*, North-Holland, Amsterdam, 1958. Republished by Dover, New York, 1981. The treatment is most interesting and at a high level of difficulty. Only the bare bones of the arguments are presented, leaving the reader with lots of opportunities to fill in the details.
4. Nayfeh, A. H., *Introduction to Perturbation Techniques*, Wiley, New York, 1981. The discussion is thorough, including a very clear and extensive treatment of turning-point problems. The pace is leisurely and the material easy to read. There are lots of problems and illustrative examples.
5. Olver, F. W. J., *Asymptotics and Special Functions*, Academic Press, New York, 1974. The section on the Liouville–Green transformation is extensive and at a high level of difficulty. You are rewarded for your efforts by a discussion of some of the more subtle aspects of the subject.
6. Jeffreys, H., and B. Jeffreys, *Methods of Mathematical Physics*, Third Edition, Cambridge University Press, Cambridge, 1966. This is a classic text with a section on asymptotic expansions as applied to second-order linear differential equations. The material is clearly presented and is both interesting and relatively easy to follow.
7. Tricomi, F. G., *Integral Equations*, Interscience, New York, 1957. Reprinted by Dover, New York, 1985. This is the most readable and comprehensive text on integral equations.
8. Lovitt, W. V., *Linear Integral Equations*, McGraw-Hill, New York, 1924. Reprinted by Dover, New York, 1950. It is amazing that this book is still in print and that its discussion is timely. It has many illustrative examples and easily serves as both an introductory and intermediate text on integral equations.

PART II

NONLINEAR SYSTEMS

CHAPTER 7

THE SIMPLE HARMONIC OSCILLATOR AND THE LOGISTIC EQUATION

In our attempt to analyze a variety of phenomena, we build models that contain the essential elements of the problem under consideration. One of the most universal of these, which is both linear and easy to analyze, is the simple harmonic oscillator (SHO). It often originates in a study of Newton's equations in one dimension in the form

$$\frac{d^2x}{dt_1^2} = -\frac{dV}{dx} \tag{7.1}$$

where potential $V(x)$ has a minimum at $x = 0$. If we expand the potential about the origin and retain only quadratic terms, we have

$$V(x) = V(0) + \frac{x^2}{2!}\frac{d^2V(0)}{dx^2}$$

Then Equation (7.1) takes the form

$$\frac{d^2x}{dt_1^2} = -\omega_0^2 x \tag{7.2}$$

whose solution is $x = A\cos(\omega_0 t_1 + \phi_0)$. Quantity A is called the amplitude and is equal to the maximum displacement; the square of the angular frequency ω_0 is equal to the second derivative of the potential; the initial phase is denoted as ϕ_0. Often, oscillatory motion is analyzed by means of a

Fourier-series expansion. One sees that in such an expansion, the solution to Equation (7.2) contains only a single harmonic, leading us to call the associated motion "simple harmonic." Two of its characteristics are a constant amplitude and an amplitude-independent angular velocity. It is our experience that Equation (7.2) occurs in diverse areas, for example, in a variety of models in mechanics where we consider vibrations, in electricity and magnetism in a discussion of circuits and electromagnetic waves, and in atomic and nuclear physics where we consider spectra. In addition, the SHO provides the basic framework for the study of weakly nonlinear oscillatory motion. It is with these points in mind that we discuss it in some detail. We begin by introducing some of the notation and basic techniques that we will be using throughout our discussion of oscillatory motion. We postpone, until Chapter 8, our orientation to nonlinear oscillatory motion. At that juncture we provide an overview of our treatment of the subject.

7.1 THE SIMPLE HARMONIC OSCILLATOR (SHO)

Our basic form of the oscillator equation is given by Equation (7.2). It is usually convenient to work with dimensionless variables. Thus, we measure all displacements $x(t_1)$ in terms of the amplitude A of the oscillation. We write $x(t) = x(t_1)/A$ and also introduce a dimensionless time variable t:

$$t = \omega_0 t_1$$

Our equation then becomes:

$$\frac{d^2x}{dt^2} + x = 0$$

or

$$\ddot{x} + x = 0 \tag{7.3}$$

We introduce the dimensionless velocity, denoted by symbol y, such that

$$y = \dot{x} = \frac{dx}{dt} \tag{7.4}$$

and thus write our basic equation also as two coupled first-order equations:

$$\frac{dx}{dt} = y \tag{7.5a}$$

$$\frac{dy}{dt} = -x \tag{7.5b}$$

We will go back and forth between these representations, i.e., the coupled first-order equations and the second-order equation.

We are interested in finding the energy of the oscillator. To accomplish this, we multiply Equation (7.5a) by x, Equation (7.5b) by y, and add:

$$x\dot{x} + y\dot{y} = \frac{1}{2}\frac{d}{dt}\left(x^2 + y^2\right) = 0 \tag{7.6}$$

This equation can be integrated to yield a constant, which we call E, and identify with the total energy of the system.

We discuss both conservative systems in which the total energy is constant and systems in which there is damping and a resultant change of the total energy in time.

7.2 TRANSFORMATION OF OUR EQUATIONS

We often work in polar coordinates, designated by variables r and θ. The transformation between Cartesian and polar coordinates is

$$x = r\cos\theta \tag{7.7a}$$

$$y = -r\sin\theta \tag{7.7b}$$

We have exchanged our Cartesian variables x and y for polar coordinates r and θ. We have to find the derivatives of these relationships. They are

$$\dot{x} = \dot{r}\cos\theta - r\dot{\theta}\sin\theta \tag{7.8a}$$

$$\dot{y} = -\dot{r}\sin\theta - r\dot{\theta}\cos\theta \tag{7.8b}$$

$$\begin{bmatrix}\dot{x}\\ \dot{y}\end{bmatrix} = A\begin{bmatrix}\dot{r}\\ r\dot{\theta}\end{bmatrix}$$

$$A = \begin{bmatrix}\cos\theta & -\sin\theta\\ -\sin\theta & -\cos\theta\end{bmatrix} \tag{7.9a}$$

The inverse transformation is

$$\begin{bmatrix}\dot{r}\\ r\dot{\theta}\end{bmatrix} = A^{-1}\begin{bmatrix}\dot{x}\\ \dot{y}\end{bmatrix}; \qquad A^{-1} = \begin{bmatrix}\cos\theta & -\sin\theta\\ -\sin\theta & -\cos\theta\end{bmatrix} \tag{7.9b}$$

We then always have the relationships

$$x\dot{x} + y\dot{y} = r\dot{r} \tag{7.10a}$$

$$y\dot{x} - x\dot{y} = r^2\dot{\theta} \tag{7.10b}$$

Angle θ is measured from the x axis and increases in the clockwise direction in the x–y plane, as illustrated in Figure 7.1.

EXERCISE 2.1 Consider Equations (7.5a) and (7.5b) that describe simple harmonic motion. Introduce polar coordinates and conclude that the radius

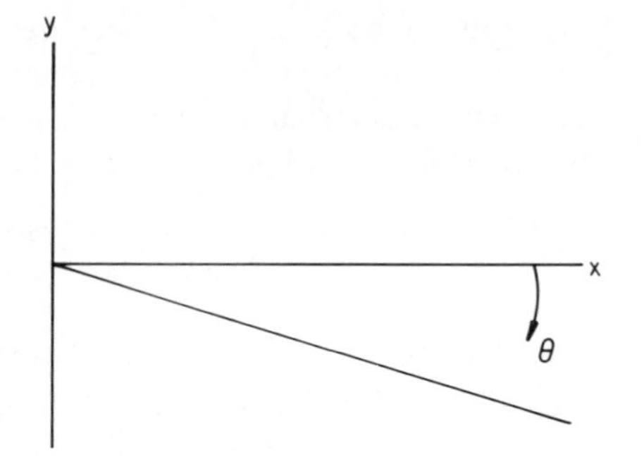

Figure 7.1 The angle θ increases in a clockwise direction. This is compatible with the sign convention given by Equations (7.7a) and (7.7b) that relates (x, y) and (r, θ).

and the angular velocity are constant. If we use the sign convention for the angular velocity given by Equation (7.10b), we have $\dot{\theta} = +1$. ■

7.3 PERTURBATIONS

We are interested in systems that are close to simple harmonic. Thus, we consider the perturbed oscillator given by

$$\frac{d^2x}{dt^2} + x = \varepsilon f\left(x, \frac{dx}{dt}\right) = \varepsilon f(x, y) \tag{7.11}$$

where the perturbing term $\varepsilon f(x, y)$ is always small. This means that during the entire time development of the system, for each interval of time comparable to a period of the unperturbed oscillator, there is only a small change in the frequency and amplitude. However, there can be a sizable cumulative effect. For example, we can have $f(x, y) = y$, indicating a slightly damped linear oscillator:

$$\ddot{x} + x = \varepsilon \dot{x} = \varepsilon y; \qquad |\varepsilon| \ll 1$$

The sign of ε determines whether we have positive damping, $\varepsilon < 0$ (i.e., decay of oscillations) or negative damping, $\varepsilon > 0$ (i.e., growth of oscillations). Note that the damping is slight if

$$\frac{|x(t + T) - x(t)|}{|x(t)|} \ll 1$$

where T is the period of the unperturbed oscillator.

Example 3.1 We can express the damped linear oscillator as two coupled first-order equations. Thus, given

$$\frac{d^2x}{dt^2} + x = -\mu \frac{dx}{dt}; \quad |\mu| \ll 1$$

we can write

$$\dot{x} = y \tag{7.12a}$$

$$\dot{y} = -x - \mu y \tag{7.12b}$$

$$\begin{bmatrix} \dot{x} \\ \dot{y} \end{bmatrix} = \begin{bmatrix} 0 & 1 \\ -1 & -\mu \end{bmatrix} \begin{bmatrix} x \\ y \end{bmatrix} \tag{7.13}$$

We can then analyze the motion of the system by studying the properties of the matrix in Equation (7.13). ■

7.4 LINEAR TRANSFORMATIONS

We sometimes find that our oscillator equations are not presented in their simplest form because the coordinates that we are using are not the appropriate ones. Then we have to perform a transformation to the correct coordinate system. In order to explain this, we review a few properties of 2×2 matrices. (See Chapter 1.) For example, under what circumstances does the set of equations

$$\frac{dx}{dt} = ax + by \tag{7.14a}$$

$$\frac{dy}{dt} = cx + dy \tag{7.14b}$$

with all coefficients real, have solutions that correspond to simple harmonic motion? Is there damping? What happens if we change the relationships among the coefficients?

There are a few ways that we can answer these questions. For example, we can differentiate Equation (7.14a) and then use Equation (7.14b) to obtain:

$$\begin{aligned} \ddot{x} &= a\dot{x} + b\dot{y} = a\dot{x} + b(cx + dy) \\ &= a\dot{x} + bcx + d(\dot{x} - ax) \\ &= (a + d)\dot{x} + (bc - ad)x \end{aligned}$$

or

$$\ddot{x} - (a + d)\dot{x} + (ad - bc)x = 0 \tag{7.15}$$

We then know from our experience with the SHO that we can solve Equation (7.15) by the trial solution:

$$x = \exp \lambda t$$

We can find the values of λ:

$$\lambda = \frac{a+d}{2} \pm \frac{1}{2}\sqrt{(a+d)^2 - 4(ad-bc)} \tag{7.16a}$$

$$= \frac{a+d}{2} \pm \frac{1}{2}\sqrt{(a-d)^2 + 4bc} \tag{7.16b}$$

In order to have oscillatory solutions, we need λ to be complex (or pure imaginary). Thus, we see from Equation (7.16b) that both b and c must be different from zero and of opposite signs, and

$$-4bc > (a-d)^2 \qquad \text{or} \qquad 4(ad-bc) > (a+d)^2 \tag{7.16c}$$

The type of damping is determined by the sign of $a + d$:

if $a + d < 0$, we have positive damping;
if $a + d > 0$, we have negative damping; and
if $a + d = 0$, we have no damping.

Alternatively, we can solve Equations (7.14a) and (7.14b) by means of matrix theory and seek a solution in terms of the associated matrix M:

$$\frac{d}{dt}\begin{bmatrix} x \\ y \end{bmatrix} = \begin{bmatrix} \dot{x} \\ \dot{y} \end{bmatrix} = M\begin{bmatrix} x \\ y \end{bmatrix} \tag{7.17}$$

$$M = \begin{bmatrix} a & b \\ c & d \end{bmatrix} \tag{7.18}$$

We consider only situations in which all the matrix elements are real and elements b and c are both different from zero.

We make the following comments and observations:

(a) Determinant Δ has the value

$$\Delta = ad - bc$$

(b) The determinant is equal to the product of the eigenvalues. We designate the eigenvalues by the symbols λ_1 and λ_2. Thus, the determinant is written

$$\Delta = \lambda_1 \lambda_2$$

(c) The trace of matrix M is equal to the sum of the eigenvalues and is written

$$\text{Trace}\, M = \text{Tr}\, M = a + d = \lambda_1 + \lambda_2$$

(d) We are interested in knowing the connection between the matrix elements and the eigenvalues. For example, under what conditions are the eigenvalues real or complex? Transcribing the inequality given by Equation (7.16c) into matrix notation, we conclude that our solution is oscillatory if

$$4\Delta > (\operatorname{Tr} M)^2 \tag{7.19}$$

We also know from our study of matrices that if

$$\operatorname{Tr} M = 0 \qquad \text{and} \qquad \Delta > 0$$

the eigenvalues are purely imaginary. If the eigenvalues are complex, the sign of the trace indicates the sign of the real part of the eigenvalues. Thus, if $\Delta > 0$ and $\operatorname{Tr} M > 0$, both eigenvalues have positive real parts, and if $\operatorname{Tr} M < 0$, our eigenvalues are complex with negative real parts.

(e) We know that the trace and determinant are invariant under a similarity transformation S.

$$\operatorname{Tr} M = \operatorname{Tr}(SMS^{-1}); \qquad \det M = \det(SMS^{-1})$$

(f) If the eigenvalues are complex, we can always find an appropriate transformation so that the sum of the eigenvalues, and, hence, twice their real part, is equal to the matrix elements that occupy either the lower right-hand corner or the upper left-hand corner of the matrix.

This is demonstrated in Exercises 4.1 and 4.2. We note that the transformed matrix is antisymmetric, if the eigenvalues are purely imaginary.

EXERCISE 4.1 Matrix M, associated with Equations (7.17) and (7.18), is given:

$$M = \begin{bmatrix} a & b \\ c & d \end{bmatrix}; \qquad \Delta = ad - bc$$

having complex eigenvalues. Show that matrix S:

$$S = \begin{bmatrix} 1 & 0 \\ \dfrac{a}{\sqrt{\Delta}} & \dfrac{b}{\sqrt{\Delta}} \end{bmatrix}; \qquad S^{-1} = \begin{bmatrix} 1 & 0 \\ \dfrac{-a}{b} & \dfrac{\sqrt{\Delta}}{b} \end{bmatrix} \tag{7.20}$$

is such that

$$SMS^{-1} = A = \begin{bmatrix} 0 & \sqrt{\Delta} \\ -\sqrt{\Delta} & a + d \end{bmatrix}$$

Let us consider the matrix differential equation given by Equation (7.17). We have learned how to find the similarity transformation S that brings M into its special form. In the process of transforming M, we have rotated the vector (x, y) into a new vector (v, w). The relationship is given by

$$\begin{bmatrix} x \\ y \end{bmatrix} = S^{-1}\begin{bmatrix} v \\ w \end{bmatrix}; \qquad \begin{bmatrix} v \\ w \end{bmatrix} = S\begin{bmatrix} x \\ y \end{bmatrix}$$

Thus, Equation (7.17) becomes, after multiplying through by the transformation S,

$$\left(\frac{d}{dt}\right)S\begin{bmatrix} x \\ y \end{bmatrix} = SMS^{-1}S\begin{bmatrix} x \\ y \end{bmatrix} \qquad \text{or} \qquad \frac{d}{dt}\begin{bmatrix} v \\ w \end{bmatrix} = A\begin{bmatrix} v \\ w \end{bmatrix}$$

Now we solve for variables (v, w), and after we are finished, we transform back to the pair (x, y). ■

EXERCISE 4.2 Given the same matrix M discussed in Exercise 4.1, show that matrix P:

$$P = \begin{bmatrix} -\dfrac{c}{\sqrt{\Delta}} & -\dfrac{d}{\sqrt{\Delta}} \\ 0 & 1 \end{bmatrix}; \qquad P^{-1} = \begin{bmatrix} -\dfrac{\sqrt{\Delta}}{c} & -\dfrac{d}{c} \\ 0 & 1 \end{bmatrix} \tag{7.21}$$

is such that

$$PMP^{-1} = B = \begin{bmatrix} a + d & \sqrt{\Delta} \\ -\sqrt{\Delta} & 0 \end{bmatrix}$$

■

To summarize: if M has real elements, a positive determinant, both off-diagonal elements different from zero, and

$$4\Delta > (\operatorname{Tr} M)^2$$

it will have pure imaginary or complex eigenvalues and can always be brought to the form

$$A = \begin{bmatrix} 0 & \sqrt{\Delta} \\ -\sqrt{\Delta} & \lambda_1 + \lambda_2 \end{bmatrix} \tag{7.22}$$

or the form

$$B = \begin{bmatrix} \lambda_1 + \lambda_2 & \sqrt{\Delta} \\ -\sqrt{\Delta} & 0 \end{bmatrix} \tag{7.23}$$

by an appropriate similarity matrix.

EXERCISE 4.3 We conclude our analysis by discussing another transformation that brings a given matrix M into a convenient form. We observe that matrices S and P, given by Equations (7.20) and (7.21), respectively, have the quantities $\pm\sqrt{\Delta}$ as their off-diagonal matrix elements. In addition, the term $(a+d)/2$ is not symmetrically placed. In the absence of damping, $\Delta = ad - bc$ is equal to the square of the frequency of oscillation. When damping is present, we have a shift, so that

$$\omega_0 = \sqrt{ad-bc} \rightarrow \omega = \sqrt{\frac{4(ad-bc)-(a+d)^2}{4}} \tag{7.24}$$

In light of these observations, it is often convenient to find a matrix J such that

$$JMJ^{-1} = C$$

where

$$C = \begin{bmatrix} \dfrac{a+d}{2} & \sqrt{\dfrac{4(ad-bc)-(a+d)^2}{4}} \\ -\sqrt{\dfrac{4(ad-bc)-(a+d)^2}{4}} & \dfrac{a+d}{2} \end{bmatrix} \tag{7.25}$$

Notice that the trace of $C = a + d$ and det $C = ad - bc$, so that we can, in principle, find J. The exercise is then, given Equation (7.15) with an oscillatory solution of the form,

$$\begin{aligned} x(t) &= \exp\frac{(a+d)t}{2}\cos\left\{\left[\sqrt{(ad-bc)} - \frac{(a+d)^2}{4}\right]t + \phi\right\} \\ &\equiv (\exp \mu t)\cos(\omega t + \phi) \end{aligned}$$

to find a matrix J such that

$$JMJ^{-1} = \begin{bmatrix} \mu & \omega \\ -\omega & \mu \end{bmatrix} = C \tag{7.26}$$

where $\omega \neq \omega_0$ unless $a + d = 0$.

Since the normalization of J does not affect the result, it might be convenient to pick

$$J = \begin{bmatrix} 1 & \alpha \\ \beta & \gamma \end{bmatrix}$$

and then solve for α, β, γ. ■

EXERCISE 4.4 Given the oscillator equation

$$\frac{d^2x}{dt^2} + x = \varepsilon \frac{dx}{dt}; \qquad |\varepsilon| \ll 1$$

how do the eigenvalues depend on the sign of ε?

First, write

$$\frac{dx}{dt} = y; \qquad \frac{dy}{dt} = -x + \varepsilon y$$

Then introduce matrix M:

$$\begin{bmatrix} \dot{x} \\ \dot{y} \end{bmatrix} = M \begin{bmatrix} x \\ y \end{bmatrix}; \qquad M = \begin{bmatrix} 0 & 1 \\ -1 & \varepsilon \end{bmatrix}$$

It is clear that

if $\varepsilon > 0$, the system is negatively damped; and
if $\varepsilon < 0$, the system is positively damped.

Solve the eigenvalue equation and show that we need $\varepsilon^2 < 4$ in order to have oscillatory solutions.

We say that the system is lightly or slightly damped if $|\varepsilon| \ll 1$. ■

7.5 THE LOGISTIC EQUATION

7.5.1 The Exponential

We encounter the equation for exponential growth in a variety of contexts, for example, in the growth of a population $N(t)$ that has no predators and an unlimited supply of food. The growth rate is then determined by a multiplicative factor σ that is some constant multiplied by the difference between the birth and death rates, perhaps as follows:

$$\frac{dN}{dt} = \sigma N(t) \tag{7.27}$$

We are familiar with this equation and its solution. We pause here to discuss it from a point of view that will be used in later chapters. We assume that at some arbitrary time, say $t = 0$, we count and find that the population has the value $N(0)$. (Hereafter, whenever we refer to the initial value of something, we denote its argument by "0.") We want to know whether it increases or decreases as time progresses. This depends on the sign of parameter σ. That is to say, if $\sigma > 0$, the population increases in time, and if $\sigma < 0$, the population decreases. Thus, we can determine the future population if we know $N(0)$ and the sign and magnitude of parameter σ.

7.5.2 The Logistic Equation

We now alter the exponential model slightly by introducing the concept of saturation. There might be limited space or food supplies available to the species. Then there is a new factor that limits the growth of the population, namely, the size of the population itself. The most popular model for this situation is called the logistic model:

$$\frac{dN}{dt} = \sigma N\left(1 - \frac{N}{K}\right) \tag{7.28}$$

where we have indicated the limited resources by altering the growth coefficient from

$$\sigma \to \sigma\left(1 - \frac{N}{K}\right)$$

Constant K is called the carrying capacity. This small change in our equation alters the growth of the species in a fundamental way. Suppose we have a population that is not changing:

$$\frac{dN}{dt} = 0 \qquad \Rightarrow \qquad N \equiv \overline{N}$$

This can occur for two values of N, namely, $\overline{N} = 0$ and $\overline{N} = K$. These are called the equilibrium values, steady-state values, or equilibrium points. (We introduce a bar over the variable to indicate that it is at its equilibrium value.) Note that in our linear system described by Equation (7.27), we had only one equilibrium point, $N = 0$.

Now we ask what will happen if the system is at one of these equilibrium values and it is slightly disturbed by a change in its value so that we now have

$$N \neq 0 \qquad \text{or} \qquad K$$

This leads us to the concept of examining the stability of the equilibrium point. Assume that a system is initially located at an equilibrium point. We make a small change in the initial condition and watch how the system responds. If it returns to the equilibrium point, we say that the equilibrium point is stable. If it moves away from both the initial value and the equilibrium point, we say that the equilibrium point is unstable. If it stays at the new initial value, we say that the equilibrium point is neutrally stable. Thus, we characterize the equilibrium point in terms of how the system behaves when it is slightly disturbed from this location. This concept is central to our study of nonlinear systems. Keep in mind that we are only asking for a local judgment. That is to say, we only ask for the behavior of the system under "small" deviations. For example, consider the "stability of the origin." We write

$$N(t) = \overline{N} + n(t) = 0 + n(t)$$

We require $|n(t)| \ll K$, since K is the scale of the final size of the population. Equation (7.28) becomes

$$\frac{dn}{dt} = \sigma n(t)\left[1 - \frac{n(t)}{K}\right] \approx \sigma n(t) \tag{7.29}$$

We see that the origin is stable or unstable, depending on the sign of σ.

Let us now examine the stability of point $\overline{N} = K$. We write

$$N(t) = K + n(t); \qquad |n(t)| \ll K$$

Thus, our initial condition is a disturbance of the maximum population. Our equation becomes

$$\begin{aligned} \frac{dn}{dt} &= \sigma(n + K)\left(1 - \frac{K}{K} - \frac{n}{K}\right) \\ &= -\sigma(n + K)\frac{n}{K} \\ &\approx -\sigma n(t) \end{aligned} \tag{7.30}$$

We see that the quantities $1 - n(t)/K$ and $K + n$ are both positive; thus, Equation (7.29) is of the same form as Equation (7.30). If we now compare Equations (7.29) and (7.30), we see that they indicate that the stability of the equilibrium points are of opposite sign. That is to say, if motion about the origin is stable, then the motion about the carrying capacity K is unstable and vice versa. Alternatively, we say that the origin is unstable if $\sigma > 0$, since it

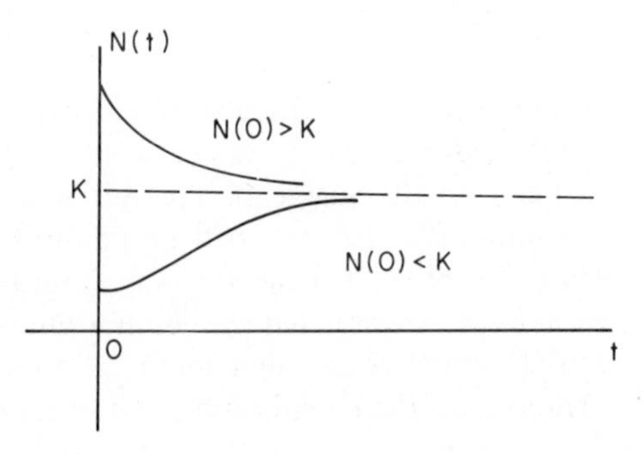

Figure 7.2 Observe the smooth monotonic approach to the carrying capacity $N = K$.

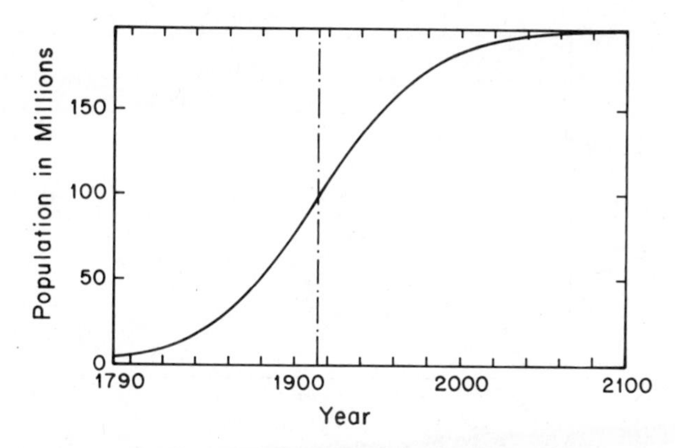

Figure 7.3 We plot a graph due to Pearl and Reed, as shown in the book by Lotka, page 68, that fits the United States population data from 1790 to 1910. (With the permission of Dover Publications, Inc.)

repels all populations that are less than the carrying capacity K. If parameter $\sigma < 0$, the origin is an attractor and K is a repeller.

We consider a situation in which we initially have a population at some value $N(0)$ that is small enough to be in the region of exponential growth and parameter $\sigma > 0$. We then disturb it by some small perturbation and observe that it grows monotonically and approaches the limiting value K. This comes about because it is always repelled by the origin and always attracted to the equilibrium point at $N = K$. This behavior is due to the interplay between the linear and nonlinear terms. When the amplitude is small, the linear terms dominate and the amplitude grows until it becomes large enough to be controlled by the nonlinear terms that lead it toward its asymptotic value. In addition, if the initial population is greater than the carrying capacity, it decays almost exponentially with the carrying capacity as its limiting value. In this region, both the linear and the nonlinear terms lead to an attraction of the population toward the carrying capacity value $N = K$. See Figure 7.2.

Equation (7.28) is used extensively as a model for population growth. See, for example, the book by Lotka in which you find comparisons with data. See Figures 7.3 and 7.4.

EXERCISE 5.1 Given the logistic equation in the form

$$\frac{dN}{dt} = \sigma N\left(1 - \frac{N}{K}\right); \qquad \sigma, K > 0$$

The initial condition is such that $N(0) \neq K, 0$.

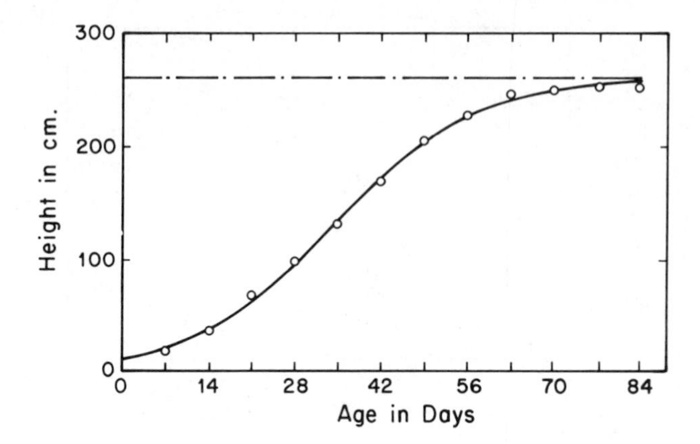

Figure 7.4 This figure illustrates the growth of sunflower seedlings according to the data of Reed and Holland, taken from the book by Lotka, page 74. (With the permission of Dover Publications, Inc.)

(a) Solve the logistic equation for $N(t)$. The solution is

$$N(t) = \frac{KN(0)\exp \sigma t}{K + N(0)(\exp \sigma t - 1)}$$

(b) Plot $N(t)$ vs. t for $N(0) = K/4$, $K/2$, and $3K/4$.

Observe that $N(t)$ monotonically approaches the limiting value K independent of the initial condition. (See Figure 7.2.) ■

A system of logistic-type equations, devised by Ross for the spread of malaria, yields good agreement between the theory and the data. (See the books by Lotka and D'Ancona, which have much relevant material and many references to the original experimental and theoretical papers.) A fundamental paper of historical interest on this subject is due to William Feller, "On the Logistic Law of Growth and Its Empirical Verification in Biology," *Acta Biotheoretica*, 5 (1940).

We often bring ourselves to the conclusion that the logistic equation adequately or correctly models a variety of growth situations in part because of the simplicity and plausibility of its derivation. We then appeal to the data to demonstrate agreement between theory and experiment. Professor Feller shows that a variety of models, including the logistic equation, yields satisfactory agreement with the data under consideration and thus the logistic model cannot be singled out as preferable by this comparison alone. Thus, we have to be aware that comparison of available data at a given level of precision is not always sufficiently reliable to discriminate between competing models. The paper gives a nice insight into some of the difficulties one faces when appealing to experiments as a confirmation of a hypothesis.

CHAPTER 8

ASPECTS OF HARMONIC MOTION AND THE CONCEPT OF SECULAR TERMS

We begin by reviewing some aspects of linear oscillatory motion. In this way, we will prepare ourselves for those new ideas that emerge when we discuss nonlinear oscillatory motion as well as to be able to distinguish between those aspects that are associated with linear and nonlinear parts.

Our major goal is to discuss the long-time behavior of weakly nonlinear oscillatory systems. These are systems that are generated from the harmonic oscillator by small perturbations. We restrict our attention to weakly nonlinear systems, and this means that in the lowest order of approximation, the system is always viewed as a simple harmonic oscillator. The nonlinearity enters as a correction term.

We want to describe the behavior of these systems for long times and we find that straightforward or ordinary perturbation theory is inadequate since it generates so-called secular terms or nonuniform approximations.

Since most of our experience is based upon an analysis of linear systems, it is necessary for us to adopt new ways of looking at things as we try to understand the new phenomena that emerge from the nonlinearities. In order to accomplish this transformation in perspective, on occasion, we refer ahead to material that is covered in a subsequent chapter. This makes sense if we keep in mind that Chapters 8 through 15 can be viewed as one long connected unit devoted to a treatment of various aspects of oscillatory motion. It also enables us to introduce and briefly discuss a new idea and then to take it up again at a later stage in the development of the subject.

We begin by discussing perturbation techniques and their application to linear systems that can be easily solved. Then we examine these methods again when we investigate the associated nonlinear phenomena. In the course of this development, we encounter two ideas or concepts that frequently occur in our analysis, namely, the "Flashing Clock" and secular terms.

8.1 THE FLASHING CLOCK

I first read about the "flashing clock" concept in a text by Minorsky. However, I assume that it was known and used by Poincaré. This is also known as the "stroboscopic" method.

Let us consider an harmonic oscillator with angular frequency ω equal to 1.01. We want to use the fact that $\omega \approx 1$. This is accomplished by comparing our system with $\omega = 1.01$ with a system that has the angular frequency $\omega_0 = 1$. Thus, we have System 1:

$$\frac{d^2x}{dt^2} + 1 \cdot x = 0; \qquad x(0) = A, \quad \dot{x}(0) = 0 \tag{8.1}$$

The solution is $x(t) = A\cos t$. Now consider System 2:

$$\frac{d^2s}{dt^2} + (1.01)^2 s = 0; \qquad s(0) = A, \quad \dot{s}(0) = 0 \tag{8.2}$$

The solution is $s(t) = A\cos 1.01t$.

We use the variable x when referring to System 1 and variable s when referring to System 2. It is often convenient at this point to introduce a graph of the velocity vs. the displacement. (This is called a "phase plane" plot and is discussed at length in Chapter 9.)

The s system yields an ellipse as its phase portrait. You can then rescale the velocity appropriately to transform the ellipse into a circle. See Figures 8.1a to 8.1c.

The initial conditions are not a problem, so we impose the same conditions on each system.

Both systems are periodic, but their periods are not identical. Thus, if they start in phase, after a sufficiently long period of time, they will be out of phase. We see this clearly if we construct phase plane plots for each. We will have

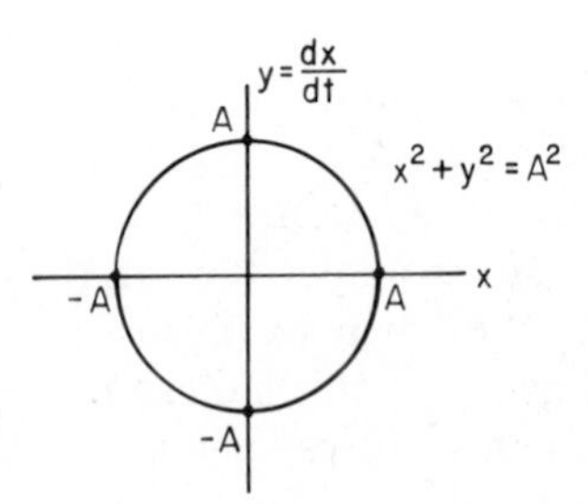

Figure 8.1*a* The phase plane plot of the solution to Equation (8.1) is a circle.

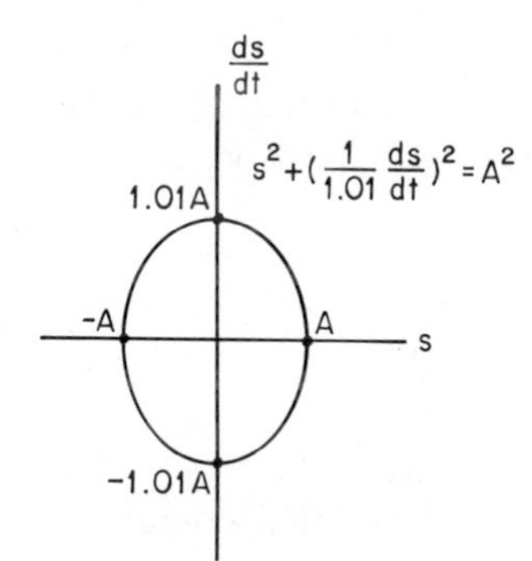

Figure 8.1*b* The phase plane plot of the solution to Equation (8.2) is an ellipse.

circles with the same radius:

$$x^2 + \left(\frac{dx}{dt}\right)^2 = A^2 \tag{8.3a}$$

and

$$s^2 + \left(\frac{1}{1.01}\frac{ds}{dt}\right)^2 = A^2 \tag{8.3b}$$

If we flash a light every second (unit of time) on the systems as they traverse the trajectory, we would find that System 1, with $\omega_0 = 1$, appears to be stationary, whereas System 2, with $\omega = 1.01$, would drift slowly in a clockwise direction.

If we keep on flashing the light, the point describing System 2 moves around the circle and eventually comes back and coincides with the stationary point describing System 1. Thus, we would observe the periodicity of System 2. But this periodic behavior might not be apparent on reasonable time scales because we are flashing the light at the wrong frequency. That does not mean

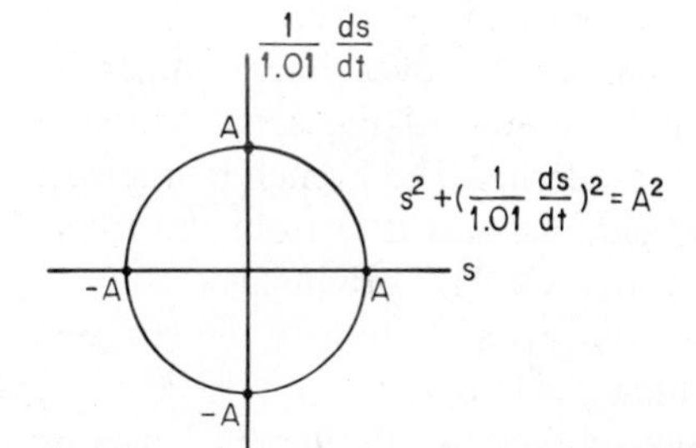

Figure 8.1*c* Upon rescaling the velocity, we obtain a circle for the phase plane plot of the solution to Equation (8.2).

that it is not periodic; you just cannot tell. So in order to recapture the periodic nature of the motion, you try to find the correct frequency to flash the light so as to observe the phase point of the system as a fixed or stationary point.

8.2 SECULAR TERMS

We discuss this concept in some detail, so if it is not clear at first, be patient. Secular terms are aperiodic functions, or aperiodic terms, that contain the product of a monotonically increasing function and a periodic function. For example, $t \sin t$ or $t \cos t$ are secular terms as is $t^2 \sin t$. They often are present in the solution to a problem. We will see that this is indeed the case in the solution to Equation (8.8) given by Equation (8.9). But they can also occur as a direct consequence of the perturbation scheme that is used.

We will generally be studying problems whose solutions are unknown but that are close to solvable problems with known solutions. We then wish to use the known solutions as the basis of a perturbation expansion by constructing an approximate solution in the form of an expansion in terms of a small parameter ε.

(Note that when we use the symbol "ε," it always means a small parameter. In general, it can be positive or negative. If we require a specific sign for ε, we will specify it.)

Our approximation should be such that small terms remain small for all time. That is to say, if x_0 is the principal term and εx_1 is the correction to it, we want $|\varepsilon x_1| \ll |x_0|$ for all time.

When this is accomplished, the resulting solution is said to be "uniformly valid in time." If the principal solution x_0 is a periodic function and the correction term εx_1 is a secular term, after a sufficiently long time, the required inequality will fail. It is with this in mind that we attempt to develop perturbation expansions that are free of spurious secular terms.

Note that in some problems, it is difficult to know in advance if secular terms are part of the structure of the solution or a manifestation of the particular perturbation expansion employed. (For example, consider the periodic function $\sin(1 + \varepsilon)t$. An expansion in powers of ε yields $\sin t + \varepsilon t \cos t +$ h.o.t. masking its periodic structure.) Secular terms received considerable attention in the study of perturbation problems in astronomy, where the cumulative effect of the perturbations were only sizable over times comparable to a century. Considerable efforts were directed toward the construction of perturbation expansions that were free of these terms. The French mathematicians were leaders in some of these investigations, and it is from the french word "siècle," meaning century, that the word "secular" originates.

Let us now look at some simple linear problems and see how the concepts associated with secular terms and the flashing clock arise.

We examine, in Sections 8.3 and 8.4, the solutions to the forced harmonic oscillator and the altered simple harmonic oscillator.

Each problem is analyzed in a manner so as to reveal the underlying weakness of the traditional perturbative method and to guide us toward identifying an approach that works.

8.3 THE FORCED HARMONIC OSCILLATOR

Our basic equation is

$$\ddot{x} + x = \varepsilon f(t); \qquad |\varepsilon| \ll 1 \tag{8.4}$$

which corresponds to a simple harmonic oscillator that has a known forcing term, $\varepsilon f(t)$. Note that we restrict our attention to systems with a small forcing term since this maintains our connection with weakly perturbed systems. However, the solution of the equation is valid for a forcing function of arbitrary strength. This is an inhomogeneous linear second-order differential equation that can, for example, be solved by the method of variation of parameters that we discussed in Chapter 6, Section 6.2.

The solution to Equation (8.4),

$$x(t) = A\cos t + B\sin t + \varepsilon\int_0^t f(s)\sin(t-s)\,ds \tag{8.5}$$

consists of two parts:

(a) the homogeneous solution and
(b) an integral that contains $\sin(t-s)$ and our forcing term $f(s)$.

(Verify that this is indeed a solution of the equation.)

It is important to stress that the problem is solved if $f(t)$ is known. That is to say, we just evaluate the integral in Equation (8.5). Let us consider some examples. If $f(t)$ is of the form $\cos\omega t$ or $\sin\omega t$, then it is convenient to solve for $x(t)$ using the trial solution method. This is discussed in Section 0.3, Example 3.1, and Exercises 3.1 and 3.2 of Chapter 0. With this in mind, it might be useful to do Examples 3.1–3.5 using both methods. It is immediately clear that the trial solution method involves significantly less algebra than the integral method. However, the latter method is of interest because of its generality and the clear way that it identifies the origin of the secular behavior of the solution.

Example 3.1

$$\ddot{x} + x = \varepsilon f(t) = \varepsilon\cos t; \qquad x(0) = A, \quad \dot{x}(0) = B \tag{8.6}$$

We evaluate the integral

$$\int_0^t \cos s\,\sin(t-s)\,ds$$

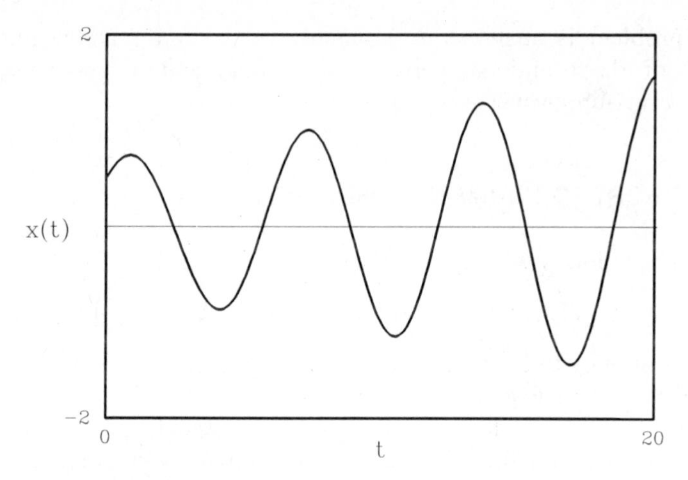

Figure 8.2 $x(t) = 0.5(\cos t + \sin t) + 0.05t \sin t$.

and find that the solution to the differential equation is

$$x(t) = A \cos t + B \sin t + \frac{\varepsilon t}{2} \sin t \qquad (8.7)$$

Notice that there are lots of cancellations from terms in the integrals.

Observe that $x(t)$ is not periodic. It consists of a periodic and an aperiodic piece. *This is important*. See Figure 8.2. ■

Example 3.2 Consider

$$\ddot{x} + x = \varepsilon \sin t; \qquad x(0) = A, \quad \dot{x}(0) = B \qquad (8.8)$$

This is essentially the same as Example 3.1 and the solution is

$$x(t) = A \cos t + B\left(1 + \frac{\varepsilon}{2B}\right) \sin t - \frac{\varepsilon t}{2} \cos t \qquad (8.9)$$

(Note that we have added a correction term to the constant B in order to satisfy the initial conditions.) ■

Example 3.3 Consider

$$\ddot{x} + x = \varepsilon C \cos at; \qquad x(0) = A, \quad \dot{x}(0) = B \qquad (8.10)$$

with parameter $a \neq 1$. In this case, our forcing term $\varepsilon f(t)$ is mismatched to the natural frequency of the oscillator. The solution is

$$x(t) = A\left(1 - \frac{\varepsilon C}{A(1-a^2)}\right)\cos t + B\sin t + \frac{\varepsilon C}{1-a^2}\cos at \tag{8.11}$$

Consider Equation (8.11). Let $a = 1 + \delta$, where $|\delta| \ll 1$. Expand in powers of $1 - a$ and reorganize the equation to obtain

$$x(t) = A\cos t + B\sin t + \frac{\varepsilon C}{-2\delta}\left[\cos(1+\delta)t - \cos t\right] + \text{h.o.t.} \tag{8.12}$$

Then if the quantity $|(\delta t)| \ll 1$, we can expand $\cos[(1+\delta)t]$ and approximate $\cos\delta t$ by 1 and $\sin\delta t$ by δt and obtain Equation (8.7). Thus, we expect Equation (8.11) to look like Equation (8.7) for times t such that $|(\delta t)| \ll 1$. ■

Example 3.4 Consider

$$\ddot{x} + x = \varepsilon M\cos^3 t; \qquad x(0) = A, \quad \dot{x}(0) = B \tag{8.13}$$

We can solve this problem by the same methods as we used to solve the previous examples. Let us proceed as follows:

Write

$$\cos^3 t = \tfrac{1}{4}(3\cos t + \cos 3t) \tag{8.14}$$

Then our forcing term consists of two pieces, and since the equation is linear, the solution can be found for each piece separately and then joined together. Thus, we have

$$x(t) = A\left(1 + \frac{\varepsilon M}{32A}\right)\cos t + B\sin t + \frac{3\varepsilon}{8}M(t\sin t) - \frac{\varepsilon}{32}M\cos 3t \tag{8.15}$$

Notice that the solution contains a periodic piece that comes from the term $\cos 3t$ and an aperiodic, or secular, piece $t\sin t$ that can be traced to the presence of the $\cos t$ term in the decomposition of $\cos^3 t$. ■

Example 3.5 Consider

$$\ddot{x} + x = \varepsilon N\cos^2 t; \qquad x(0) = A, \quad \dot{x}(0) = B \tag{8.16}$$

We solve this problem by writing

$$\cos^2 t = \tfrac{1}{2}(1 + \cos 2t)$$

and then proceeding as before to obtain

$$x(t) = A\left(1 - \frac{\varepsilon N}{3A}\right)\cos t + B \sin t + \frac{\varepsilon N}{2} - \frac{\varepsilon N}{6}\cos 2t \tag{8.17}$$

There is a shift in the "zero point" since the forcing term has a zero frequency component. (We encounter this when we analyze the behavior of an oscillating mass at the end of a vertical spring in the gravitational field. All that gravity does is to shift the zero point of the motion.)

Notice that the solution is periodic since the forcing term has an even power of $\cos t$ and, hence, when decomposed into components, lacks a term proportional to $\cos t$. This contrasts with the situation when $f(t)$ is an odd power of $\cos t$, which upon decomposition always gives a term linear in $\cos t$. ■

Now let us summarize our findings. In Example 3.1, we found that our solution was not periodic. It contained a periodic piece and a secular term that was the product of a periodic term and a factor that increased linearly in time. In Example 3.3, the solution is periodic if the quantity a in the frequency of the forcing term $\cos at$ is $\neq 1$. If a is rational, there is an interval of time T such that $x(t + T) = x(t)$. For example, if $a = 3$, our solutions coincide after intervals of time equal to 2π. If, for example, $a = \frac{2}{3}$, then we have to wait for intervals of time equal to 6π to get coincidence. But it is clear that our solution is periodic in direct contrast to that of Example 3.1. Finally, if parameter a is irrational, then $\cos t$ and $\cos at$ are in phase only at $t = 0$ and we say that Equation (8.11) represents "almost periodic" motion.

8.4 THE ALTERED SIMPLE HARMONIC OSCILLATOR

Often, we find it instructive to analyze a solvable model by a variety of approximation techniques, in part, to help us develop intuition for the study of more complex problems and, in part, to help us become more familiar with the techniques themselves. It is with these thoughts in mind that we consider a situation where we slightly alter the spring constant so the basic simple harmonic oscillator equation becomes

$$\frac{m\,d^2x}{dt_1^2} + k_1 x = 0; \qquad k_1 = k(1 + \varepsilon), \quad x(0) = \alpha, \quad \dot{x}(0) = \beta \tag{8.18}$$

Introduce

$$\omega_0^2 = k/m; \qquad \omega^2 = \frac{k_1}{m} = (1 + \varepsilon)\,\omega_0^2$$

so that Equation (8.18) can be written as

$$\frac{d^2x}{dt_1^2} + (1 + \varepsilon)\,\omega_0^2 x = 0 \tag{8.19}$$

It is convenient to introduce the dimensionless time t:

$$t = \omega_0 t_1; \qquad \frac{d}{dt_1} = \omega_0 \frac{d}{dt} \tag{8.20}$$

Then Equation (8.19) becomes

$$\frac{d^2x}{dt^2} + (1 + \varepsilon)x = 0 \tag{8.21a}$$

or

$$\frac{d^2x}{dt^2} + x = -\varepsilon x; \qquad x(0) = A, \quad \dot{x}(0) = B \tag{8.21b}$$

We refer to Equation (8.21b) as "the altered simple harmonic oscillator." Observe that in the unperturbed simple harmonic oscillator,

$$\frac{d^2x}{dt^2} + x = 0$$

the "acceleration" term d^2x/dt^2 is equal to the negative of the displacement x. Noting this, take Equation (8.21a) and divide by $1 + \varepsilon$ to obtain

$$\frac{1}{1 + \varepsilon}\,\frac{d^2x}{dt^2} + x = 0 \tag{8.22a}$$

Since $|\varepsilon| \ll 1$, we write

$$\frac{1}{1 + \varepsilon} = 1 - \varepsilon + O(\varepsilon^2) \approx 1 - \varepsilon$$

and obtain

$$(1 - \varepsilon)\,\frac{d^2x}{dt^2} + x = 0 + O(\varepsilon^2) \tag{8.22b}$$

or

$$\frac{d^2x}{dt^2} + x = +\varepsilon \frac{d^2x}{dt^2} + O(\varepsilon^2) \tag{8.22c}$$

Upon comparing Equations (8.21b) and (8.22c), we conclude that to the lowest order in our small parameter ε, the two ways of posing the problem are equivalent. Thus, we can either think of our perturbation as a correction to the displacement or to the acceleration.

Example 4.1 We know that the exact solution to Equation (8.21b) is

$$x(t) = A\cos(\sqrt{1+\varepsilon}\, t) + \frac{B}{\sqrt{1+\varepsilon}} \sin(\sqrt{1+\varepsilon}\, t) \tag{8.23}$$

With the exact solution as a guide, we attempt to solve Equation (8.21b) by a "reasonable approximation." In this way, we hope to gain some insight into how an approximation is constructed. We also note that we can always solve this problem exactly by the method of variation of parameters. That is to say, we can write

$$x(t) = A\cos t + B\sin t - \varepsilon \int_0^t x(t')\sin(t - t')\, dt' \tag{8.24}$$

Our first guess for an approximate solution of Equation (8.21b) is to think of expanding displacement $x(t)$ in powers of small parameter ε:

$$x(t) = x_0(t) + \varepsilon x_1(t) + O(\varepsilon^2) \tag{8.25}$$

The term εx_1 is a small correction to x_0 if

$$\frac{|x_1(t)|}{|x_0(t)|} = O(1) \tag{8.26}$$

or, equivalently, if the ratio of absolute values is bounded for all t.

Then we seek a solution by substituting the power series into Equation (8.21b) and equating like powers of ε:

$$\varepsilon^0: \qquad \ddot{x}_0 + x_0 = 0 \tag{8.27}$$

The zero order, or lowest-order, equation is that of the simple harmonic oscillator and the solution is

$$\begin{aligned} x_0(t) &= A_0 \cos t + B_0 \sin t \\ x_0(0) &= A_0; \qquad \dot{x}_0(0) = B_0 \end{aligned} \tag{8.28}$$

We seek the first-order correction. Thus,

$$\varepsilon: \qquad \ddot{x}_1 + x_1 = -x_0 = -A_0 \cos t - B_0 \sin t \tag{8.29}$$

The l.h.s. of Equation (8.29) is the differential operator associated with the simple harmonic oscillator, and the r.h.s. consists of known functions. Solving, we obtain

$$x_1(t) = A_1 \cos t + B_1 \left(1 - \frac{B_0}{2B_1}\right) \sin t - \frac{A_0}{2} t \sin t + \frac{B_0}{2} t \cos t \tag{8.30}$$

We have $x_1(0) = A_1$ and $\dot{x}_1(0) = B_1$. Thus, the initial conditions for Equation (8.21b)

$$x(0) = A \quad \text{and} \quad \dot{x}(0) = B \quad \text{become} \quad A = A_0 + \varepsilon A_1; \quad B = B_0 + \varepsilon B_1 \tag{8.31}$$

But, there is something strange about the solution given by Equation (8.30). Our original problem clearly has a periodic solution because all we have done is altered the spring constant. On the other hand, our method of solution to Equation (8.21b) yields a correction term εx_1 that has a secular component. It cannot then satisfy the inequality given by Equation (8.26) for all times. We conclude that our expansion procedure is unsatisfactory, especially for long times. ■

Let us continue our analysis of Equation (8.21b). The expansion procedure seems so plausible and powerful, yet, as we have seen, it yields unexpected difficulties. We gain another "grain of insight" by examining Equation (8.29) by the integral equation iteration scheme. In this method, we write a formal solution as

$$x_1(t) = A_1 \cos t + B_1 \sin t - \varepsilon \int_0^t (A_0 \cos s + B_0 \sin s) \sin(t - s)\, ds \tag{8.32}$$

We see that the upper limit of the integration is t, and, hence, if we have a constant term within the integral, it leads to a term in the displacement that is of the form $t \cos t$ or $t \sin t$, indicating aperiodic behavior. Thus, if this is the form of our expansion, it is only valid for a brief interval of time and probably masks the periodic nature of our solution. So our immediate task is to arrange our expansion so as to avoid this difficulty. This procedure leads to the expression "suppression of secular terms."

To guide us toward a correct perturbation expansion, notice that Equation (8.21b) has its exact solution given by Equation (8.23). If we look at our system by means of our "flashing clock" after an interval

$$T = \frac{2\pi}{\sqrt{1 + \varepsilon}} \tag{8.33}$$

we have $x(t + T) = x(t)$. If we continue to look at the system at intervals spaced by the correct period T, the system takes on the same values or appears as a fixed point.

Now look at the system at intervals associated with the unperturbed oscillator $T_0 = 2\pi$. After the first observation, we see that there is a drift proportional to ε since

$$\begin{aligned} x(T_0 = 2\pi) - x(t = 0) &= A\cos(\sqrt{1 + \varepsilon}\, 2\pi) + \frac{B}{\sqrt{1 + \varepsilon}} \sin(\sqrt{1 + \varepsilon}\, 2\pi) - A \\ &= B\pi\varepsilon + O(\varepsilon^2) \end{aligned}$$

If we observe the system after an interval equal to $2T_0$, the displacement is approximately $2B\pi\varepsilon$. If we continue in this manner, the system appears to continue to drift at a rate proportional to small parameter ε. So we conclude that we are observing the system at incorrect intervals to observe its periodicity.

A fundamental aspect of our "incorrect" approach to the problem involves the linking, or interconnection, between the small parameter ε and the length of time the system is viewed as measured by a clock flashing at intervals equal to the unperturbed period. This leads to a perturbation expansion that is "time limited" in its validity. To see this, we rework the analysis of the altered simple harmonic oscillator from a slightly different approach.

Example 4.2 Consider Equation (8.21b) and write its solution in the form:

$$x(t) = A\cos t - \varepsilon \int_0^t x(t') \sin(t - t')\, dt'; \qquad x(0) = A, \quad \dot{x}(0) = 0 \tag{8.34}$$

where we have chosen initial conditions that simplify the algebra. Our first step is to replace the unknown function $x(t')$ that appears under the integral sign by its expression as given in terms of the integral. (At this point, it may be worthwhile to review this technique, which is given in Chapter 6, Section 6.6.) Write Equation (8.34) as

$$x(t) = A\cos t - \varepsilon \int_0^t \left[A\cos t' - \varepsilon \int_0^{t'} x(t'') \sin(t' - t'')\, dt'' \right] \sin(t - t')\, dt' \tag{8.35}$$

Verify that the solution to Equation (8.21b) given by Equation (8.35) is indeed a solution. This involves differentiating the r.h.s. of Equation (8.35) twice and concluding that the resulting solution is in the form given by Equation (8.34).

We could continue our substitution procedure as long as we wish, always keeping our equation exact. However, let us stop here and rewrite the r.h.s. of Equation (8.35) as

$$x(t) = A\cos t - \varepsilon \int_0^t A\cos t' \sin(t - t')\, dt' + \varepsilon^2 \int_0^t \int_0^{t'} x(t'') \sin(t' - t'') \sin(t - t')\, dt''\, dt' \tag{8.36}$$

It appears as if Equation (8.36) has three ordered contributions on its r.h.s.:

1. the known homogeneous solution;
2. a term, containing known quantities, that is proportional to the small parameter ε; and
3. an unknown piece that is proportional to ε^2.

Although Equation (8.36) is a correct solution, it still gives a misleading impression. To see this, we evaluate the term in Equation (8.35) that is proportional to ε:

$$-\varepsilon A \int_0^t \cos t' \sin(t - t')\, dt' = -\varepsilon A \sin t \left(\frac{t}{2} + \frac{1}{4} \sin 2t \right) + \frac{\varepsilon A}{4} (\cos t)(1 - \cos 2t) \tag{8.37a}$$

This equation can be simplified further to yield:

$$= -\frac{\varepsilon A t}{2} \sin t \qquad \blacksquare \tag{8.37b}$$

We observe that Equation (8.37b) is a secular term. This means that if we wait a sufficiently long period of time, this supposed correction term will be comparable to or bigger than the homogeneous solution. But this is undesirable since we insist that it be a correction term for all time. It is important to emphasize that Equation (8.36) is still an exact solution to Equation (8.21b). We have not done anything wrong. There is just a confusion regarding what terms are small and for how long they are small. First, note that the neglected term with the small coefficient ε^2 is not small or negligible for long times. Thus, it can only be correctly neglected for short times. Furthermore, if it is evaluated, it yields a contribution that cancels the *secular* term given by Equation (8.37b). Thus, our exact solution retains its periodicity.

In Examples 4.1 and 4.2, we have developed a solution through a substitution procedure that uses the homogeneous solution as the basis for the iteration. The clock, or natural time scale, of the homogeneous solution is based on the unperturbed oscillator and not on the time scale of the perturbed system. Thus, the method we have used is trying to develop a periodic solution by using an incorrect time-measuring scheme. By viewing this slightly differently, Equation (8.21b) has a periodic solution and we have developed a solution that intermixes periodic and aperiodic terms in its perturbation expansion.

Example 4.3 Consider Equation (8.21a):

$$\frac{d^2x}{dt^2} + (1 + \varepsilon)x = 0 \tag{8.21a}$$

We want to introduce a transformation of the time variable to a new time that incorporates the lowest-order correction to the frequency. We know that

$$\omega = \sqrt{1 + \varepsilon}\,\omega_0 = \left(1 + \frac{\varepsilon}{2}\right)\omega_0 + O(\varepsilon^2) \tag{8.38}$$

This motivates us to introduce the new time:

$$t' \equiv \left(1 + \frac{\varepsilon}{2}\right)t$$

Then Equation (8.21a) becomes

$$\left(1 + \frac{\varepsilon}{2}\right)^2 \frac{d^2x}{dt'^2} + (1 + \varepsilon)x = 0 + O(\varepsilon^2) \tag{8.39}$$

Let us find the solution to the lowest order in the small parameter ε. Equation (8.39) becomes

$$(1 + \varepsilon)\frac{d^2x}{dt'^2} + (1 + \varepsilon)x = 0 \tag{8.40}$$

The solution is

$$\begin{aligned} x(t') &= A' \cos t' + B' \sin t' \\ &= A' \cos\left[\left(1 + \frac{\varepsilon}{2}\right)t\right] + B' \sin\left[\left(1 + \frac{\varepsilon}{2}\right)t\right] \end{aligned} \tag{8.41}$$

Now we observe that if we flash our light at intervals $T' = 2\pi/(1 + \varepsilon/2)$, we see the system as almost stationary for times until the next order term in the expansion is required. The length of time involved depends upon the accuracy of our measuring apparatus. Let us say that we want to continue viewing our system. Then we would take the next term in the expansion of our frequency as given by Equation (8.38):

$$\omega = \sqrt{1+\varepsilon}\,\omega_0 = \left(1 + \frac{\varepsilon}{2} - \frac{\varepsilon^2}{8}\right)\omega_0 + O(\varepsilon^3) \tag{8.42}$$

We then introduce the new time:

$$t'' = \left(1 + \frac{\varepsilon}{2} - \frac{\varepsilon^2}{8}\right)t$$

and have

$$\left(1 + \frac{\varepsilon}{2} - \frac{\varepsilon^2}{8}\right)^2 \frac{d^2x}{dt''^2} + (1+\varepsilon)x = 0 + O(\varepsilon^3) \tag{8.43}$$

Keeping terms through $O(\varepsilon^2)$, we have

$$\left(1 + \frac{\varepsilon}{2} - \frac{\varepsilon^2}{8}\right)^2 = 1 + \varepsilon + O(\varepsilon^3) \tag{8.44}$$

Then our solution to Equation (8.42) is

$$x(t'') = A\cos t'' + B\sin t'' = A\cos\left[\left(1 + \frac{\varepsilon}{2} - \frac{\varepsilon^2}{8}\right)t\right] + B\sin\left[\left(1 + \frac{\varepsilon}{2} - \frac{\varepsilon^2}{8}\right)t\right] \tag{8.45}$$

This approximate solution follows the exact solution for a longer time than did our previous approximation. This means that if we give a slightly finer tuning to our frequency, we can observe our system as stationary for a longer time.

So, by adjusting the frequency of our flashing clock, we were able to suppress the secular terms and see the system as a fixed point. Remember that since we are free to choose the rate at which we flash our clock, we have the potential to adjust it closer and closer to the correct frequency. This adjustment process yields the higher-order correction terms to the frequency. Then we can generate an expansion for the frequency in powers of ε:

$$\omega = \omega_0 + \varepsilon\omega_1 + \varepsilon^2\omega_2 + \varepsilon^3\omega_3 + O(\varepsilon^4)$$

where the terms containing ω_1, ω_2, and ω_3 are the higher-order corrections to the unperturbed, or natural, frequency ω_0. ■

We have learned at least two lessons in this Chapter. If you force an SHO at its natural frequency, you correctly obtain a solution that contains a secular term corresponding to the increase in the amplitude or energy of the system as time progresses.

If you consider a SHO with an altered spring constant, then you may obtain, in a perturbation expansion, a secular term rather than a periodic solution. The origin of the disease is clear. It originates in our decision to use the unperturbed frequency as the basic time signal, or flashing clock, that generates the perturbation expansion. Then the true clock and our clock become seriously mismatched over a significant interval of time and we obtain an incorrect view of the behavior of our system.

We will see that these observations are sufficient to guide us toward the development of a consistent perturbation expansion for conservative oscillatory systems that are close to the SHO.

We discuss these ideas at greater length in Chapter 10.

CHAPTER 9

EQUILIBRIUM POINTS AND THE PHASE PLANE

In this chapter, we continue our development of the tools that enable us to treat the qualitative and quantitative aspects of weakly nonlinear systems.

In later chapters, we develop analytical skills, but for the present we are content to learn how to generate a sketch that adequately describes the motion of our system. The approach we use involves the combination of equilibrium points and the phase plane.

We begin by locating the equilibrium points of our system, making a transformation of variables, and analyzing the motion about these points. The trajectory of the system is described by a curve in the phase plane.

9.1 EQUILIBRIUM POINTS

If we have a first-order differential equation

$$\frac{dx}{dt} = f(x) \tag{9.1}$$

where x is a scalar, the equilibrium points are those values of x for which $f(x) = 0$.

We designate such points as $\bar{x}$. So $f(\bar{x}) = 0$. In order to study small-amplitude motion, we expand $f(x)$ about the equilibrium points, retaining the linear terms. The equilibrium points are classified as stable or unstable, depending on the motion in the vicinity of the points. The equilibrium point is said to be neutrally stable if it acts as neither an attractor nor a repeller of small-amplitude motion in its vicinity.

Thus, we write

$$x = \bar{x} + z; \qquad |z| \ll 1 \tag{9.2}$$

So

$$\frac{dx}{dt} = \frac{dz}{dt} = f(\bar{x} + z) = f(\bar{x}) + zf'(\bar{x}) + \text{h.o.t.}$$

or

$$\frac{dz}{dt} \sim zf'(\bar{x}); \qquad z \sim z_0 \exp[f'(\bar{x})t] \tag{9.3}$$

Then:

if the real part of $f'(\bar{x}) > 0$, we have an unstable point;
if the real part of $f'(\bar{x}) < 0$, we have a stable point; and
if the real part of $f'(\bar{x}) = 0$, we say that the equilibrium point is neutrally stable.

It is important to note that this is a statement about local stability, not global stability.

EXERCISE 1.1 (Review of the logistic equation.)

$$\frac{dx}{dt} = \sigma x\left(1 - \frac{x}{K}\right); \qquad \sigma, K > 0$$

Show that the equilibrium points are $x = 0, K$. Also demonstrate that motion about the origin is unstable. We say that the origin repels. Show that motion about $x = K$ is stable. We say that $x = K$ is an attractor. ■

It is straightforward to extend this analysis to coupled first-order differential equations of the form:

$$\frac{dx}{dt} = ax + by + f(x, y) \tag{9.4a}$$

$$\frac{dy}{dt} = cx + dy + g(x, y) \tag{9.4b}$$

where a, b, c, and d are constants, and f and g are polynomials in x and y that have no linear or constant terms. This means that the first terms can begin as xy, x^2, or y^2. We identify the equilibrium points as those values of x and y for which we have, simultaneously,

$$\frac{dx}{dt} = \frac{dy}{dt} = 0 \tag{9.5}$$

We first observe that the origin (0, 0) is an equilibrium point. If there are other points, they occur where

$$ax + by + f(x, y) = 0$$

and

$$cx + dy + g(x, y) = 0$$

After we have found the equilibrium points, we expand, retaining the linear terms. Then the local stability analysis is given in terms of the eigenvalues of the resulting matrix equation.

Example 1.1 (Review of oscillator equations.) We are interested in studying weakly nonlinear problems and we require the linear part of the system of equations to describe harmonic motion. This includes the possibility of damped motion.

Thus, we first consider a set of coupled linear equations of the form

$$\frac{dx}{dt} = ax + by \tag{9.6a}$$

$$\frac{dy}{dt} = cx + dy \tag{9.6b}$$

We can, of course, write Equations (9.6a) and (9.6b) in matrix form:

$$\frac{d}{dt}\begin{bmatrix} x \\ y \end{bmatrix} = A\begin{bmatrix} x \\ y \end{bmatrix}; \qquad A = \begin{bmatrix} a & b \\ c & d \end{bmatrix}$$

The stability of the equilibrium points can then be cast in terms of eigenvalues of matrix A. We recall from the discussion in Chapter 7 that if the motion is oscillatory, the eigenvalues of the matrix A must be complex, with the imaginary part indicating the angular frequency and the real part indicating the damping. (Question: What happens if det $A = 0$?) ■

Example 1.2 Consider the set of coupled equations:

$$\frac{dx}{dt} = ax - bx^2y \tag{9.7a}$$

$$\frac{dy}{dt} = -cy + dx^2y \tag{9.7b}$$

with a, b, c, and $d > 0$.

The equilibrium points $(\bar{x}, \bar{y})$ are $(0, 0)$,

$$\left(+\sqrt{\frac{c}{d}}, +\frac{a}{b}\sqrt{\frac{d}{c}}\right),$$

$$\left(-\sqrt{\frac{c}{d}}, -\frac{a}{b}\sqrt{\frac{d}{c}}\right)$$

We are interested in the small amplitude or linear motion in the neighborhood of the equilibrium points. If we linearize our equations about point $(0, 0)$, the equations decouple and lead to exponential decay and exponential growth.

Let us turn our attention to the other equilibrium points. We observe that in our particular case, since they are reflections of one another,

$$\bar{x} \to -\bar{x}; \qquad \bar{y} \to -\bar{y}$$

we can study the motion about one of them. We make a transformation, or shift of origin, to the equilibrium point:

$$(\bar{x}, \bar{y}) = \left(+\sqrt{\frac{c}{d}}, +\frac{a}{b}\sqrt{\frac{d}{c}}\right)$$

We then introduce small displacements u and v:

$$x = \bar{x} + u; \qquad y = \bar{y} + v; \qquad |u|, |v| \ll 1$$

Equations (9.7a) and (9.7b) become

$$\frac{du}{dt} = -au - \frac{bc}{d}v + \text{n.l.t.} \tag{9.8a}$$

$$\frac{dv}{dt} = +2\frac{ad}{b}u + \text{n.l.t.} \tag{9.8b}$$

We can think of these linear equations as a matrix equation of the form

$$\frac{d}{dt}\begin{bmatrix} u \\ v \end{bmatrix} = M\begin{bmatrix} u \\ v \end{bmatrix}; \qquad M = \begin{bmatrix} -a & -\dfrac{bc}{dt} \\ +2\dfrac{ad}{b} & 0 \end{bmatrix}$$

We then see that

$$\text{Tr}\, M = -a; \qquad \det M = 2ac$$

indicating that we have damped oscillatory motion if $8c > a$. (See Equation 7.19.) It is important to observe that if $a = 0$, corresponding to no damping, we also have no oscillations. Why? Because if $a = 0$, we have $\det M = 0$.

If our parameters take on the values $a = b = c = d = 1$, we have

$$(\bar{x}, \bar{y}) = (0,0); \qquad (1,1); \quad (-1,-1)$$

We expand about $(1,1)$; $x = 1 + u$, $y = 1 + v$, and our linearized equations are

$$\frac{du}{dt} = -u - v \tag{9.9a}$$

$$\frac{dv}{dt} = 2u \tag{9.9b}$$

These equations describe an oscillator with a damping factor $= -\frac{1}{2}$ and a frequency of oscillation

$$\omega = \frac{\sqrt{7}}{2}$$

■

EXERCISE 1.2 Given the set of coupled first-order equations:

$$\frac{dx}{dt} = A + x^2y - (B+1)x \tag{9.10a}$$

$$\frac{dy}{dt} = Bx - x^2y \tag{9.10b}$$

with A and B both constants that are greater than zero. (This is known as the Brusselator model and is used to describe oscillating chemical reactions.)

(i) Find the equilibrium point $(\bar{x}, \bar{y})$.

(ii) Expand about the equilibrium point

$$x = \bar{x} + \alpha; \qquad y = \bar{y} + \beta$$

Retain the linear terms.

(iii) Develop a matrix equation of the form

$$\frac{d}{dt}\begin{bmatrix}\alpha\\ \beta\end{bmatrix} = M\begin{bmatrix}\alpha\\ \beta\end{bmatrix}; \qquad M = \begin{bmatrix}a & b\\ c & d\end{bmatrix}$$

What are the values of a, b, c, and d?

(iv) Introduce the transformation P:

$$P = \begin{bmatrix} -\frac{c}{\sqrt{\Delta}} & -\frac{d}{\sqrt{\Delta}} \\ 0 & 1 \end{bmatrix}$$

Find P^{-1} such that $PP^{-1} = 1$. Then find $J = PMP^{-1}$.

(v) Now view parameter A as fixed and parameter B as variable. Discuss what should be the relationship between B and A so that one obtains:
(a) undamped oscillatory motion;
(b) positively damped oscillations; and
(c) negatively damped oscillations. ■

Example 1.3 Consider the set of coupled equations of the form:

$$\frac{dN_1}{dt} = \varepsilon_1 N_1 - \gamma_1 N_1 N_2 \tag{9.11a}$$

$$\frac{dN_2}{dt} = -\varepsilon_2 N_2 + \gamma_2 N_1 N_2 \tag{9.11b}$$

where the coefficients are all positive constants.

These are the famous Volterra–Lotka equations that occur in the study of predator–prey interactions in ecological systems. They are supposed to represent an idealization of the interaction of two isolated species, a predator and a prey. The prey satisfies its appetite by feeding on a source outside of the predator–prey interaction. (It may be convenient to think of rabbits and foxes. The rabbits eat grass and the foxes do not eat grass; rather, foxes eat rabbits.) Coefficient ε_1 of species N_1 represents a multiplicative factor times the birth rate minus the death rate. The sign in front of that term is positive, indicating that if there were no members of the second species, the first species would grow exponentially. When the species interact or encounter one another, there is a certain fixed probability that it results in the loss of a member of the first species. The situation is similar for the second species. If left alone, their numbers will fall off exponentially since the growth coefficient in their case is negative. That is to say, the predators need prey to survive. Their numbers will increase in direct proportion to the loss of the first species with a different coefficient representing a conversion efficiency. See Figure 9.1.

It is a beautiful model that has played an important role in the development of mathematical biology. It is interesting to note that the development of these equations due to Volterra (see the text by D'Ancona) was greatly influenced by the developments that were taking place in statistical mechanics. Most particularly, Volterra's discussion of the so-called method of encounters is reminiscent of the methods of Boltzmann associated with his study of the approach to equilibrium of an ideal gas, the celebrated "H Theorem." (This material is

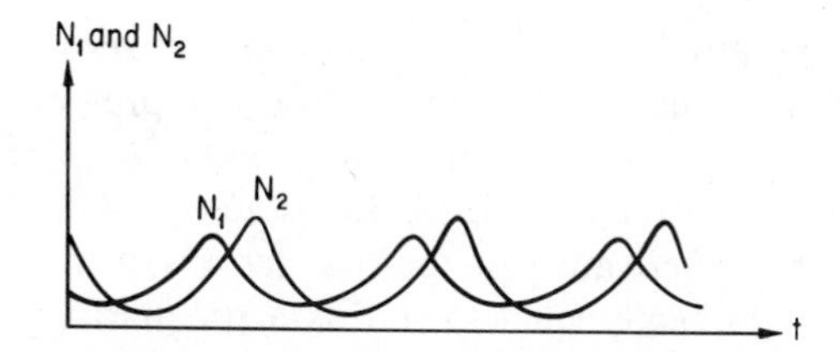

Figure 9.1 The solutions $N_1(t)$ and $N_2(t)$ of the Volterra–Lotka equations are plotted. Observe that the behavior is periodic. (Reprinted from the text by Andronov, Vitt, and Khaiken, page 145, with the permission of Dover Publications, Inc.)

traditionally studied in a course on statistical physics.) The development due to Lotka is more focused on the study of chemical kinetics and rate equations. His book is still in print and readily available. Thus, it can be consulted to obtain his perspective.

We now turn to an analysis of the stability of the equilibrium points and the associated small amplitude or linear motion about them. We observe that the origin

$$\overline{N}_1 = \overline{N}_2 = 0$$

and

$$\overline{N}_1 = \frac{\varepsilon_2}{\gamma_2}; \qquad \overline{N}_2 = \frac{\varepsilon_1}{\gamma_1}$$

are equilibrium points.

We locate our system at the origin $(0, 0)$ and disturb it slightly and see what happens. The equations become decoupled, N_1 grows exponentially, and N_2 decays toward zero.

We are now interested in studying the small oscillations of the two species about the equilibrium point:

$$\overline{N}_1 = \frac{\varepsilon_2}{\gamma_2}; \qquad \overline{N}_2 = \frac{\varepsilon_1}{\gamma_1}$$

We shift the origin to this equilibrium point and expand about it

$$N_1 = \overline{N}_1 + n_1; \qquad N_2 = \overline{N}_2 + n_2; \qquad |n_1|, |n_2| \ll 1$$

(Note that n_1 and n_2 are really dimensionless population densities.)

Our equation for species n_1 becomes

$$\frac{dn_1}{dt} = \left(\varepsilon_1 \overline{N}_1 - \gamma_1 \overline{N}_1 \overline{N}_2\right) + \left(\varepsilon_1 - \gamma_1 \overline{N}_2\right) n_1 - \gamma_1 \overline{N}_1 n_2 - \gamma_1 n_1 n_2 \quad (9.12)$$

We notice that the terms in parenthesis are equal to zero. The first group of terms in parenthesis must equal zero because it represents the shift of the origin of the coordinates to the equilibrium point. (In other words it equals zero because the r.h.s. of Equation (9.11a) is zero at $(\overline{N}_1, \overline{N}_2)$.) The second group is equal to the derivative of the first term with respect to $\overline{N}_1$ at the equilibrium point. Its being equal to zero is a peculiarity of this model and indicates neutral stability and purely imaginary eigenvalues. We simplify Equation (9.12) and write it as

$$\frac{dn_1}{dt} = -\left(\frac{\gamma_1 \varepsilon_2}{\gamma_2}\right) n_2 - \gamma_1 n_1 n_2 \tag{9.13a}$$

We can immediately write the equation for variable n_2 by noting the symmetry of the coefficients in Equations (9.11a) and (9.11b):

$$\varepsilon_1 \to -\varepsilon_2; \qquad \gamma_1 \to -\gamma_2$$

Thus, we have

$$\frac{dn_2}{dt} = \left(\frac{\gamma_2 \varepsilon_1}{\gamma_1}\right) n_1 + \gamma_2 n_1 n_2 \tag{9.13b}$$

We now study the small oscillations about the equilibrium point. Thus, we neglect quadratic terms since both n_1 and n_2 are small compared to one and write Equations (9.13a) and (9.13b) as

$$\frac{dn_1}{dt} = -\left(\frac{\gamma_1 \varepsilon_2}{\gamma_2}\right) n_2 \tag{9.14a}$$

$$\frac{dn_2}{dt} = +\left(\frac{\gamma_2 \varepsilon_1}{\gamma_1}\right) n_1 \tag{9.14b}$$

These are just the equations of simple harmonic motion. We can, of course, differentiate Equation (9.14a), link it with Equation (9.14b), and write

$$\frac{d^2 n_1}{dt^2} + \varepsilon_1 \varepsilon_2 n_1 = 0 \tag{9.15}$$

Thus we see that the system undergoes small-amplitude simple harmonic oscillations about the equilibrium point, with angular frequency

$$\omega = \sqrt{\varepsilon_1 \varepsilon_2}$$

We can return to Equations (9.14a) and (9.14b) and write them in a more symmetric form by introducing the scaled variables:

$$\xi = \sqrt{\frac{\gamma_2 \varepsilon_1}{\gamma_1}}\, n_1; \qquad \eta = \sqrt{\frac{\gamma_1 \varepsilon_2}{\gamma_2}}\, n_2 \tag{9.16}$$

We then have

$$\frac{d\xi}{dt} = -\sqrt{\varepsilon_1 \varepsilon_2}\, \eta \tag{9.17a}$$

$$\frac{d\eta}{dt} = +\sqrt{\varepsilon_1 \varepsilon_2}\, \xi \tag{9.17b}$$

These equations can be thought of as the matrix equation

$$\frac{d}{dt}\begin{bmatrix} \xi \\ \eta \end{bmatrix} = A \begin{bmatrix} \xi \\ \eta \end{bmatrix}; \qquad A = \begin{bmatrix} 0 & -\sqrt{\varepsilon_1 \varepsilon_2} \\ +\sqrt{\varepsilon_1 \varepsilon_2} & 0 \end{bmatrix} \tag{9.18}$$

The eigenvalues of matrix A are pure imaginary, indicating that the equilibrium point is neutrally stable and that locally we have simple harmonic motion. However, in order to ascertain that we have periodic motion, it is necessary to show that the motion is bounded. One has to be certain that nonlinear terms that appear in higher order do not lead to unbounded motion. This is discussed in Example 1.4. In this regard it is valuable to refer to the discussion given in the text by Bender and Orszag, pages 183–185. ■

Example 1.4 Consider the set of coupled equations:

$$\frac{dx}{dt} = -y + x\sqrt{x^2 + y^2} \tag{9.19a}$$

$$\frac{dy}{dt} = +x + y\sqrt{x^2 + y^2} \tag{9.19b}$$

Observe that the origin $(0, 0)$ is neutrally stable and that the linearized equations are those of the simple harmonic oscillator. If we multiply Equation (9.19a) by x and Equation (9.19b) by y and add, we have

$$\frac{x\,dx}{dt} + \frac{y\,dy}{dt} = \frac{1}{2}\frac{d}{dt}\left(x^2 + y^2\right) = \left(x^2 + y^2\right)^{3/2} \tag{9.20a}$$

or if

$$r = \sqrt{x^2 + y^2} \tag{9.20b}$$

$$r\frac{dr}{dt} = r^3 \tag{9.20c}$$

Equation (9.20c) is solved to yield:

$$r = \frac{r_0}{1 - r_0 t} \tag{9.21}$$

where r_0 is the initial radius, and we observe that radius $r(t)$ tends toward infinity when $t \to 1/r_0$. Also observe that even though the differential equation has well-behaved coefficients, it is possible for a spontaneous singularity to occur for a finite value of independent variable t. This property of nonlinear differential equations is absent in linear systems. ■

We conclude this discussion by noting that the formulation of the Volterra–Lotka model given by Equations (9.11a) and (9.11b) has some deficiencies; specifically:

(1) The frequency of small oscillations is independent of the interaction coefficients γ_1 and γ_2.
(2) The prey are allowed an inexhaustible supply of food, thus permitting them to grow exponentially in the absence of predation.
(3) The predators have an insatiable appetite, viz., they increase in direct proportion to the number of prey. This seems unrealistic especially if the number of prey is either very scarce or abundant. More realistic Volterra–Lotka models that address these points are discussed in many texts on ecological modeling. For example see the books by May and D'Ancona.

9.2 THE PHASE PLANE

We introduce the concept of the phase plane so as to obtain a graphical picture of the behavior of the system. We primarily consider equations that can be transformed into

$$\frac{d^2x}{dt^2} + x = \varepsilon f\left(x, \frac{dx}{dt}\right) = \varepsilon f(x, y); \qquad y = \frac{dx}{dt} \tag{9.22}$$

When we consider such equations, we agree on a convention to have the perturbative term on the right-hand side. (Note: We also use the convention $dx/dt = -y$. See Section 9.3.)

Unless specified to the contrary, we restrict our attention to functions $f(x, y)$ that do not explicitly depend upon time. Our system is then said to be autonomous. The solution of the associated differential equation remains valid when we make a translation of our time variable t by a constant t_0. (Systems in which the forcing function f depends upon time are called nonautonomous. They can exhibit motions that are forbidden for our trajectories because the

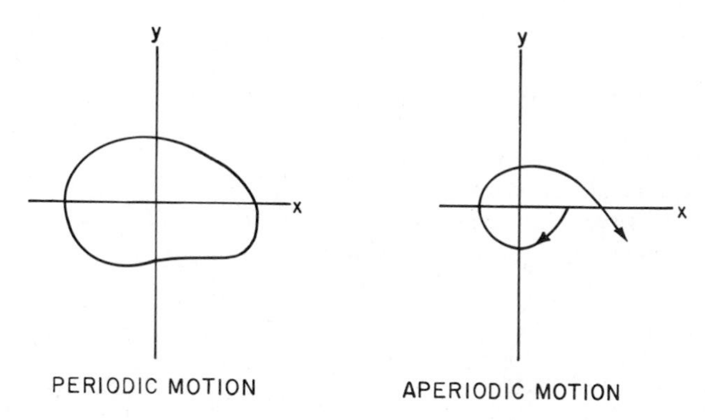

Figure 9.2 Periodic motion is characterized by closed orbits, whereas aperiodic motion yields open trajectories.

system is governed by different forces at different times.) When $\varepsilon = 0$, we have simple harmonic motion with angular frequency $\omega_0 = 1$. We want to see the change in this unperturbed motion when we introduce the perturbing term $\varepsilon f(x, y)$. Since we have a second-order differential equation, we can specify as initial conditions position $x(0)$ and initial velocity $\dot{x}(0) = y(0)$. Position x and velocity y are convenient coordinates to describe the motion of the system. The system is characterized by the motion of point (x, y) as it evolves into a trajectory in the plane formed by the x and y axes.

This plane is called the phase plane. If the motion is periodic, then the trajectory in the phase plane is a closed curve. See Figure 9.2.

Note that trajectories cannot cross since the future of any system is specified uniquely by its present state and the equations of motion. If trajectories crossed at point (a, b), then two different future motions would be possible, contrary to the equations specifying the problem. See Figure 9.3.

Example 2.1 Consider the simple harmonic oscillator described by the equation

$$\frac{d^2x}{dt^2} + x = 0$$

with the initial conditions

$$x(0) = A \qquad \text{and} \qquad \frac{dx(0)}{dt} \equiv y(0) = 0$$

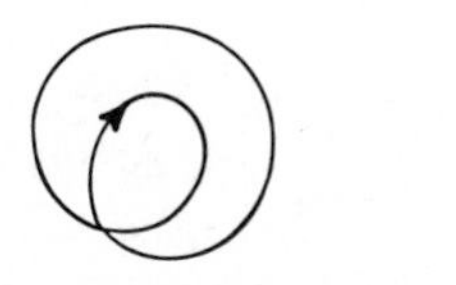

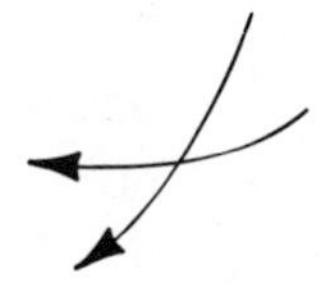

Figure 9.3 We consider situations where $f(x, y)$ is independent of time, leading to the impossibility of phase trajectories of the type given in the figure.

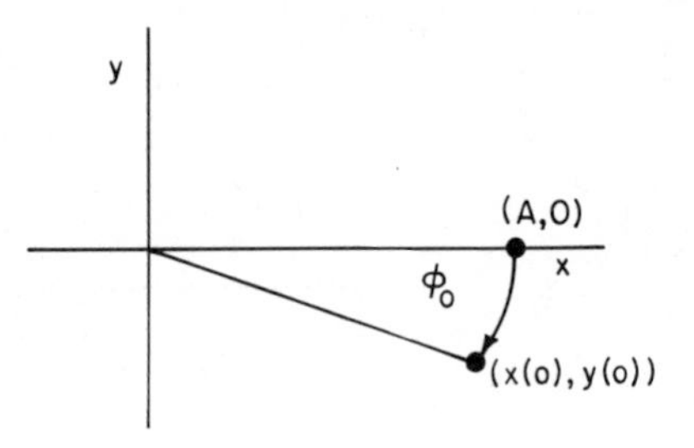

Figure 9.4 The initial conditions play a passive role in conservative systems, since by a translation of the time variable, we can easily make a comparison between systems of equal energy. However, the situation becomes tricky when discussing systems for which energy is not conserved.

The solution can be written

$$x = A \cos t; \qquad y = -A \sin t; \qquad y \equiv \frac{dx}{dt}$$

We now want to plot its trajectory in the phase plane. We note that the solution has the property that

$$x^2 + y^2 = A^2$$

The phase plane plot is then a circle centered at the origin $(0, 0)$ with radius A. Initially, the system is at point $(A, 0)$. We have chosen simple initial conditions such that the system has its maximum amplitude A at the origin of time, $t = 0$. If we picked different initial conditions:

$$x(0) = A \cos \phi_0; \qquad y(0) = -A \sin \phi_0$$

we would have the same trajectory, but our point would initially be at an angle $-\phi_0$. See Figure 9.4.

Since the solutions of the equations do not depend on the origin of time, we almost always choose to start observing our system at time $t = 0$ such that the phase point is at coordinate $(A, 0)$. Clearly, in actual problems, we are faced with a variety of initial conditions, and we have to introduce transformations in order to cast the problem in the appropriate form. In addition, we often rescale our velocity so as to transform ellipses into circles. This involves the definition of a new time, as discussed in Example 2.2. ■

Example 2.2 Consider

$$\frac{d^2x}{dt_1^2} + \omega_0^2 x = 0; \qquad \omega_0 \neq 1 \tag{9.23}$$

If we write

$$\frac{dx}{dt_1} = y \tag{9.24a}$$

$$\frac{dy}{dt_1} = -\omega_0^2 x \tag{9.24b}$$

then our phase plane plot is an ellipse defined by

$$x^2 + \left(\frac{1}{\omega_0}\frac{dx}{dt_1}\right)^2 = A^2$$

It is much more convenient to have the trajectory as a circle. To accomplish this, we rescale our velocity y by adjusting our clock to run on dimensionless time:

$$t = \omega_0 t_1; \qquad \frac{d}{dt_1} = \omega_0 \frac{d}{dt}$$

We then write Equation (9.23) as

$$\frac{d^2x}{dt^2} + x = 0 \tag{9.25}$$

which is the standard form of our equation for the SHO.

If we plot $y = dx/dt$ vs. x, we obtain a circle for our phase-plane trajectory.

■

Example 2.3 Consider the linearized Volterra–Lotka equations given by Equations (9.14a) and (9.14b):

$$\frac{dn_1}{dt} = -\left(\frac{\gamma_1\varepsilon_2}{\gamma_2}\right)n_2 \tag{9.14a}$$

$$\frac{dn_2}{dt} = +\left(\frac{\gamma_2\varepsilon_1}{\gamma_1}\right)n_1 \tag{9.14b}$$

We multiply Equation (9.14a) by $(\gamma_2/\gamma_1)\varepsilon_1 n_1$ and Equation (9.14b) by $(\gamma_1/\gamma_2)\varepsilon_2 n_2$ and add them to obtain

$$\left(\frac{\gamma_2\varepsilon_1}{\gamma_1}n_1\right)\frac{dn_1}{dt} + \left(\frac{\gamma_1\varepsilon_2}{\gamma_2}n_2\right)\frac{dn_2}{dt} = 0$$

or

$$\frac{\gamma_2\varepsilon_1}{\gamma_1}n_1^2 + \frac{\gamma_1\varepsilon_2}{\gamma_2}n_2^2 = \text{a constant} \tag{9.26}$$

We introduce a phase plane formed by the n_1 and n_2 coordinate axes. Then each coordinate in the plane indicates the size of the fluctuation in the population densities n_1 and n_2. The trajectory describing the system, given by Equation (9.26), is an ellipse. It is possible, however, to introduce scaled variables ξ and η, defined by Equation (9.16), and then our phase-plane equation becomes a circle:

$$\xi^2 + \eta^2 = \text{a constant}$$

■

Example 2.4 We are given the damped harmonic oscillator described by the equation

$$\frac{d^2x}{dt_1^2} + \omega_0^2 x = \varepsilon \frac{dx}{dt_1}; \qquad x(0) = A, \quad \dot{x}(0) = 0 \tag{9.27a}$$

where we think of the damping term as a perturbation, $|\varepsilon| \ll 1$. We want to plot the trajectory in the phase plane in such a manner that the unperturbed problem yields a circle. Introduce the proper time

$$t = \omega_0 t_1$$

and write Equation (9.27a) as

$$\frac{d^2x}{dt^2} + x = \frac{\varepsilon}{\omega_0} \frac{dx}{dt} \tag{9.27b}$$

Now introduce the phase-plane variables x and y, and write Equation (9.27b) as

$$\frac{dx}{dt} = y \tag{9.28a}$$

$$\frac{dy}{dt} = -x + \frac{\varepsilon}{\omega_0} y \tag{9.28b}$$

We perform a transformation to polar coordinates as follows: Let

$$x = r\cos\theta \tag{9.29a}$$

and

$$y = -r\sin\theta \tag{9.29b}$$

Differentiating, we have

$$\frac{dx}{dt} = \frac{dr}{dt}\cos\theta - r\frac{d\theta}{dt}\sin\theta \tag{9.30a}$$

$$\frac{dy}{dt} = -\frac{dr}{dt}\sin\theta - r\frac{d\theta}{dt}\cos\theta \tag{9.30b}$$

Note that it is possible to choose relationships between the Cartesian coordinates (x, y) and the polar coordinates (r, θ) other than those given by Equations (9.29a) and (9.29b). This is discussed in Section 9.3.

This is a convenient transformation because it establishes the velocity and position as equivalent variables or coordinates for describing the trajectory in the phase plane. In addition, when we transform to polar coordinates, we can describe the phase trajectory of the SHO as a circle with radius r such that $dr/dt = 0$. (Refer to Chapter 7.) We want to obtain equations for dr/dt and $d\theta/dt$. From Equations (9.30a) and (9.30b), we see that we can obtain dr/dt by multiplying Equation (9.28a) by x and Equation (9.28b) by y and adding. This yields

$$x\frac{dx}{dt} + y\frac{dy}{dt} = r\frac{dr}{dt} = \frac{\varepsilon}{\omega_0}r^2\sin^2\theta$$

or

$$\frac{dr}{dt} = \frac{\varepsilon}{2\omega_0}r(1 - \cos 2\theta) \tag{9.31a}$$

Similarly, to obtain $d\theta/dt$, we multiply Equation (9.28a) by y and Equation (9.28b) by $-x$ and add:

$$-x\frac{dy}{dt} + y\frac{dx}{dt} = r^2\frac{d\theta}{dt}$$

$$= +r^2 + \frac{\varepsilon r^2}{2\omega_0}\sin 2\theta$$

or

$$\frac{d\theta}{dt} = +1 + \frac{\varepsilon}{2\omega_0}\sin 2\theta \tag{9.31b}$$

Our radial velocity dr/dt has a slowly varying component:

$$\frac{dr(\text{smooth})}{dt} = \frac{\varepsilon}{2\omega_0}r$$

Angular velocity $d\theta/dt$ has a smoothly varying component $= +1$.

Initially, the system is at $(A, 0)$. If there is positive damping, the smoothed radius traces a decreasing spiral in the phase plane. Most often, we are only interested in the smoothed parts of the motion as they provide an easily obtainable view of the system. It is, in general, reasonably difficult to develop analytical techniques that properly take into account all the wiggly aspects of

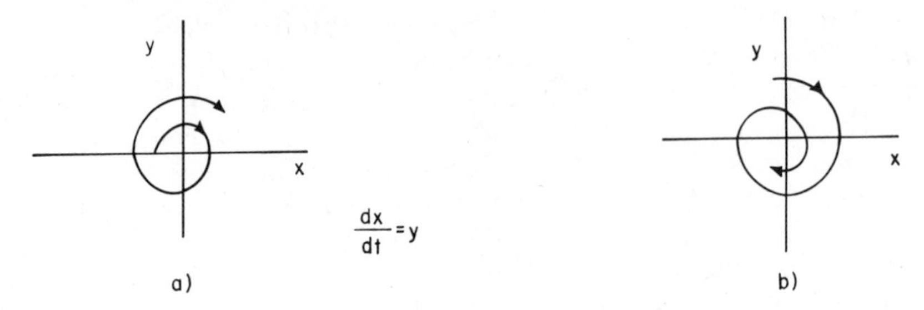

Figure 9.5 We plot spirals that develop in time in a clockwise direction according to the sign convention, $dx/dt = y$. In Figure 9.5a, the amplitude is increasing in time, and in Figure 9.5b, the system is positively damped. Note that the "sense" of the motion is maintained.

the trajectory. Note that since ε is a small parameter, we say that

$$\frac{dr}{dt} = O(\varepsilon) \qquad \text{and} \qquad \frac{d\theta}{dt} = +1 + O(\varepsilon)$$

and, thus, we clearly observe that the motion is clockwise and a slowly decreasing or increasing spiral, depending on the sign of parameter ε. See Figure 9.5. ■

9.3 SIGN CONVENTIONS

We now try to make certain that we understand the sign or direction of the arrows on the phase-plane trajectory. There is no standard convention; some books change their convention within the text without giving a hint that they have done it. So beware of sign problems.

Our trajectory in the phase plane should always be drawn with an arrow indicating the direction of time. There is no confusion if the trajectory is closed. However, if we have a spiral trajectory and we omit the arrow, we cannot tell whether it represents a growing or a diminishing spiral. This depends upon a convention and we adopt the convention that yields clockwise motion. How did we do this? It follows from the relation between variables x and y. We originally defined velocity y as

$$y = \frac{dx}{dt} \tag{9.32a}$$

Thus,

$$x(t + \delta t) - x(t) = y(t)(\delta t) \tag{9.32b}$$

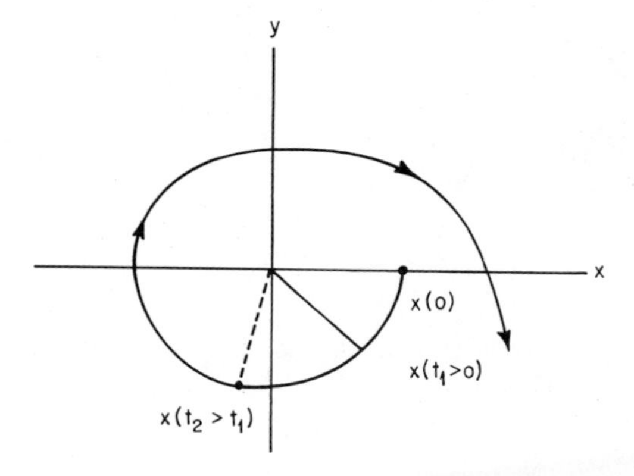

Figure 9.6 Here is a typical example of how to unravel a sign-convention problem. One examines the appropriate equation at times t_1 and $t_2 > t_1$ and makes certain that the spiral opens.

If $x(t)$ has its maximum value at $t = 0$, then it initially decreases, and thus y changes from its initial value of $y = 0$ to a negative value. This is clear if we write

$$x(t + \delta t) < x(t)$$

for t slightly greater than 0. The system maintains its direction, so we can put an arrow on the trajectory, as shown in Figure 9.6.

It should be clear that the direction of the spiral is unchanged if we change the sign of the damping coefficient. The sign is entirely determined from the relationship between the two phase-plane variables x and y. Let us choose a different relationship between dx/dt and y than that given by Equation (9.32a). Specifically, let us now define velocity y' as the negative of dx/dt:

$$y' = -\frac{dx}{dt} \tag{9.33}$$

Then, for example, the damped oscillator equation would be written as

$$\frac{d^2x}{dt^2} + x = \varepsilon\frac{dx}{dt} = -\varepsilon y' \tag{9.34}$$

or, equivalently,

$$\frac{dx}{dt} = -y' \tag{9.35a}$$

$$\frac{dy'}{dt} = +x + \varepsilon y' \tag{9.35b}$$

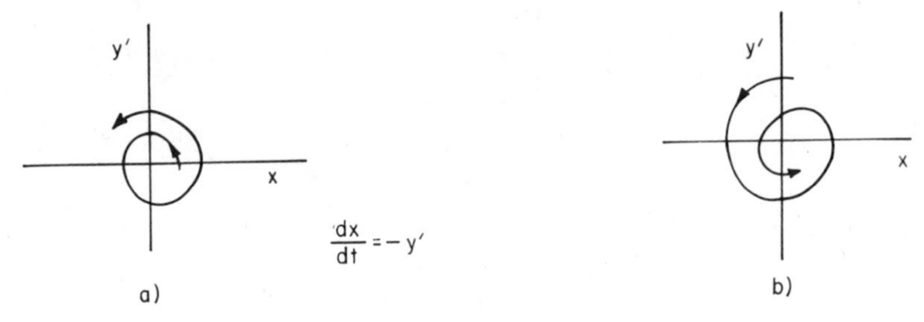

Figure 9.7 We plot spirals that develop in time in a counterclockwise direction according to the convention $dx/dt = -y'$.

If we introduce the phase-plane variables (x, y'), we would observe the trajectory as a counterclockwise spiral.

The motion in the phase plane would be the reflected trajectory of that given in Figure 9.5, since now we have our arrow of time going counterclockwise. See Figure 9.7.

Let us expand on our discussion of the choices for the relationship between dx/dt and the variable we call y. We discuss two sign conventions.

(*Note to the reader*: a word of explanation is in order regarding my emphasis on a possible confusion that arises from the various sign conventions. Many of my colleagues think that I am the one who is confused! They say that one just chooses a convention and sticks to it. But I think that this is not the case. Yes, I have great trouble getting the signs right in an expression. But I am not alone in this matter. As evidence, I point to the confusion that students have in using right- and left-handed coordinate systems; and the plight of those who try to master the so-called "right-hand rule." Of course, there is also the confusion that arises from misprints, and I know that some (hopefully not too many) are present in this text. With these words said, I give the reader a choice. Either skim over the elaborated discussion of sign conventions or read it with an appropriate smile.)

Convention I

$$y \equiv \frac{dx}{dt} \tag{9.36}$$

This corresponds to clockwise motion in the phase plane. (This is the convention used in Figure 9.5.) We choose a relationship between Cartesian coordinates (x, y) and polar coordinates (r, θ), such that angle θ increases in time.

Then we have

$$x = r\cos\theta \tag{9.37a}$$

$$y = -r\sin\theta \tag{9.37b}$$

$$\theta = +t + \phi \tag{9.37c}$$

Differentiating, we get

$$\frac{dx}{dt} = \frac{dr}{dt}\cos\theta - r\frac{d\theta}{dt}\sin\theta \tag{9.38a}$$

$$\frac{dy}{dt} = -\frac{dr}{dt}\sin\theta - r\frac{d\theta}{dt}\cos\theta \tag{9.38b}$$

Then we also have

$$x\frac{dx}{dt} + y\frac{dy}{dt} = r\frac{dr}{dt} \tag{9.39a}$$

$$-x\frac{dy}{dt} + y\frac{dx}{dt} = r^2\frac{d\theta}{dt} \tag{9.39b}$$

Observe that angle θ is measured from the x axis so that it increases in a clockwise direction. Introduce θ in the phase-plane coordinates (x, y). Plot the trajectory in the phase plane and observe that it is a clockwise winding spiral as time t increases.

Now consider the following.

Convention II

$$y \equiv -\frac{dx}{dt} \tag{9.40}$$

(This is the convention used in Figure 9.7.) We still want to have polar angle θ increase with time. Thus, our motion is counterclockwise and we write

$$x = r\cos\theta \tag{9.41a}$$

$$y = r\sin\theta \tag{9.41b}$$

$$\theta = +t + \phi \tag{9.41c}$$

Differentiating, we have

$$\frac{dx}{dt} = \frac{dr}{dt}\cos\theta - r\frac{d\theta}{dt}\sin\theta \tag{9.42a}$$

$$\frac{dy}{dt} = \frac{dr}{dt}\sin\theta + r\frac{d\theta}{dt}\cos\theta \tag{9.42b}$$

We also have

$$x \frac{dx}{dt} + y \frac{dy}{dt} = r \frac{dr}{dt} \tag{9.43a}$$

$$x \frac{dy}{dt} - y \frac{dx}{dt} = r^2 \frac{d\theta}{dt} \tag{9.43b}$$

It is useful to compare each of the appropriate pairs in Equations (9.36) through (9.43b). Notice that some signs are the same. For example, if we have a situation where the radius is increasing in time, both conventions must yield $dr/dt > 0$.

To summarize the sign conventions: in Convention I, velocity dx/dt is equal to the quantity we call y. In Convention II, velocity dx/dt is equal to $-y$. In both conventions, we have chosen our relationship between polar angle θ and time, so that the angle increases as time increases. We have a slight preference for sign *Convention II* as given by Equation (9.40). It often leads to simpler *sign* relationships. The convention to use is purely a matter of taste and convenience.

EXERCISE 3.1 Given the equation:

$$\frac{d^2x}{dt^2} + x = \varepsilon \frac{dx}{dt}; \qquad x(0) = A, \quad \dot{x}(0) = \frac{\varepsilon A}{2} \tag{9.44}$$

Solution:

$$x(t) = A \exp \frac{\varepsilon t}{2} \cos\left(\sqrt{1 - \frac{\varepsilon^2}{4}}\, t\right)$$

Use sign Conventions I and II and write Equation (9.44) as a set of coupled first-order equations. Introduce polar coordinates and plot the trajectories in the phase plane for the case $\varepsilon > 0$. Make certain that you obtain an increasing spiral for both sign conventions. ■

We now have some background and can begin to use these ideas to study nonlinear systems. We see that any autonomous system described by Equation (9.22) that has a trajectory that is a closed curve in the phase plane describes periodic motion. Or, alternatively, if the motion is periodic, with period T, then

$$x(T + t) = x(t); \qquad \frac{dx(T + t)}{dt} = \frac{dx(t)}{dt}$$

If the trajectory does not close, then we can study the long-term behavior of the system by following the progress of the trajectory. We find that nonlinear systems can exhibit behavior that is impossible for linear systems and that the phase plane gives us a useful analytical and graphical tool.

CHAPTER 10

CONSERVATIVE SYSTEMS

We are going to introduce some of the central ideas associated with the development of an asymptotic method for handling conservative, weakly nonlinear oscillatory systems. The oldest systematic method is due to Lindstedt and Poincaré. It is easy to understand, but it is limited to systems that have periodic motion. We discuss it in this chapter and then in later chapters develop the more powerful and more universally applicable methods.

10.1 REVIEW OF THE BASIC IDEAS

(1) We begin with a system that is close to a simple harmonic oscillator (SHO). This means that we have a small parameter that characterizes how close we are to the basic system.

(2) The SHO has a natural frequency ω_0 and thus a natural time scale appears in the problem.

(3) The solutions of the SHO have constant amplitude and constant frequency, independent of amplitude, whereas the solutions to the perturbed problem are not of this form, but instead have the possibility of slowly varying amplitude and (or) an amplitude-dependent frequency. We find that on a time scale of one period of the unperturbed oscillator, the perturbed system has the possibility of small changes. We can think of observing our system with camera flashes that occur at the natural frequency ω_0 and record only the change after each of these intervals, i.e., by means of a flashing clock.

(4) We have to avoid secular terms. These appear because we have a small effect that acts over a long time and we have to organize our calculation so as to clearly identify terms that are small for all times and thus cause no trouble and terms that are only small for short times and have to be handled with care. So we introduce methods that reveal the underlying structure of the calculations and enable us to separate the terms into these two general classes.

How do we do this? If the system is periodic, we adjust the flashing of our light to be closer to the correct frequency rather than the unperturbed frequency. We are free to do this and we use this freedom to suppress or cancel the secular terms.

(5) If the system is not periodic but has damping, we use our freedom to track the average energy over a cycle of the unperturbed period and watch how it changes. The crucial quantity to watch is the energy, or amplitude, rather than the displacement. (This is discussed in Chapter 11.)

(6) We also sometimes work in Cartesian and sometimes in polar coordinates. We then have to be careful to understand how we transform back and forth between the coordinate systems. It is often easier to work in polar coordinates if the amplitude changes, since we then can concentrate on the change in energy per cycle and are not distracted by the oscillating parts.

10.2 PROPERTIES OF CONSERVATIVE SYSTEMS

We are going to consider conservative systems that differ slightly from simple harmonic oscillators through a small change in the potential. The motion remains periodic and we expect the new period to be close to the old one.

We begin with a potential of the form $V(x) = \frac{1}{2}x^2 +$ "a small correction term." The total energy E of the system consists of the kinetic energy K_E and the potential energy P_E. If we set the system in motion, it oscillates, exchanging kinetic and potential energy. The motion is confined to the region determined by its turning points, i.e., the points where the velocity is equal to zero and, hence, the displacement is maximum. Its motion is consequently bounded and periodic. See Figure 10.1.

Let us see how we come to this conclusion. We write our equation for the total energy as

$$\left(\frac{dx}{dt}\right)^2 = 2[E - V(x)] \tag{10.1a}$$

We then have

$$\frac{dx}{dt} = \pm\sqrt{2[E - V(x)]} \tag{10.1b}$$

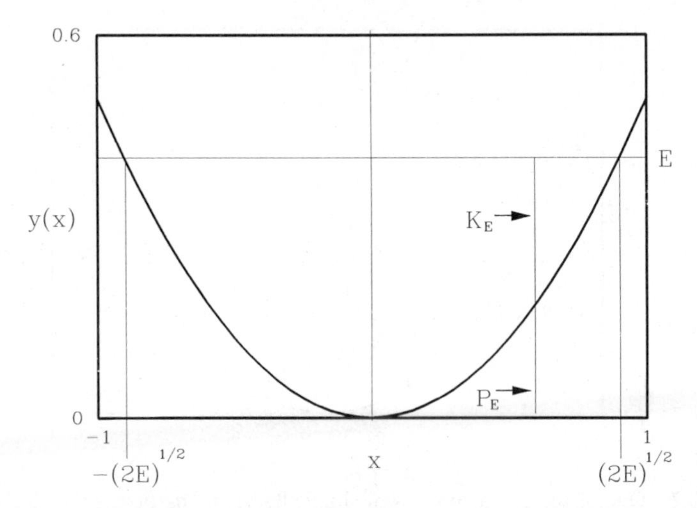

Figure 10.1 It is possible to prove that one-dimensional bounded motion in a conservative force field is periodic. For example, see the text by Landau and Lifshitz. The total energy is indicated as E, the potential energy as P_E, and the kinetic energy as K_E.

or

$$\pm \int \frac{dx}{\sqrt{2[E - V(x)]}} = \int dt \tag{10.1c}$$

We integrate x over the region between turning points x_1 and x_2 given by the values of x for which $E = V$. This then gives us, for the time integral, the corresponding time. The system now turns around and goes back to its origin. These integrals are just numbers. Then it is clear that each half cycle is equal and the sum is the total period of the motion.

$$\int_{x_1}^{x_2} \frac{dx}{\sqrt{2[E - V(x)]}} = \int_0^{T/2} dt = \frac{T}{2} \tag{10.2}$$

If we have the quadratic potential $V(x) = \frac{1}{2}x^2$, the integrals are easily evaluated as follows:

$$\int_{-\sqrt{2E}}^{+\sqrt{2E}} \frac{dx}{\sqrt{2(E - x^2/2)}} = \int_{-1}^{1} \frac{dy}{\sqrt{1 - y^2}} = \pi = \tfrac{1}{2}\,\text{period} \tag{10.3}$$

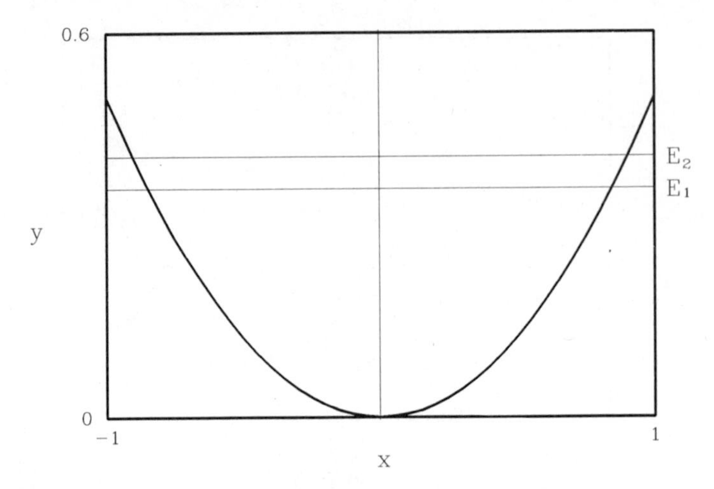

Figure 10.2 One of the characteristics of simple harmonic motion is that the period is independent of the amplitude or equivalently the energy. $y(x) = \frac{1}{2}x^2$.

To summarize:

1. Motion in a potential well is periodic.
2. Motion in a quadratic potential yields a period, $T = 2\pi$, that is independent of the energy and the amplitude. Thus, in Figure 10.2, System 1, with energy E_1, has the same period as System 2, with energy E_2. Suppose we have a flashing clock adjusted to flash at intervals equal precisely to the period of System 1, then we will see the motion as fixed in space, or as a fixed point. Since the light catches the system where it has the same position and velocity every time, it appears at rest.

   Now, since System 1 has the same period as System 2, it will also appear as a fixed point if we flash the light at the frequency that we have found for System 1. See Figure 10.2.
3. Now we try the same experiment for a system described by a potential of the form

$$V(x) = \tfrac{1}{2}x^2 + \varepsilon\left(\tfrac{1}{4}\right)x^4; \qquad 0 < \varepsilon \ll 1 \tag{10.4}$$

   See Figure 10.3.

   This corresponds to the Duffing oscillator, a system governed by the equation of motion:

$$\frac{d^2x}{dt^2} + x = -\varepsilon x^3; \qquad 0 < \varepsilon \ll 1 \tag{10.5}$$

   (Note that we have already prepared our problem by introducing dimensionless variables x and t.)

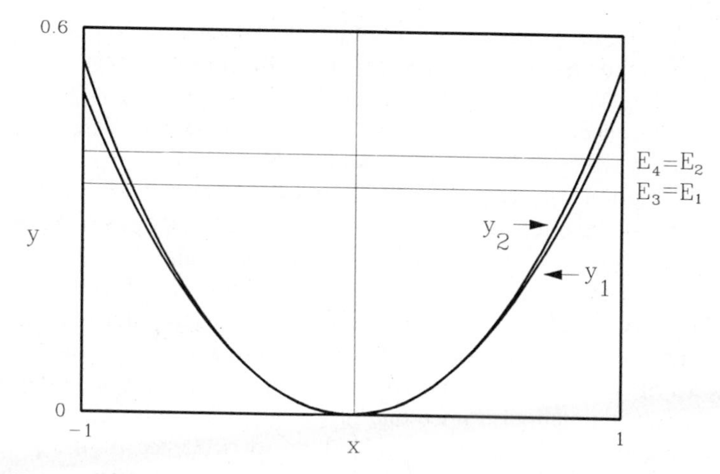

Figure 10.3 The potential for the Duffing oscillator has a small quartic term that leads to an amplitude-dependent frequency. It is precisely because the potential differs only slightly from $\frac{1}{2}x^2$ for the region of interest that we are confident that we have the capacity to develop a perturbation expansion in the small parameter ε. $y_1 = \frac{1}{2}x^2$; $y_2 = \frac{1}{2}x^2 + \varepsilon(\frac{1}{4})x^4$.

Our discussions of conservative nonlinear systems are centered on a discussion of the Duffing oscillator. It is really the simplest system to analyze and its characteristics are typical of a wide class of systems. By concentrating almost extensively on its properties, we will be able to compare and contrast various perturbation schemes.

We can think of our system as consisting of a mass m and an altered spring, so that the Hooke's law force is changed:

$$-kx \to -k(x + \gamma x^3); \qquad \gamma = \frac{\varepsilon}{k}$$

The system described by this equation executes oscillations of fixed amplitude since the total energy

$$E = \frac{1}{2}\left(\frac{dx}{dt}\right)^2 + \frac{1}{2}x^2 + \varepsilon\left(\frac{1}{4}\right)x^4 \tag{10.6}$$

is conserved. We can write this conclusion as $dE/dt = 0$. Now consider a Duffing oscillator, described as System 3, with energy E_3, and as System 4, with energy E_4. Then, referring to Figure 10.2, we arrange our systems so that $E_3 = E_1$ and $E_4 = E_2$. Assume that we can establish by some method of analysis that the period depends upon the amplitude, so that Systems 3 and 4 have different periods. (This is done in Exercise 2.2 and in the equations leading up to Equation (10.35).)

Perhaps we can visualize the amplitude dependence as follows. First, we adjust the frequency of our flashing clock to correspond to that of the unperturbed oscillator with energy E_1, denoted as System 1. When we view System 3, with energy $E_3 = E_1$, we notice that it is not seen as a fixed point, but appears to drift slowly. We then carefully adjust the timing of our flashes to stop the motion. Everything is fine as long as we do not observe the system for too long a time interval. Even though the system has a precise period, it is unknown to us. After we are satisfied with our value for the frequency associated with System 3, we try it on System 4, with $E_4 > E_3$. We find that it does not work, since the period depends upon the amplitude, or the total energy, and so we have to make a different adjustment of our flashing light. This is a critical difference between motion in a quadratic potential and in other symmetric potentials. Since one of our goals is to develop a consistent perturbation expansion, we have to find the dependence of the period or frequency on the amplitude of motion in a variety of potentials.

EXERCISE 2.1 Consider the altered SHO given by the equation

$$\frac{d^2x}{dt^2} + x = -\varepsilon x; \qquad x(0) = A, \quad \dot{x}(0) = 0$$

(a) Find the potential $V(x)$.
(b) What are the turning points of the motion?
(c) Plot $V(x)$ vs. x for a system that has total energy E. Indicate the turning points on the plot. ■

Consider Equation (10.2) for a case in which the potential $V(x)$ is a symmetric function. We outline in Exercises 2.2–2.4 an expansion scheme that enables us to find the period of oscillation as a power series in the small parameter ε.

EXERCISE 2.2 Consider a particle with total energy E and maximum displacement A moving in a symmetric potential

$$V(x) = \tfrac{1}{2}x^2 + \tfrac{1}{4}\varepsilon x^4; \qquad |\varepsilon| \ll 1$$

We can evaluate the energy when the velocity is equal to zero and write

$$E = \tfrac{1}{2}A^2 + \tfrac{1}{4}\varepsilon A^4$$

where $\pm A$ are the turning points. Then Equation (10.2) takes the form

$$\frac{T}{2} = \int_{-A}^{A} \frac{1}{\left(A^2 + \frac{1}{2}\varepsilon A^4 - x^2 - \frac{1}{2}\varepsilon x^4\right)^{1/2}}\,dx$$

Since $A^2 - x^2$ is a factor of the polynomial, we can write

$$T = 4\int_0^A \frac{1}{\left[(A^2 - x^2)\left(1 + \frac{1}{2}\varepsilon A^2 + \frac{1}{2}\varepsilon x^2\right)\right]^{1/2}}\,dx$$

Introduce $x = A\sin\theta$ and obtain

$$T = 4\int_0^{\pi/2} \frac{1}{\left[1 + \frac{1}{2}\varepsilon A^2 + \frac{1}{2}\varepsilon A^2 \sin^2\theta\right]^{1/2}}\,d\theta$$

Expand the denominator and conclude that

$$T = 2\pi\left[1 - \tfrac{3}{8}\varepsilon A^2 + O(\varepsilon^2)\right]$$

Continue the expansion and show that the next term is $(57/256)\varepsilon^2 A^4$. ■

EXERCISE 2.3 The factorization technique can be used whenever the potential is a symmetric function of x. For example, consider a particle with total energy E and maximum displacement A moving in a potential of the form

$$V(x) = \tfrac{1}{2}x^2 + \tfrac{1}{6}\varepsilon x^6; \qquad |\varepsilon| \ll 1$$

Use the factorization method to write the appropriate denominator as

$$(A^2 - x^2)\left(1 + \tfrac{1}{3}\varepsilon A^4 + \tfrac{1}{3}\varepsilon A^2 x^2 + \tfrac{1}{3}\varepsilon x^4\right)$$

Show that the period of oscillation is

$$T = 2\pi\left[1 - \tfrac{5}{16}\varepsilon A^4 + O(\varepsilon^2)\right]$$ ■

EXERCISE 2.4 Consider a potential that is of the form

$$V(x) = \tfrac{1}{2}x^2 + \tfrac{1}{4}\varepsilon x^4 + \tfrac{1}{6}\varepsilon x^6$$

Show that the lowest-order correction to the frequency of oscillation is simply the sum of the terms found in Exercises 2.2 and 2.3. The frequency corrections cannot interfere, in the lowest order, because they enter the expansion as different powers of the amplitude. ■

Comment: If $V(x)$ contains odd powers of x, the motion is no longer symmetric about the origin. Then $A^2 - x^2$ is no longer a factor of the denominator, and an obvious change of variables is not available to simplify the integrand.

10.3 ORIENTATION TO THE SPIRIT OF THE CALCULATIONS

We now introduce the calculational techniques that are effective for the study of conservative systems. We concentrate our attention on the Duffing oscillator, given by Equation (10.5), since it is a typical nonlinear conservative system. It has precisely a cubic nonlinearity, rather than a nonlinearity that we approximate as a cubic. (For example, a simple pendulum, executing small-amplitude oscillations, can be described in an approximation that contains a term that is cubic in the amplitude.) We are concerned with the following aspects of the system:

(1) Is the motion periodic? Yes, because the system is one-dimensional and executes bounded motion in a conservative field of force.
(2) Is the period of oscillation changed by the perturbation? Yes. We will find this out by calculation, but it is expected since we have a slightly stiffer spring.
(3) What is the shape of the orbit in the phase plane? We guess that it will be close to a circle.

10.3.1 Review

We begin our study of the problem by plotting the trajectory in the phase plane. There are two aspects to the motion. There is the change in frequency of oscillation and the change in shape of the orbit.

It is our expectation that since the perturbation is small, the change in the frequency will be correspondingly small. However, you have to be careful regarding what this means. Let us look at a phase-plane plot of our Duffing oscillator equation. See Figure 10.4.

In Figure 10.4, we indicated five points, labeled a–e, on the trajectory. We think of our system as starting at time $t = 0$ and at point a.

We now wait for an interval of time equal to the period of the unperturbed oscillator, $T_0 = 2\pi$, and we find that the system is not at point a but at some

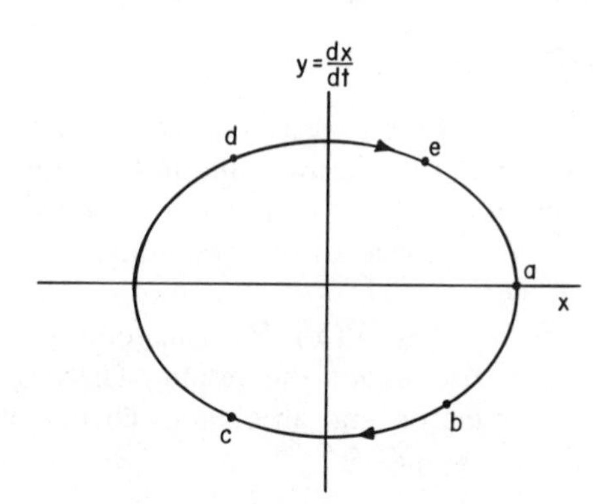

Figure 10.4

other point, e.g., at b or e, depending upon whether the frequency of the Duffing oscillator is greater or less than the corresponding simple harmonic oscillator. (We will find out, through our calculations, that the Duffing oscillator has a greater frequency than the corresponding SHO. Thus, it would be found at point b rather than point e.) Initially, we restrict our viewing of the system to intervals separated by periods of time that are equal to the period of the unperturbed oscillator.

If the perturbation changed the period of motion and if we were to continue to observe the system by flashes at intervals separated by the unperturbed period, the system would appear to drift and we would find it at point c, etc. Thus, the periodic nature of the motion would be masked by the method of observation.

10.3.2 Organization of the Calculation

We want to organize our calculations in such a manner as to exhibit or maintain the periodic nature of the solution. We do this by an expansion that is uniformly valid in time. That means that if we say that two terms are small, one relative to the other, this relationship remains true or is maintained for all time.

Let us make these ideas clearer by considering the following equation:

$$\frac{d^2x}{dt^2} + (1+\varepsilon)^2 x = 0; \qquad x(0) = A, \quad \dot{x}(0) = 0 \tag{10.7}$$

The solution is

$$x(t) = A\cos[(1+\varepsilon)t] \tag{10.8a}$$

What is most important to observe is that the displacement $x(t)$ is a periodic function. Thus, we are thinking of a problem that has yielded a small correction to the natural frequency of oscillation.

Now, let us assume that instead of finding the exact solution given by Equation (10.8a), we found an approximate solution by a perturbation expansion. It might come to us in the form

$$x(t) = A\cos t - (\varepsilon A t)\sin t + O(\varepsilon^2 t^2) \tag{10.8b}$$

This approximate solution is correct. However, it can be misleading because it can lead one to assume incorrectly that it is an expansion in a small parameter ε rather than in the small parameter εt. When we developed our perturbation expansion, we may have started at time $t = 0$ and then worked forward. Thus, there is a coupling between the time t and the parameter ε. This leads to the possible incorrect conclusion that we have aperiodic rather than periodic motion due to the presence of terms in our perturbation expansion such as $(\varepsilon A t)\sin t$.

We now return to our Duffing oscillator, given by Equation (10.5), and we will be certain that our perturbation expansion is organized so as to maintain in each term the periodic nature of the solution even though we will be simultaneously searching for the correct period or frequency. Of course, the precise frequency is unknown to us and thus we will have to continuously adjust our flashing rate in order to catch the motion as a fixed point.

Let us make a correspondence between this approach and our adjustments in our calendar. We insert a leap year at fixed intervals in order to correct for our error in the length of the day per year. But this needs further corrections since the ratio of the length of a day to the length of the year is not a rational number. Our technique with the calendar is just to continue to make corrections in order to obtain as adequate a match as fits our needs. We will do the same thing. We will find the first correction to our frequency, readjust our flasher, and find the next correction, etc. Our goal is to obtain a perturbation expansion for $x(t)$ in terms of functions that are periodic with the correct frequency ω rather than the unperturbed frequency ω_0.

However, our expansion cannot be valid for all time since we are not able to find the precise or exact frequency of oscillation. This means that eventually our approximation to the phase-plane orbit will "open" and the finite time validity of our approximation will be evident.

This is the central idea of our improved perturbation expansion. Thus, we do not seek an expansion in the form

$$x(t) = x_0(t) + \varepsilon x_1(t) + \varepsilon^2 x_2(t) + \text{h.o.t.} \tag{10.9a}$$

where εx_1 and $\varepsilon^2 x_2$ are small corrections to x_0. Rather, we seek an expansion in the form

$$x(\tau) = x_0(\tau) + \varepsilon x_1(\tau) + \varepsilon^2 x_2(\tau) + \text{h.o.t.} \tag{10.9b}$$

where $\varepsilon x_1(\tau)$ and $\varepsilon^2 x_2(\tau)$ are small corrections to $x_0(\tau)$. We have transformed the time variable in our problem to $\tau = \omega t$. This transformation introduces the correct, or exact, frequency ω, an unknown quantity that we seek at the same time as we develop the perturbation calculation. It is important to observe that $t \neq \tau$ whenever the period of the motion is altered by the perturbation.

Our basic assumptions are as follows:

(1) The frequency is only slightly altered by the small perturbation, and it can be expressed as a power series in the small parameter ε:

$$\omega = \omega_0 + \varepsilon\omega_1 + O(\varepsilon^2); \qquad \omega_0 = 1 \tag{10.10}$$

We find the correction terms sequentially.

(2) The displacement, $x(\tau)$, and the velocity of the periodic oscillation, $dx/d\tau$, differ only slightly in functional form from the unperturbed

displacement and the unperturbed velocity, $x(t)$ and $dx(t)/dt$, respectively. Since our equation is second order in time, we have to find a good approximation for both the displacement and the velocity.

10.4 THE POINCARÉ – LINDSTEDT METHOD

The Poincaré–Lindstedt method provides us with a consistent perturbation scheme that identifies the frequency correction that is required to suppress secular generating terms and gives us an approximation for the displacement $x(t)$. It is limited to conservative systems and is thus of less generality than either the method of averaging or the method of multiple time scales. (These more general methods are discussed in Chapters 12–14.) However, it is a rigorous method that gives us insight into the behavior of conservative nonlinear oscillatory systems. We choose to develop the perturbation expansion by converting our differential equation to an integral equation and then solving by a method of successive approximations. The detailed steps are explained through the analysis of the Duffing oscillator in which we find an expression for the displacement $x(t)$ with an error $O(\varepsilon^2)$ for time $t = O(1)$. (Note that the technique is also outlined in Exercise 4.2 through an analysis of the altered simple harmonic oscillator. This problem is quite simple and should be referred to if you get confused by the calculations we go through for the Duffing oscillator.) The analysis is straightforward, but algebraically complicated. The trial solution method can also be used and is generally recommended for actual calculations since it entails a minimum of algebra. The reason we emphasize the integral equation method is that it gives us insight into the source of the secular generating terms and is useful for the development of an error analysis as discussed in Chapter 15. It is also a general technique for solving inhomogeneous differential equations.

Our original equation is

$$\frac{d^2x}{dt^2} + x = -\varepsilon x^3; \qquad 0 < \varepsilon \ll 1 \tag{10.11}$$

with initial conditions

$$x(0) = A; \qquad \frac{dx(0)}{dt} = 0 \tag{10.12}$$

We know from our energy analysis that the total energy E is

$$E = \frac{1}{2}\left(\frac{dx}{dt}\right)^2 + \frac{1}{2}x^2 + \varepsilon\left(\frac{1}{4}\right)x^4 \tag{10.13}$$

Thus, our particular initial conditions are such that the maximum displacement is $x_{max} = A$.

The analysis proceeds in a sequence of steps:

(1) Our unperturbed frequency ω_0 is equal to 1 and we seek the frequency ω of our Duffing oscillator as a series expansion in powers of the small parameter ε:

$$\omega = 1 + \varepsilon\omega_1 + O(\varepsilon^2) \tag{10.14}$$

where $\varepsilon\omega_1$ is a small correction to 1.

(2) Introduce the "proper" time that is connected with the period of the motion:

$$\tau \equiv \omega t \tag{10.15}$$

and expand our displacement in powers of ε:

$$x(\tau) = x_0(\tau) + \varepsilon x_1(\tau) + \text{h.o.t.} \tag{10.16}$$

We also express our differential operators in terms of the proper time:

$$\frac{d}{dt} = \omega\frac{d}{d\tau}; \qquad \frac{d^2}{dt^2} = \omega^2\frac{d^2}{d\tau^2}$$

(3) Equation (10.11) becomes

$$\frac{d^2x}{d\tau^2} + \frac{1}{\omega^2}x = \frac{-\varepsilon}{\omega^2}x^3 \tag{10.17a}$$

We can also write Equation (10.17a) as

$$\frac{d^2}{d\tau^2}(x_0 + \varepsilon x_1) + \frac{1}{(1 + \varepsilon\omega_1)^2}(x_0 + \varepsilon x_1)$$

$$= -\frac{\varepsilon}{(1 + \varepsilon\omega_1)^2}(x_0 + \varepsilon x_1)^3 + \text{h.o.t.} \tag{10.17b}$$

(4) We seek a solution to Equation (10.17b) by solving our equations successively in powers of the small parameter ε through $O(\varepsilon)$ for times $t = O(1)$ or $t \ll 1/\varepsilon$. We impose this restriction on the length of time for which we require our solution to be valid because of the linking of the small parameter ε with the time t within the argument of the trigonometric functions. For example, consider a displacement of the form $x(t) = \cos[(1 + \varepsilon\omega_1 + \varepsilon^2\omega_2)t]$. If we are interested in finding the displacement correct to $O(\varepsilon)$ for $t = O(1)$, *we need not*

include the $\varepsilon^2\omega_2$ term. However, if we continue our time of observation until $t = O(1/\varepsilon)$, *we need* to include the $\varepsilon^2\omega_2$ term in order to retain the same accuracy.

With these observations, we proceed to construct the solution to Equation (10.17b). We are going to retain only terms of order ε, so we can write equation (10.17b) as

$$\left(\frac{d^2x_0}{d\tau^2} + x_0\right) + \varepsilon\left(\frac{d^2x_1}{d\tau^2} + x_1\right) = +2\varepsilon\omega_1 x_0 - \varepsilon x_0^3 + O(\varepsilon^2) \quad (10.17c)$$

Observe that since the parameter ε multiplies the terms on the right-hand side of Equation (10.17c), one uses the unperturbed solution for the displacement.

We then solve order by order in ε.

$$O(\varepsilon^0): \qquad \frac{d^2x_0}{d\tau^2} + x_0 = 0 \quad (10.18)$$

The solution is

$$x_0(\tau) = A_0 \cos \tau \quad (10.19a)$$

where the initial conditions require

$$x_0(0) = A_0; \qquad \dot{x}_0(0) = 0 \quad (10.19b)$$

(5) Keep in mind that we are going to find only the corrections that are of lowest order in ε. We now write our equation for the first amplitude correction x_1 as

$$\frac{d^2x_1}{d\tau^2} + x_1 = 2\omega_1 x_0 - x_0^3 \quad (10.20)$$

We solve this equation by the method of variation of parameters, yielding the solution in the form

$$x_1(\tau) = A_1 \cos \tau + B_1 \sin \tau + \int_0^{\tau} \sin(\tau - \tau')\left[2\omega_1 x_0(\tau') - x_0^3(\tau')\right] d\tau' \quad (10.21a)$$

The initial conditions can be used to find A_1:

$$x_1(0) = A_1; \qquad \frac{dx_1(0)}{d\tau} = 0 \Rightarrow B_1 = 0 \quad (10.21b)$$

We have encountered this type of equation and its solution previously.

(See, for example, Equation (8.5) and the discussion that follows it.) By taking the derivative of Equation (10.21a), one sees that

$$\frac{dx_1}{d\tau} = -A_1 \sin \tau + \int_0^{\tau} \cos(\tau - \tau')\left[2\omega_1 x_0(\tau') - x_0^3(\tau')\right] d\tau' \quad (10.21c)$$

The particular initial conditions we have imposed cannot change the structure of the solution $x(\tau)$ since any other initial conditions yield a solution that can be put into this form by an appropriate translation of the time variable. (This, however, can involve extensive algebraic manipulation.)

Note that we were given an initial condition on the correct displacement and the correct velocity. Since we are solving for the displacement and velocity in successive orders of perturbation, we impose the initial conditions in each order. Thus, referring to Equation (10.12), we write

$$A = A_0 + \varepsilon A_1 + O(\varepsilon^2) \quad (10.22)$$

This is equivalent to writing

$$x(0) = A; \qquad x_0(0) = A_0; \qquad x_1(0) = A_1; \qquad \text{etc.} \quad (10.23)$$

where the amplitude A is equal to the maximum displacement.

We can also satisfy our initial conditions by choosing

$$x_0(0) = A; \qquad \dot{x}_0(0) = 0; \qquad x_i(0) = 0; \qquad \dot{x}_i(0) = 0; \qquad i > 0 \quad (10.24)$$

In the procedure given by Equation (10.24), the homogeneous solution of $x_i(t)$ and its derivative are set equal to the negative of the particular solution and its derivative, respectively, at time $t = 0$. This way of satisfying the initial conditions is the basis of Exercises 4.1 and 5.1. (Both methods yield the same results and are analyzed in some detail in Chapter 14).

Let us pause before we find the correction to the natural frequency. We would like to emphasize the nature of our perturbation expansion. We have organized the calculation so that in each order of the perturbation expansion, the l.h.s. always contains the differential operator associated with the unperturbed oscillator.

$$\text{l.h.s.} = \left(\frac{d^2}{d\tau^2} + 1\right) x_i; \qquad i = 0, 1, 2 \ldots$$

The r.h.s. is always a combination of terms that have been evaluated in finding the lower-order corrections to the frequency and displacement. Thus, the structure of the first-order calculation is typical of all of the higher-order calculations. We are most interested in finding the frequency corrections since they enable us to view the system as a fixed point. We expect that the amplitude corrections will be less important since they are really amplitude corrections to the higher harmonics. Thus, if perchance, we do not calculate them, we will be missing only the wiggles on the phase-plane trajectory.

We, therefore, concentrate primarily on finding the correction to the fundamental frequency. Note that there is no amplitude correction for conservative systems. Keep in mind that throughout our calculation we will insist on retaining the periodicity of the solution in each order, and we use this requirement to guide us toward a choice of ω_1.

10.4.1 Computation of x_1

In order to solve Equation (10.21a), we expand the term x_0^3. We have

$$\cos^3\tau = \tfrac{1}{4}(3\cos\tau + \cos 3\tau) \tag{10.25}$$

Thus, Equation (10.21a) can be written

$$x_1(\tau) = A_1\cos\tau + \int_0^\tau \sin(\tau - \tau')\left[\left(-\tfrac{3}{4}A_0^3 + 2A_0\omega_1\right)\cos\tau' - \tfrac{1}{4}A_0^3\cos 3\tau'\right] d\tau' \tag{10.26}$$

Similarly, we can write Equation (10.21c) as

$$\frac{dx_1}{d\tau} = -A_1\sin\tau + \int_0^\tau \cos(\tau - \tau')\left[\left(-\tfrac{3}{4}A_0^3 + 2A_0\omega_1\right)\cos\tau' - \tfrac{1}{4}A_0^3\cos 3\tau'\right] d\tau' \tag{10.27}$$

It is important to note that the range of integration is from 0 to τ. We will encounter four different types of integrals:

$$I_1 = \int_0^\tau \cos^2\tau'\, d\tau' = \tfrac{1}{2}\tau + \tfrac{1}{4}\sin 2\tau \tag{10.28}$$

$$I_2 = \int_0^\tau \cos\tau'\cos 3\tau'\, d\tau' = \tfrac{1}{2}\int_0^\tau (\cos 2\tau' + \cos 4\tau')\, d\tau' = \tfrac{1}{4}\sin 2\tau + \tfrac{1}{8}\sin 4\tau \tag{10.29}$$

$$I_3 = \int_0^\tau \sin\tau'\cos 3\tau'\, d\tau' = \tfrac{1}{2}\int_0^\tau (\sin 4\tau' - \sin 2\tau')\, d\tau' = -\tfrac{1}{8} - \tfrac{1}{8}\cos 4\tau + \tfrac{1}{4}\cos 2\tau \tag{10.30}$$

$$I_4 = \int_0^\tau \sin\tau'\cos\tau'\, d\tau' = \tfrac{1}{4}(1 - \cos 2\tau) \tag{10.31}$$

If we examine the structure of the terms, we see that

$$I_1(2\pi) = \pi$$
$$I_2(2\pi) = I_3(2\pi) = I_4(2\pi) = 0$$

Now we require both the displacement $x_1(\tau)$ and the velocity $dx_1/d\tau$ to be periodic with period $T = 2\pi$. (This follows since we are using the unknown but exact frequency ω as the basis for our time scale τ and thereby adjust the τ period so that it is equal to 2π.) Thus, we choose our free parameter ω_1 to cancel the terms associated with the integral I_1. We then have

$$-\tfrac{3}{4}A_0^3 + 2\omega_1 A_0 = 0 \qquad \text{or} \qquad \omega_1 = \tfrac{3}{8}A_0^2$$

Our expressions for the displacement and velocity become

$$x_1(\tau) = A_1 \cos\tau - \tfrac{1}{4}A_0^3(I_2 \sin\tau - I_3 \cos\tau) \tag{10.32}$$

$$\frac{dx_1}{d\tau} = -A_1 \sin\tau - \tfrac{1}{4}A_0^3(I_2 \cos\tau + I_3 \sin\tau) \tag{10.33}$$

These can be greatly simplified using the addition formulas for products of sines and cosines. Alternatively, one can use the complex representation for sines and cosines. After some straightforward but perhaps tedious algebra, we find

$$x_1(\tau) = A_1 \cos\tau + \frac{A_0^3}{32}\cos 3\tau - \frac{A_0^3}{32}\cos\tau \tag{10.34}$$

and the frequency is

$$\omega = \left(1 + \varepsilon\tfrac{3}{8}A_0^2\right) + O(\varepsilon^2) \tag{10.35}$$

We recall that

$$\tau = \omega t; \qquad x(\tau) = x_0(\tau) + \varepsilon x_1(\tau) + \text{h.o.t.}$$
$$x(0) = A; \qquad \dot{x}(0) = 0$$

So we have, after imposing the initial conditions and using Equation (10.19a) and (10.34),

$$x(t) = \left[A_0 + \varepsilon\left(A_1 - \frac{A_0^3}{32}\right)\right]\cos\omega t + \frac{\varepsilon A_0^3}{32}\cos 3\omega t + O(\varepsilon^2)$$

$$\text{for } t = O(1) \tag{10.36}$$

Our initial conditions are satisfied if we write

$$A = A_0 + \varepsilon A_1 + O(\varepsilon^2)$$

Since we are interested only in finding our solutions correct through $O(\varepsilon^2)$ and times $t = O(1)$, we write $\omega_0 = 1$ for the frequency ω in the $\cos 3\omega t$ term since

$$\cos 3\omega t \sim \cos 3\omega_0 t + O(\varepsilon) \qquad \text{for } t = O(1)$$

Thus, we write

$$\frac{\varepsilon A_0^3}{32} \cos 3\omega t = \frac{\varepsilon A^3}{32} \cos 3t$$

and, similarly,

$$\frac{\varepsilon A_0^3}{32} \cos \omega t = \frac{\varepsilon A^3}{32} \cos t$$

with an error $O(\varepsilon^2)$ for $t = O(1)$. Our final equation becomes

$$x(t) = A \cos\left\{\left[1 + \varepsilon\left(\frac{3}{8}\right)A^2\right]t\right\} + \frac{\varepsilon A^3}{32}(\cos 3t - \cos t) + O(\varepsilon^2) \qquad \text{for } t = O(1) \quad (10.37)$$

We know that the exact solution of the Duffing oscillator is periodic. Thus, a phase-plane plot of the exact velocity versus the exact displacement would be a closed trajectory. Our perturbation expansion is a power-series expansion in terms of ε for the displacement and velocity leading to a phase-plane plot of the "approximate velocity" versus the "appropriate displacement." Consequently, after a sufficiently long interval of time, there is a lack of closure of the orbit, indicating that we have exceeded the time validity of our expansion. Consider Figures 10.5a and 10.5b. We choose $x(0) = 1$ and $\varepsilon = 0.3$, and we let the time run for 10 units, where one period of the unperturbed system is equal to 2π units.

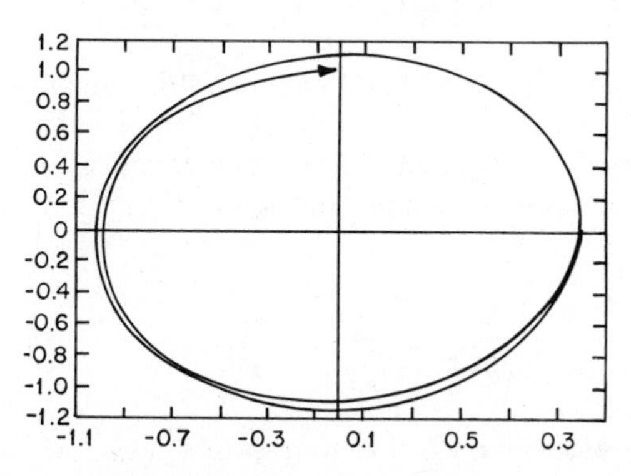

Figure 10.5*a* This is a phase-plane plot of the Duffing oscillator based on the solution given by Equation (10.37).

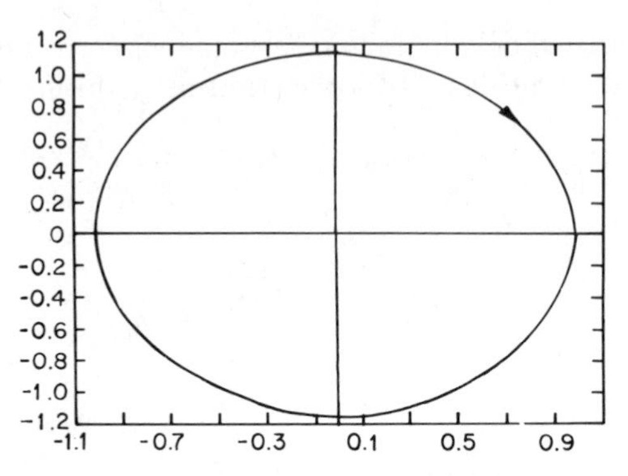

Figure 10.5*b* This phase-plane plot of the Duffing oscillator was obtained by including terms proportional to ε^0, ε^1, and ε^2.

Note that the correction to the natural frequency enables us to see the system as a fixed point for times $t = O(1)$ or $\varepsilon t \ll 1$. If we observe our system for longer times, we will see it begin to drift, and, hence, we have to calculate the next correction term for our frequency. This necessity for frequency updating is familiar when we observe the motion of the hands on a watch. Initially, only the second hand appears to move, then the minute hand appears to move, and, finally, the hour hand. Actually, each hand is moving all of the time. But the ratio of their small parameters is 1/60, so we have to wait a long time on the time scale of the second or minute hand to observe the motion of the hour hand.

EXERCISE 4.1 Repeat the calculation of the frequency correction for the Duffing oscillator with the initial conditions given by Equation (10.12) using a perturbation expansion in which the following initial conditions are satisfied by

$$x_0(0) = A; \qquad \dot{x}_0(0) = 0; \qquad \dot{x}_i(0) = x_i(0) = 0; \qquad i \geq 1 \quad (10.38)$$

This leads to equations analogous to Equations (10.21a) through (10.36). ■

Example 4.1 Consider Equation (10.20) together with the knowledge that the secular generating terms are suppressed if $\omega_1 = \frac{3}{8}A_0^2$. The relevant equation becomes

$$\frac{d^2x_1}{d\tau^2} + x_1 = -\frac{A_0^3}{4}\cos(3\tau).$$

We can find the particular solution by a trial solution of the form $x_1(\tau) = B\cos(3\tau)$ and find $B = A_0^3/32$. Then, upon satisfying the initial conditions, we immediately obtain Equation (10.34). We note that it is useful to know a variety of techniques for solving inhomogeneous differential equations of this form. The trial solution method is often preferred because of its calculational simplicity. ■

Convince yourself that Equations (10.36) and (10.37) are *equivalent* for times $t = O(1)$.

EXERCISE 4.2 Consider the altered SHO

$$\ddot{x} + x = -\varepsilon x; \qquad x(0) = A; \qquad \dot{x}(0) = 0 \tag{10.39}$$

Find the first and second-order frequency corrections.

(a) After introducing the proper time and the perturbation expansion for $x(\tau)$, obtain an equation analogous to Equation (10.21a).
(b) Conclude that we require $\omega_1 = \frac{1}{2}$ in order to cancel the secular terms.
(c) Develop the expansion to the second order and find that $\omega_2 = -\frac{1}{8}$. It is interesting to note in this calculation that the x_1 terms yield zero because we already fixed the lowest-order frequency correction. We return to this point in Chapter 14. ■

10.5 ANOTHER VIEWPOINT

We outline here an alternative way to realize the freedom in adjustment of our flashing clock for conservative systems. It utilizes an expansion in two adjustable parameters in such a way as to eliminate the secular generating terms by an alteration of the oscillator frequency of the unperturbed solution. Let us see how it works.

Consider the Duffing oscillator in the form given by Equations (10.11) and (10.12). We rewrite the r.h.s. of Equation (10.11) by adding and subtracting a term proportional to the displacement. Then, we have

$$\frac{d^2x}{dt^2} + x = -\varepsilon\left(x^3 - \mu x\right) - \varepsilon\mu x \tag{10.40a}$$

We have introduced the constant μ that will be determined by the requirement that our expansion be free from secular terms. It is important to observe that the parameter μ is $O(1)$ and is not another small parameter.

Our equation is unchanged, but we now choose to look at the isolated linear term on the r.h.s. as part of the unperturbed equation. To emphasize this, we put it on the l.h.s.:

$$\frac{d^2x}{dt^2} + (1 + \varepsilon\mu)x = -\varepsilon\left(x^3 - \mu x\right) \tag{10.40b}$$

We now consider as our unperturbed problem Equation (10.40b) with the r.h.s. = 0. Thus, to the lowest order, we write

$$\frac{d^2x}{dt^2} + (1 + \varepsilon\mu)x = 0 \tag{10.41}$$

Equation (10.41) can now be thought of as representing an oscillator whose natural frequency is not equal to 1. Thus we introduce a new time t' so as to rescale the frequency to $\omega_0 = 1$:

$$t' = \sqrt{1 + \varepsilon\mu}\, t; \qquad \frac{d^2}{dt^2} = (1 + \varepsilon\mu)\frac{d^2}{dt'^2} \tag{10.42}$$

Equation (10.40b) becomes

$$\frac{d^2x}{dt'^2} + x = -\frac{\varepsilon(x^3 - \mu x)}{1 + \varepsilon\mu} \tag{10.43}$$

We expand the r.h.s. of Equation (10.43), retaining terms $O(\varepsilon)$, and note that the denominator term $1 + \varepsilon\mu$ does not contribute.

$$\frac{d^2x}{dt'^2} + x = -\varepsilon(x^3 - \mu x) + O(\varepsilon^2) \tag{10.44}$$

We now develop our perturbation expansion as we have done previously. We write

$$x(t') = x_0(\varepsilon, \varepsilon\mu, t') + \varepsilon x_1(\varepsilon, \varepsilon\mu, t') + O(\varepsilon^2) \tag{10.45}$$

We include as separate terms in the arguments both ε and $\varepsilon\mu$ in order to emphasize that we have an expansion in two parameters. We will shortly see their interrelationship. We have

$$O(\varepsilon^0): \qquad \frac{d^2x_0}{dt'^2} + x_0 = 0 \tag{10.46}$$

The solution is

$$x_0 = A_0 \cos t' = A_0 \cos\left(\sqrt{1 + \varepsilon\mu}\, t\right) \tag{10.47}$$

$$\begin{aligned} O(\varepsilon^1): \qquad \frac{d^2x_1}{dt'^2} + x_1 &= -\left(x_0^3 - \mu x_0\right) \\ &= -A_0^3\cos^3 t' + \mu A_0 \cos t' \\ &= -A_0 \cos t' \left(\tfrac{3}{4}A_0^2 - \mu\right) - \tfrac{1}{4}A_0^3 \cos 3t' \end{aligned} \tag{10.48}$$

We know from our previous calculations of the need to cancel or suppress the terms that have $\cos t'$ dependence since these lead to secular terms. This leads to a choice for our free parameter μ. Let

$$\mu = \tfrac{3}{4}A_0^2 \tag{10.49}$$

With this choice, Equation (10.48) takes the form

$$\frac{d^2x_1}{dt'^2} + x_1 = -\tfrac{1}{4}A_0^3\cos 3t' \tag{10.50}$$

and the solution is

$$x_1(t') = A_1\cos t' + \frac{A_0^3}{32}\cos 3t' \tag{10.51}$$

Note that x_1 satisfies the initial conditions:

$$x_1(0) = A_1 + \frac{A_0^3}{32}; \qquad \frac{dx_1(0)}{dt'} = 0 \tag{10.52}$$

We combine x_1 and x_0 and express them both in terms of the true time t, retaining terms $O(\varepsilon)$:

$$x(t) = A_0\cos\left\{\left[1 + \varepsilon\tfrac{3}{8}A_0^2 + O(\varepsilon^2)\right]t\right\}$$
$$+\varepsilon A_1\cos t + \frac{\varepsilon A_0^3}{32}\cos 3t + O(\epsilon^2) \qquad \text{for } t = O(1) \tag{10.53}$$

It is important to note that we have correctly omitted a frequency correction in the argument of the $\cos 3t$ term. We have to satisfy the initial conditions correct to $O(\varepsilon)$:

$$x(0) = A_0 + \varepsilon\left(A_1 + \frac{A_0^3}{32}\right) + O(\varepsilon^2) = A \tag{10.54}$$

We choose

$$A_1 = -\frac{A_0^3}{32}$$

and then have

$$A_0 = A + O(\varepsilon^2)$$

Thus, the amplitude A is the same as the maximum displacement. The zero-velocity initial condition is satisfied by the construction of the solution.

With these choices, we see that Equation (10.53) becomes

$$x(t) = A\cos\{[1 + \varepsilon\tfrac{3}{8}A^2]t\} + \frac{\varepsilon A^3}{32}(\cos 3t - \cos t) + O(\varepsilon^2) \quad \text{for } t = O(1)$$

which is identical to Equation (10.37).

EXERCISE 5.1 Repeat the analysis of the Duffing oscillator, beginning with Equation (10.40a), using the initial conditions given by Equation (10.38).

$$x_0(0) = A; \qquad \dot{x}_0(0) = 0$$
$$x_i(0) = \dot{x}_i(0) = 0 \qquad \text{for } i \geq 1$$

This is often called the zero-zero initial conditions. ■

In this section, we have introduced an alternative way of viewing the Poincaré–Lindstedt method. It is clearly effective in any problem that has a perturbation term that is a nonlinear polynomial. It eliminates the linear term in the perturbation by incorporating it as an adjustment to the spring constant or frequency of the fundamental. Then, in the perturbation expansion, there is no term on the r.h.s. of the equation of the form $\cos t$. If we continue our expansion to the next order, we again introduce a free linear term that suppresses the secular generating terms in that order. It is incorporated into the adjusted spring constant as we did in our example. This method continues to work as we compute higher orders and generates the higher-order corrections to the fundamental frequency.

EXERCISE 5.2 It is straightforward to go through the analysis that we gave for the Duffing oscillator for the equation

$$\frac{d^2x}{dt^2} + x = -\varepsilon x^5; \qquad 0 < \varepsilon \ll 1$$
$$x(0) = A; \qquad \dot{x}(0) = 0$$

Everything proceeds in a like manner and there is a frequency correction equal to $+\varepsilon\frac{5}{16}A^4$. Find $x(t)$ with an error $O(\varepsilon^2)$ for $t = O(1)$. See Exercise 2.3.

This type of calculation is typical for problems with odd-polynomial nonlinearities. ■

10.6 CONCLUSIONS

Let us summarize our approach:

(1) We note that our differential equation represents the motion of a system executing bounded motion in a one-dimensional conservative force field; the motion is periodic.

(2) We expect the period of the motion to be altered and we are concerned that if we view the trajectory by means of a clock flashing at the unperturbed frequency, we will fail to see the system as a fixed point. Rather, it will slowly drift, thereby masking the periodic nature of the motion.

(3) Thus, we develop an expansion that requires that the periodicity be maintained in each order of the perturbation expansion.

(4) We want the new trajectory to be close to the unperturbed trajectory, so we require both the displacement $x(t)$ and the velocity dx/dt to be close to their unperturbed values.

(5) We have to be careful (or consistent) in the manner that we choose to satisfy our initial conditions. We can, for example, choose to satisfy the initial conditions *approximately* to each order of calculation in the form

$$x(0) = A = A_0 + \varepsilon A_1 + \text{h.o.t.}$$

Thus, we obtain the true amplitude A as a power-series expansion.

Alternatively, we can satisfy the initial conditions *exactly* by the lowest-order solution x_0 by introducing zero-zero initial conditions.

These options regarding the choice of initial conditions and their implications are explained at some length in Chapter 14.

(6) Finally, we emphasize that our expansion has limited time validity. This follows since our procedure is to obtain an approximation to the periodic solution and the exact frequency as a power-series expansion that intertwines the small parameter ε with the time t within the argument of trigonometric functions. Then, after a sufficiently long time, the approximation to the closed phase-plane orbit will open. This is a fundamental aspect of our general approach and occurs even though we have developed an expansion that is uniformly valid in time. These points will be discussed in later chapters and play a central role in the error analysis that is given in Chapter 15.

CHAPTER 11

NONCONSERVATIVE SYSTEMS

In Chapter 10, we concentrated our attention on studying conservative systems that were close to simple harmonic oscillators. By this we mean that over each cycle of true motion, the system looks like a SHO. Then there are correction terms to the fundamental frequency and the introduction of harmonics. When we describe the motion in terms of flashing clocks, we say that there is a fast component associated with the natural, or unperturbed, frequency and a small frequency correction that we have to include in order to see the system as a fixed point.

In this chapter, we introduce a discussion of systems that have damping and, consequently, a change in energy. In addition, we introduce the concept of a limit cycle, a new phenomenon associated with nonlinear systems.

11.1 DAMPED HARMONIC MOTION

We restrict our attention to systems that are lightly or slightly damped and by this we mean that there is very little change in the amplitude over a time scale of the order of the unperturbed period. Thus, there is never a fast component associated with the damping. This situation prevails in a variety of physical systems, e.g., the damped simple pendulum of small amplitude, an RLC circuit, etc.

Notice that this is not a small change in the displacement, but a small change in the amplitude, or envelope. This requires us to find the proper frame in which to observe this small damping term. Consider the damped linear oscillator:

$$\frac{d^2x}{dt^2} + x = -\varepsilon\frac{dx}{dt}; \qquad 0 < \varepsilon \ll 1$$

$$x(0) = A; \qquad \frac{dx(0)}{dt} = -\frac{\varepsilon A}{2} \tag{11.1}$$

There are standard methods to find the exact solution to this equation. For example, one can assume a solution of the form:

$$x(t) = \exp mt$$

and substitute it into the equation and solve for m. One then has a quadratic equation that yields

$$m = -\left(\frac{\varepsilon}{2}\right) \pm i\sqrt{1 - \frac{\varepsilon^2}{4}}$$

Thus, since our parameter ε is small, we see that the exact solution

$$x(t) = A\exp\left(-\frac{\varepsilon t}{2}\right)\cos\left(\sqrt{1 - \frac{\varepsilon^2}{4}}\,t\right) \tag{11.2}$$

is expressed as a product of a slowly decaying envelope term and an oscillatory part. Alternatively, we can say that our solution has two components:

Solution = slow variable factor * fast variable factor

In light of the nature of the solution, it is often advantageous to look at the average energy rather than the displacement. This is a reasonable approach for the linear oscillator and also proves to be quite satisfactory in the nonlinear case. Let us see how this is done by analyzing the equation for the damped linear oscillator.

Multiply both sides of Equation (11.1) by dx/dt:

$$\frac{dx}{dt}\frac{d^2x}{dt^2} + \frac{dx}{dt}x = -\varepsilon\left(\frac{dx}{dt}\right)^2 \tag{11.3}$$

The l.h.s. of Equation (11.3) is the time derivative of the energy E:

$$\frac{dE}{dt} = \frac{d}{dt}\left[\frac{1}{2}\left(\frac{dx}{dt}\right)^2 + \frac{1}{2}x^2\right] \tag{11.4}$$

We have as the initial value of the total energy, neglecting a term of $O(\varepsilon^2)$,

$$E(\text{initial}) = \tfrac{1}{2}[A(\text{initial})]^2$$

11.1.1 Simplifying Assumptions

We divide the time of the motion into intervals equal to the period of the unperturbed oscillator. Thus, we think of our time derivative as recording changes on a time scale associated with this period. It is then consistent to assume that the amplitude and the energy of the system vary very slowly during this time interval. Thus, during the entire time of the motion, we write

$$E = \tfrac{1}{2}A^2(t)$$

where A is the slowly decaying amplitude. So we are assuming that our measuring device is not sensitive enough to follow the detailed motion of the system, but, rather, it uses a clock that moves at a rate such as to restrict measurements to the time scale of the unperturbed oscillator. In this process, the rapidly varying components of the motion are missed and we restrict our interpretation of the problem to the slowly varying aspects of the motion.

Now, returning to Equation (11.3), we observe that the small parameter ε is a multiplier of the r.h.s.. Thus, we approximate the velocity dx/dt by its unperturbed form:

$$\frac{dx}{dt} = -A \sin t$$

We use the unperturbed form to indicate that even though the amplitude decays in time, on the time scale $t' = \omega_0 t$, the velocity is well approximated by $-A \sin t$. This makes sense if we write the formal solution to Equation (11.1) as

$$x(t) = A \cos t - \frac{\varepsilon A}{2} \sin t - \varepsilon \int_0^t \frac{dx(t')}{dt'} \sin(t - t')\, dt' \tag{11.5}$$

We then have

$$\frac{dx(t)}{dt} = -A \sin t - \frac{\varepsilon A}{2} \cos t - \varepsilon \int_0^t \frac{dx(t')}{dt'} \cos(t - t')\, dt' \tag{11.6}$$

Now let us see how much $dx(t)/dt$ changes in a time interval equal to 2π. We compute

$$\frac{dx(t + 2\pi)/dt - dx(t)/dt}{2\pi} = -\frac{\varepsilon}{2\pi} \int_t^{t+2\pi} \frac{dx(t')}{dt'} \cos(t - t')\, dt'$$

Note that the range of the integral is equal to 2π and that the small parameter ε is a multiplier of the r.h.s.. Thus, when we replace $dx(t')/dt'$ by its unperturbed value $-A\sin t'$, we are making an error of $O(\varepsilon)$. With this substitution, the integrals can be evaluated and our equation becomes

$$\frac{dx(t+2\pi)/dt - dx(t)/dt}{2\pi} \approx +\frac{\varepsilon A}{2}\sin t$$

So we confirm that the velocity change during a cycle is $O(\varepsilon)$. (This result is valid for short times only.)

EXERCISE 1.1 Show that

$$\frac{x(t+2\pi) - x(t)}{2\pi} \approx -\frac{\varepsilon A}{2}\cos t$$

Thus, the amplitude changes by $O(\varepsilon)$ during a time interval $t = 2\pi$. ■

With these ideas in place, we write the term $(dx/dt)^2$ occurring on the r.h.s. of Equation (11.3) as

$$\left(\frac{dx}{dt}\right)^2 = A^2\sin^2 t = \frac{A^2}{2}(1 - \cos 2t)$$

and using Equation (11.4), we have

$$\frac{dE}{dt} = -\frac{\varepsilon A^2}{2}(1 - \cos 2t) \tag{11.7a}$$

or

$$\frac{dE}{dt} = -\varepsilon E(1 - \cos 2t) \tag{11.7b}$$

We note that the term in parentheses on the r.h.s. of Equation (11.7b) consists of a constant piece and a wiggly piece. In this regard, we recall that we are really only making observations on time intervals associated with a period of the unperturbed motion. Then, the wiggly term is effectively equal to zero since the average value of $\cos 2t$ over this interval is zero. We then write Equation (11.7b) as

$$\frac{dE(\text{smooth})}{dt} = -\varepsilon E(\text{smooth})$$

The solution is

$$E(\text{smooth}) = E(\text{initial})\exp(-\varepsilon t)$$

The energy decays exponentially, and since the amplitude is proportional to the square root of the energy, it also decays exponentially:

$$A(\text{smooth}) = A(\text{initial}) \exp(-\varepsilon t/2)$$

We will see that this approach works satisfactorily when we deal with weakly damped nonlinear systems.
Note that it is straightforward to find the exact solution to Equation (11.7b). It is

$$E(t) = E(0)\exp(-\varepsilon t)\exp[(\varepsilon/2)\sin 2t] \tag{11.7c}$$

The oscillatory component is bounded:

$$\exp(-\varepsilon/2) \le \exp[(\varepsilon/2)\sin 2t)] \le \exp(\varepsilon/2)$$

and the dominant behavior is given by the decaying exponential. See Figure 11.1.

Observe that if $t = O(1)$ both $\exp(-\varepsilon t)$ and $\exp[(\varepsilon/2)\sin t]$ differ by terms of $O(\varepsilon)$ from their value at $t = 0$. In other words, for short times, they can be considered to be of comparable importance. However, when $t = O(1/\varepsilon)$, the $\exp(-\varepsilon t)$ term has yielded a cumulative effect while the oscillatory factor has only contributed wiggles. It is one of our goals in the development of approximation schemes to identify and separate terms of these types and to primarily concentrate our attention on the ones that are important over long times.

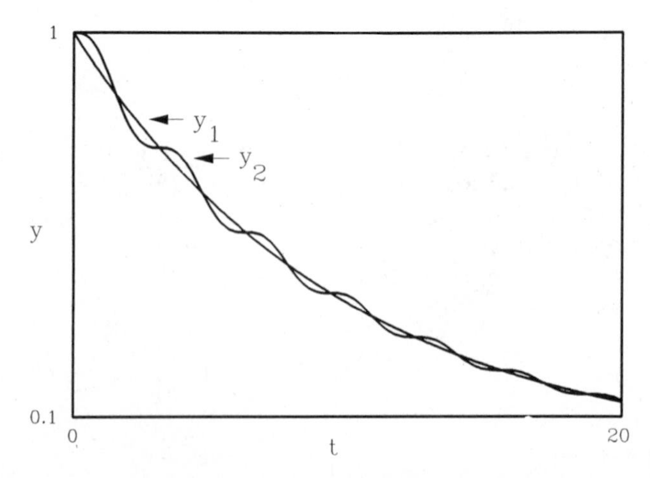

Figure 11.1 Observe how the dominant behavior is given by the smoothly varying component of the decay. $y_1 = \exp(-0.1t)$; $y_2 = \exp(-0.1t)\exp(0.05\sin 2t)$.

Example 1.1 Let us look at an oscillator with a cubic damping term:

$$\frac{d^2x}{dt^2} + x = -\varepsilon\left(\frac{dx}{dt}\right)^3; \qquad 0 < \varepsilon \ll 1$$

$$x(0) = A; \qquad \frac{dx(0)}{dt} = 0 \tag{11.8}$$

We introduce the same assumptions that we used for the damped linear oscillator. Thus, since the damping term is weak, we find it convenient to identify the total energy and then assume that it varies only slightly over an interval equal to a period of the unperturbed oscillator. We multiply each side by the velocity dx/dt and have

$$\frac{1}{2}\frac{d}{dt}\left[\left(\frac{dx}{dt}\right)^2 + x^2\right] = \frac{dE}{dt} = -\varepsilon\left(\frac{dx}{dt}\right)^4$$

We have, in the lowest order of approximation, that

$$x = A\cos t; \qquad \frac{dx}{dt} = -A\sin t$$

and

$$\left(\frac{dx}{dt}\right)^4 = A^4\sin^4 t = A^4\frac{1}{8}(3 + \cos 4t - 4\cos 2t)$$

We are interested only in the smooth part of the energy, and thus we write

$$\frac{dE(\text{smooth})}{dt} = -\frac{3\varepsilon A^4}{8} = -\frac{3\varepsilon}{2}E(\text{smooth})^2 \tag{11.9}$$

We solve Equation (11.9), denoting the initial energy by $E(0)$:

$$\frac{1}{E(0)} - \frac{1}{E} = -\frac{3}{2}\varepsilon t \tag{11.10}$$

or

$$E(\text{smooth}) = \frac{E(0)}{1 + (3\varepsilon t/2)E(0)} \tag{11.11}$$

Finally, since the amplitude is proportional to the square root of the energy, we have

$$A(\text{smooth}) = \frac{A(0)}{\sqrt{1 + (3\varepsilon t/4)A(0)^2}} \tag{11.12}$$

We see that the term $(3\varepsilon t/4)A(0)^2$ is a slow time-dependent correction to the initial amplitude $A(0)$. This gives us an idea of what is happening and yields the correct functional form for the decay of a cubic damped oscillator. Let us recapitulate: our approach to lightly damped systems is to concentrate our attention on the amplitude, or the energy, as the slowly changing variable rather than on the displacement. Furthermore, we consider only the smoothly varying components. ■

EXERCISE 1.2 Consider a nonautonomous system described by

$$\ddot{x} + x = -(4\varepsilon\cos^2 t)\dot{x}; \qquad x(0) = A, \quad \dot{x}(0) = 0, \quad \varepsilon = \frac{1}{100\pi}$$

Introduce the concept of E(smooth) and show that it decays exponentially with a factor $-\varepsilon t$. Change the initial conditions to $x(0) = 0$; $\dot{x}(0) = A$, and show that E(smooth) now has an exponential decay factor equal to $-3\varepsilon t$. Plot x_{max} versus t for $0 \leq t \leq 50\pi$. Discuss your results. ■

11.2 LIMIT CYCLES: A NONLINEAR PHENOMENON

Limit cycles are a new phenomenon that is characteristic of nonlinear systems, and they arise in a variety of contexts. We restrict our attention to stable limit cycles that are generated from weakly nonlinear oscillatory systems since this is the area of our fundamental concern. We first define limit cycles and then illustrate the basic concepts by means of a simple example due to Poincaré. We repeat ourselves a number of times so as to firmly keep in mind the fundamental aspects of the phenomena and the analysis. There are a variety of systems that exhibit limit-cycle behavior; for example:

(a) In Rayleigh's paper entitled "On Maintained Vibrations," *Philosophical Magazine*, XV (1883), he developed an approximation procedure for handling maintained vibrations that resulted from the interplay of negative and positive damping terms. He mentions that this situation occurs in a variety of systems, including organ pipes, violin strings, electromagnetic tuning forks, etc.

(b) Van der Pol, in a paper entitled "On Oscillation Hysteresis in a Simple Triode Generator," *Philosophical Magazine*, 43, 700 (1922), developed some of the seminal arguments that led to a systematic attack on limit-cycle problems. He was studying multivibrator circuits in which the resistive element altered its characteristics as a function of the current. The circuit exhibited sustained oscillations that were stable against small perturbations, leading to the concept of negative-feedback amplifiers. Sometimes, oscillators of this type are called self-excited oscillators because, if the system initially has a very small energy, it draws upon its environment so as to gain energy and increase its amplitude.

One can show the equivalence of the Rayleigh oscillator and the Van der Pol oscillator. (We discuss this in Chapter 12.)

It is easy to find a wealth of examples by consulting the appropriate texts cited in the Bibliography at the end of Chapter 15.

We note that Example 2.1, due to Poincaré, is not a perturbed simple harmonic oscillator. We analyze it in some detail because it presents in an economical fashion many of the central concepts associated with limit cycles. (We rewrite, in Exercise 2.1, the basic equations of Example 2.1, so that it becomes a perturbed simple harmonic oscillator.)

11.2.1 Definition of a Limit Cycle

A stable limit cycle is a closed curve in the phase plane that has no other closed curves arbitrarily close to it and that acts as an attractor for all trajectories from inside and outside. You can picture the trajectories as spirals that wind in or out and have a limit that they can come arbitrarily close to without actually reaching. In other words, one can think of the limit cycle as the central line in an annulus. See Figure 11.2.

Once the spiraling trajectory has entered the annulus, it cannot escape. Remember that trajectories in the phase plane cannot cross.

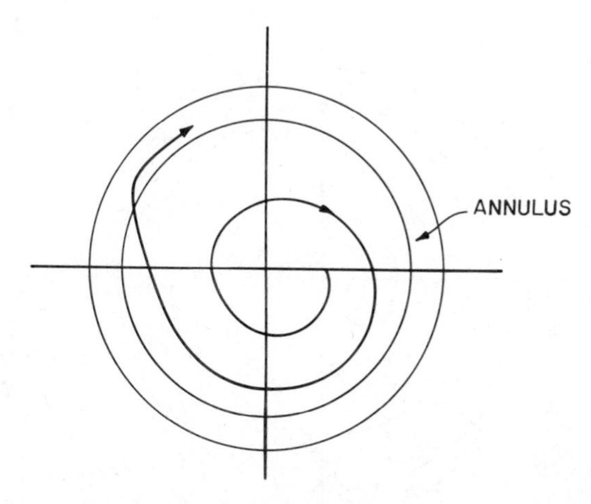

Figure 11.2 Once the spiral enters the annulus that surrounds the limit cycle it cannot escape. In addition, we know that trajectories in the phase plane cannot cross. The size of the annulus is given by the error in our approximate solution. However, since the limit-cycle solution is unknown to us, the shape of the annulus eludes our desire to picture it.

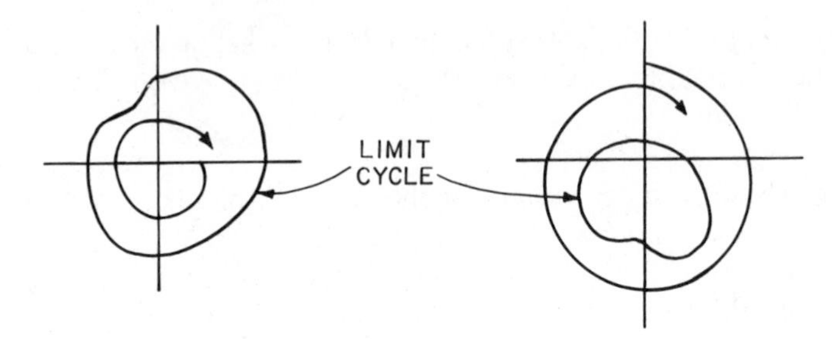

Figure 11.3

In Figure 11.3, we give two examples of possible limit cycles and the shapes of the approaching trajectories.

Note that we defined stable limit cycles. There are also unstable limit cycles that are repellers of all nearby trajectories as well as situations where there are multiple limit cycles. See Example 2.3. See Figures 11.4*a* and 11.4*b* for examples of an unstable limit cycle and of multiple limit cycles.

Hereafter, unless we say to the contrary, we consider only situations in which we have the possibility of generating a single stable limit cycle and we then omit the word "stable" when referring to a limit cycle.

With these restrictions understood, we proceed to discuss the situations we are interested in. We first outline our strategy and then go back over it point

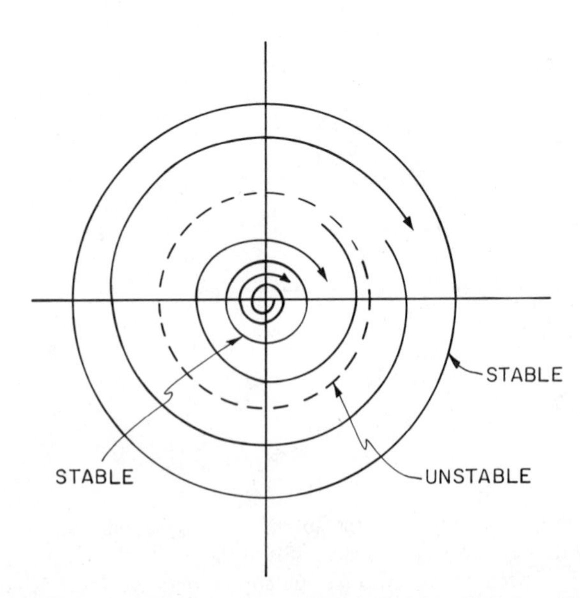

Figure 11.4*a* We depict multiple limit cycles that alternate being stable and unstable. Unstable cycles repel approaching trajectories, whereas stable ones attract them.

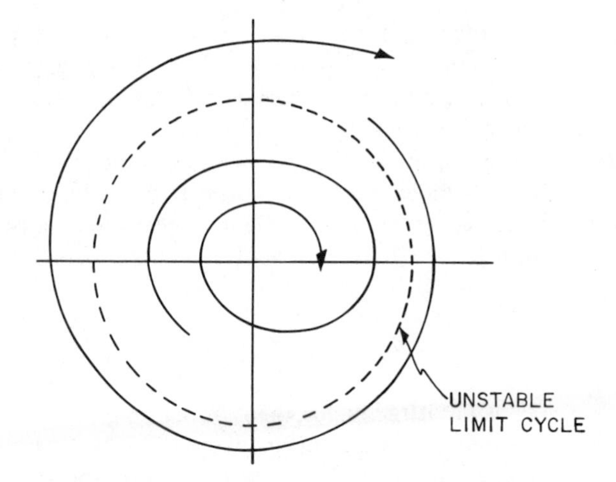

Figure 11.4*b* An example of an isolated unstable limit cycle.

by point and finish this discussion by analyzing a simple clear model that yields a limit cycle.

We are concerned with limit cycles that arise due to a perturbation of a simple harmonic oscillator. Thus, we consider a situation in which our unperturbed problem has neutral stability, i.e., purely imaginary eigenvalues. Thus, writing the basic equations of the SHO as a set of coupled first-order equations, we have

$$\frac{dx}{dt} = y \tag{11.13a}$$

$$\frac{dy}{dt} = -x \tag{11.13b}$$

We can think of the unperturbed system as described by a circle in the phase plane. We then introduce a perturbation that has linear and nonlinear velocity-dependent terms of appropriate signs.

Initially, we restrict ourselves to behavior that is generated from the origin $(0, 0)$ or zero amplitude by a small perturbation. What is the effect of the perturbation?

Since the initial amplitude is small, i.e., near the origin $(0, 0)$, the effect of the perturbation due to the linear velocity term is larger than that due to the nonlinear terms. Thus, we begin with the linear term since it introduces a real part to the eigenvalues and, hence, the associated trajectories will have changed from circles to spirals. The spirals will either be attracted or repelled from the origin, depending upon the sign of the parameter that multiplies the linear term. (Refer to Figures 9.4 to 9.6 and the associated material.)

If the spiral is attracted to the origin, we say that the perturbation introduces positive damping and that excitations decay toward the zero amplitude. In this case, the nonlinear terms do not play a significant role. On the other hand, if the spiral unwinds outward from the origin, we say that the perturbation introduces negative damping and the amplitude continues to grow until it is sufficiently large that the nonlinear terms come into play. Then, since there is a limit cycle, the net sign of the nonlinear terms must be such as to enable the system to have a finite radius of oscillation, denoted as r^*, such that

$$dr/dt = 0 \qquad \text{at } r = r^*$$

and such that it acts as an attractor for the growing amplitude of oscillation. (The limit cycle is a circle in lowest approximation. Higher-order corrections lead to harmonics.) The limit cycle is not an equilibrium point since neither dx/dt nor dy/dt is equal to zero there.

Let us say all of this once again in order to make certain that it is clear.

(1) Our unperturbed problem, given by Equations (11.13a) and (11.13b), is always the simple harmonic oscillator. It has purely imaginary eigenvalues:

$$\frac{d}{dt}\begin{bmatrix} x \\ y \end{bmatrix} = M_0 \begin{bmatrix} x \\ y \end{bmatrix}; \qquad M_0 = \begin{bmatrix} 0 & 1 \\ -1 & 0 \end{bmatrix}$$

(2) Our perturbation has a term that is linear in the velocity $\varepsilon\, dx/dt = \varepsilon y$. It may also have an amplitude term εx that leads to a frequency correction.

(3) The coefficient of the linear perturbation term in the velocity is called a bifurcation parameter since the characteristic structure of the emerging trajectories changes as it changes its sign from negative to positive values. When it is negative, the eigenvalues of the matrix associated with the linear terms have a negative real part; and when it is positive, the eigenvalues have a positive real part.

$$M = \begin{bmatrix} 0 & 1 \\ -1 & \varepsilon \end{bmatrix} \qquad \text{or} \qquad \begin{bmatrix} \varepsilon/2 & 1 \\ -1 & \varepsilon/2 \end{bmatrix}$$

(See Exercise 2.1.) We have positive damping if $\varepsilon < 0$. We have negative damping if $\varepsilon > 0$.

(4) It is necessary for the perturbation to contain nonlinear terms in order to have a limit cycle. For example, if the relative sizes and signs of the terms are correct, you can obtain a limit cycle from the interplay between a linear term and a combination of quadratic terms or cubic

terms. For example, the terms could be

$$\varepsilon(y - x^2 y) \quad \text{or} \quad \varepsilon(y - y^3) \quad \text{or} \quad (\varepsilon^2 y - xy + y^2)$$

(This last case is discussed in Section 14.5.) It is also possible to generate a limit cycle from higher-order nonlinearities.

(5) Our system always starts at or near the equilibrium point that is the origin (0, 0). Its amplitude initially comes under the influence of the linear perturbation term. If the sign of this term is negative, the amplitude decays toward the equilibrium point. If the coefficient is positive, the amplitude grows until it is large enough to come under the influence of the nonlinear terms. We require that the effective sign of the nonlinear terms be negative in order to achieve a condition where the amplitude can stop growing and reach a stable oscillatory state. This state, when it is described in terms of phase-plane variables, is called a limit cycle.

11.2.2 An Example of a Limit Cycle

Let us illustrate these remarks by a simple example due to the French mathematician Henri Poincaré, one of the pioneers in the field of nonlinear dynamics. (A variant on this model is given in Exercise 2.1.)

Example 2.1 Consider an oscillator described by the following set of coupled first-order nonlinear differential equations:

$$\frac{dx}{dt} = \alpha(x + y) - \sigma(x - y)(x^2 + y^2) \tag{11.14a}$$

$$\frac{dy}{dt} = -\alpha(x - y) - \sigma(x + y)(x^2 + y^2) \tag{11.14b}$$

with the parameters α, $\sigma > 0$. We notice that there is an equilibrium point at the origin (0, 0). Rather than analyze the stability of the origin in Cartesian coordinates, we introduce polar coordinates since this disentangles the variables and yields an immediate solution. Thus, we write

$$x = r\cos\theta; \qquad \text{and} \qquad y = r\sin\theta$$

We return to our oscillator equations and multiply Equation (11.14a) by x and Equation (11.14b) by y and add:

$$x\frac{dx}{dt} + y\frac{dy}{dt} = r\frac{dr}{dt} = \alpha r^2 - \sigma r^4 \tag{11.15a}$$

Similarly, we multiply Equation (11.14b) by x and Equation (11.14a) by y and subtract:

$$x \frac{dy}{dt} - y \frac{dx}{dt} = r^2 \frac{d\theta}{dt} = -\alpha r^2 - \sigma r^4 \tag{11.15b}$$

Thus, our equations in polar coordinates have become

$$\frac{dr}{dt} = r(\alpha - \sigma r^2) \tag{11.16a}$$

yielding $dr/dt = 0$ when $r = 0$ and when

$$r = r^* = \sqrt{\frac{\alpha}{\sigma}} \tag{11.16b}$$

and also

$$\frac{d\theta}{dt} = -\alpha - \sigma r^2 \tag{11.16c}$$

Let us see what Equations (11.16a) to (11.16c) indicate. If we begin our system at $r = 0$, it remains there, but it is in unstable equilibrium. This is clear since, in the neighborhood of the origin, the linear terms dominate and we have $dr/dt = \alpha r + \text{n.l.t.}$ Then, since $\alpha > 0$, we conclude that the origin is an unstable equilibrium point. Thus, if we consider a small initial value for r, it moves away from the point $r = 0$ and its amplitude grows until it reaches the limiting value equal to r^*. This limiting value is called the limit-cycle radius.

The angular velocity has an initial value $-\alpha$, and then it approaches the value -2α as the radius approaches the limit cycle. We can express this result in Cartesian coordinates:

$$\frac{dx}{dt} = 2\alpha y; \qquad \frac{dy}{dt} = -2\alpha x \qquad \text{when } r = r^*$$

indicating that the x and y variables undergo harmonic oscillations when the system is on the limit cycle.

Let us just concentrate on the radial motion. We now temporarily relax our condition that the parameter α be positive. We think of it as initially being negative and then slowly crossing zero to positive values. We then study the stability of the equilibrium point $r = 0$ for small amplitudes as α crosses zero:

$$\frac{dr}{dt} \approx \alpha r + \text{n.l.t.}$$

If $r < r^*$, the limit cycle is approached from inner orbits. We can show by a similar argument that if $r > r^*$, the limit cycle is approached from outer

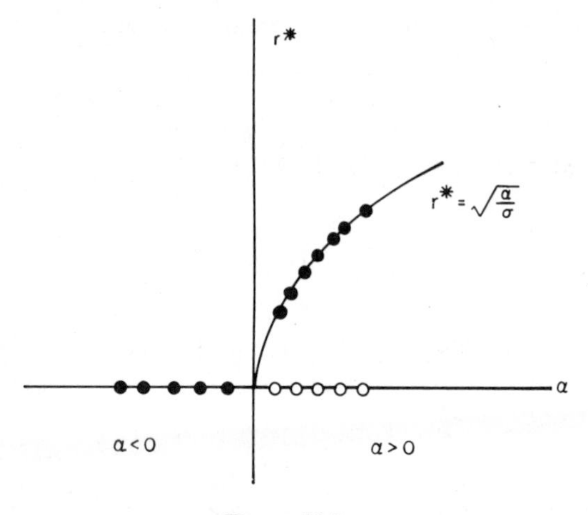

Figure 11.5

orbits. Thus, the limit cycle is thought of as an attractor. We say that the limit cycle is a stable attracting closed curve. Since trajectories in the phase plane cannot cross when

$$|r - r^*| < \delta$$

where δ is a small quantity of our choosing, we say that the system is on a limit cycle. You can think of δ as a measure of a close neighborhood of some desired value of the energy E.

We illustrate how the system behaves in Figure 11.5. We plot r^* versus α. We indicate stable equilibrium points by solid circles and unstable points by open circles. (This is standard notation.) Then we consider our system at rest and disturb it slightly and see what happens. If the equilibrium point is stable, the system returns to rest, i.e., the state where $r = 0$. If the equilibrium point is unstable, the radius of the envelope of oscillations grows. If a limit cycle is present, it attracts the system and holds it as a stable attractor. In the figure, we indicate only the radius of the oscillations.

In our example, we obtain a limit-cycle radius proportional to the square root of the bifurcation parameter α. It turns out that in a great many physical situations, this square-root behavior is characteristic of the onset of a limit cycle. This comes from the interplay of the linear and cubic velocity terms. In these situations, if you have an oscillatory system that exhibits limit-cycle behavior when a parameter crosses some critical value, the onset of the new motion is given by the square root of the appropriate parameter. We require α, $\sigma > 0$ in order to have an unstable origin and a stable limit cycle of finite radius. ■

In Chapter 12, we introduce the Van der Pol oscillator, the classic system used to study limit-cycle behavior. It is easy to understand physically and it exhibits all of the features that are inherent in nonlinear oscillatory systems, yet absent in linear systems.

Example 2.2 It is possible to show that the limit cycle that we discussed in Example 2.1 is stable. Consider Equation (11.16a):

$$\frac{dr}{dt} = r(\alpha - \sigma r^2)$$

Assume that the system starts near $r = 0$ and its amplitude grows until $r = r^*$. It is then slightly disturbed, which we indicate by introducing a small perturbation:

$$r = r^* + u; \qquad |u| \ll 1; \qquad r^* = \sqrt{\frac{\alpha}{\sigma}}$$

We have

$$\frac{du}{dt} = (r^* + u)\left[\alpha - \sigma(r^{*2} + 2r^*u + u^2)\right]$$

If we retain only the linear terms and use the value of r^*, we have

$$\frac{du}{dt} = -2\alpha u$$

indicating that small deviations from the limit-cycle radius decay exponentially for either sign of u. Thus, our limit cycle attracts deviations associated with both inner and outer orbits. ■

EXERCISE 2.1 Consider Example 2.1. Modify Equations (11.14a) and (11.14b) by the introduction of a small positive parameter ε, so that they become

$$\frac{dx}{dt} = \varepsilon x + y - \varepsilon(x - y)(x^2 + y^2)$$

$$\frac{dy}{dt} = -x + \varepsilon y - \varepsilon(x - y)(x^2 + y^2)$$

Note that εx and εy terms are both small dissipative terms. Our linearized equation is then in the form discussed in Chapter 7, Exercise 4.3. Introduce polar coordinates and find the limit cycle. Show that it is stable against small perturbations. ■

EXERCISE 2.2 There is an interesting paper by J. Shohat entitled, "On Van der Pol's and Related Nonlinear Differential Equations" (*Journal of Applied Physics, 15*, 568, 1944). In it, he discusses properties of the periodic solutions of equations of the form $d^2x/dt^2 + x = \varepsilon F(x)(dx/dt)$; $0 < \varepsilon \ll 1$, where the function $F(x)$ is a polynomial. Keep in mind that a periodic solution of this equation, with period T, has the property that $[x(t), dx(t)/dt]$ and $[x(t + T), dx(t + T)/dt]$ take on the same value at intervals spaced by T. Considering only periodic solutions, multiply the equation successively by 1, x, (dx/dt), integrate over a period T and obtain

(a) $\int_0^T x\,dt = 0$

(b) $\int_0^T \left(\frac{dx}{dt}\right)^2 dt = \int_0^T x^2\,dt$

(c) $\int_0^T F(x)\left(\frac{dx}{dt}\right)^2 dt = 0$

Give an interpretation of each of these results. Shohat's paper has an extensive discussion of perturbation methods that are applicable to this problem. This includes approximation schemes that are supposed to be valid for large values of ε. However, the last results have been seriously questioned by N. G. de Bruijn in a paper entitled "A Note On Van der Pol's Equation" (Phillips Research Report 1, 401, 1946). ■

Example 2.3 We introduce a model system that exhibits "jumps" and hysteresis, derived from an equation discussed by Minorsky on page 183 of his book entitled *Nonlinear Oscillations*. Consider a nonlinear system described by

$$\frac{dx}{dt} = -\alpha\sigma(x + y) + \beta\sigma(x + y)\sqrt{x^2 + y^2} - \sigma(x - y)(x^2 + y^2) \quad (11.17a)$$

$$\frac{dy}{dt} = +\alpha\sigma(x - y) - \beta\sigma(x - y)\sqrt{x^2 + y^2} - \sigma(x + y)(x^2 + y^2) \quad (11.17b)$$

where $\beta > 0$; $0 < \sigma \ll 1$; and α is a variable parameter. Let

$$x = r\cos\theta; \qquad y = r\sin\theta$$

Then

$$\dot{r} = -\sigma r(r^2 - \beta r + \alpha) \quad (11.18a)$$

$$\dot{\theta} = \alpha\sigma - \beta\sigma r - \sigma r^2 \quad (11.18b)$$

We have stationary values for r when $r = 0$ and when

$$r = \frac{\beta}{2} \pm \sqrt{\frac{\beta^2}{4} - \alpha} \tag{11.19}$$

(We require $r \geq 0$ since it is a radius.) In order to determine if a stationary value of r (denoted by r^*) is stable to small perturbations, we can either differentiate Equation (11.18a) with respect to r and evaluate it at $r = r^*$, or we can let

$$r = r^* + u; \qquad |u| \ll 1$$

and obtain

$$\frac{du}{dt} \approx -\sigma r^*(2r^*u - \beta u) = -\sigma r^* u(2r^* - \beta) \tag{11.20}$$

Since σ, $\beta > 0$, we have a stable limit cycle with radius r^* if $2r^* > \beta$. We also have, from Equation (11.18a), that small oscillations in the neighborhood of the origin are given by

$$\dot{r} \approx -\sigma r \alpha \tag{11.21}$$

and thus are stable if $\alpha > 0$ and unstable if $\alpha < 0$. We want to study the magnitude of the radius of the limit cycle as the parameter α varies from negative to positive values. In Figure 11.6a, we plot r^* versus α and denote stable values of r^* with solid circles and unstable values of r^* by open circles.

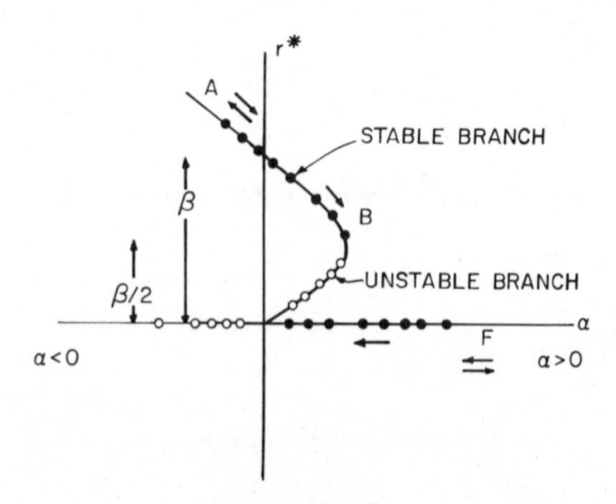

Figure 11.6a

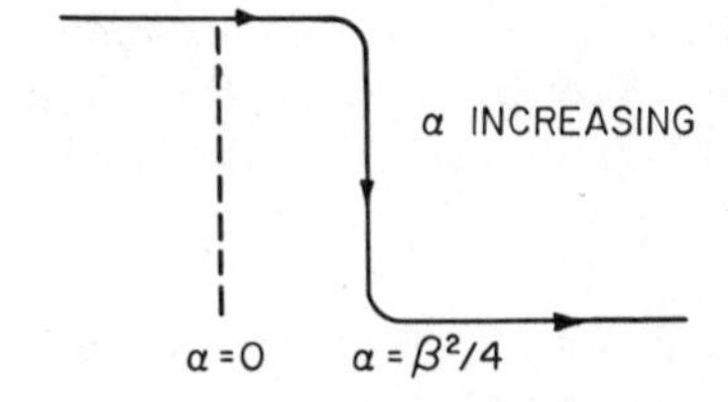

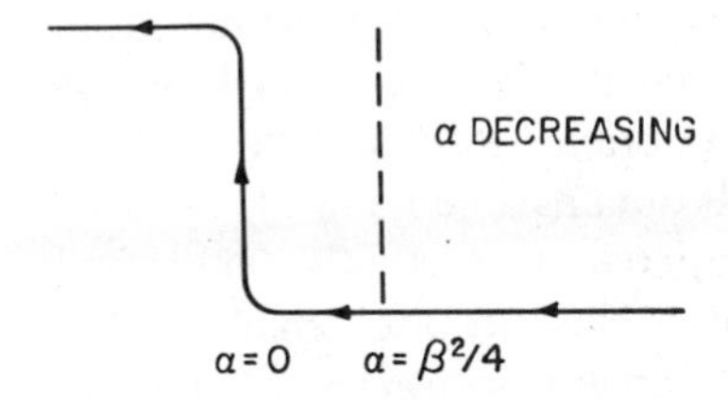

Figure 11.6b

Case I $\alpha > 0$

(i) From Equation (11.19), if $0 < \alpha < \beta^2/4$, there are two real roots.
We can see from Equation (11.20) that the larger one is stable and the smaller one is unstable.

$$r_1^* = \frac{\beta}{2} + \sqrt{\frac{\beta^2}{4} - \alpha} \tag{11.22a}$$

$$r_2^* = \frac{\beta}{2} - \sqrt{\frac{\beta^2}{4} - \alpha} \tag{11.22b}$$

(ii) $\alpha = \beta^2/4$: one real double root and the stable state is $r = 0$.

(iii) $\alpha > \beta^2/4$: no real roots.

Case II $\alpha < 0$

$$r^* = \frac{\beta}{2} + \sqrt{\frac{\beta^2}{4} + |\alpha|} \tag{11.23}$$

The origin is unstable and we have a stable limit cycle since $r^* > \beta/2$. ■

11.3 DISCUSSION OF FIGURES 11.6*a* and 11.6*b*

We want to follow the magnitude of the stable oscillations as we vary the parameter α from negative to positive values. We begin with $r \approx 0$ and $\alpha \ll 0$ and see what happens. From Equation (11.21), we see that the origin is unstable, and so the system draws energy from a reservoir and its radius increases until it reaches the limit-cycle value r^* given by Equation (11.23). If we continue to increase the parameter α past $\alpha = 0$, the radius, r^* is given by Equation (11.22a) and the system stays on the upper branch of the curve until it reaches the critical value $\alpha = \beta^2/4$. See Figure 11.6b.

At this value of α, since there are no real values of r^*, the radius collapses to $r = 0$. As the parameter α continues to increase, the fluctuations about $r = 0$ are damped and the steady state remains at $r^* = 0$. We now do the same analysis with α initially large and positive. The only stable value of r^* is $r^* = 0$. We decrease α past the value $\alpha = \beta^2/4$ and note that if our perturbation on the value of the radius is sufficiently small, the system stays on the $r^* = 0$ axis until $\alpha = 0$, at which time the lower branch loses stability and the system draws energy from the reservoir until it obtains a value of r^* given by the upper branch. We see how this system is a model that exhibits jumps and hysteresis. Note that when α is positive and very close to zero, small fluctuations enable the system to jump to the upper branch before $\alpha = 0$. Note that the lines with arrows indicate stable motion.

CHAPTER 12

THE METHOD OF AVERAGING (MOA)

Before we being the mathematical analysis, let us review two basic ideas that have been developed in the preceding chapters.

Periodic Motion

$$\frac{d^2x}{dt^2} + x = \varepsilon f(x); \qquad |\varepsilon| \ll 1$$

We expect the perturbation to change the period. The motion remains periodic, i.e., both the displacement and the velocity take on the same values after every time interval equal to the correct period of the motion. Also, the system appears as a fixed point when it is viewed with a flashing light that has been adjusted to the correct frequency. The amplitude is constant since energy is conserved.

We can develop a systematic procedure (i.e., an expansion that is free of secular terms), and in doing so, we can, in principle, find the correct frequency for a weakly nonlinear conservative oscillatory system. We note that for sufficiently long times, the phase-plane curve opens, indicating a breakdown in the usefulness of the expansion. The method is essentially that of Poincaré and Lindstedt.

Aperiodic Motion

$$\frac{d^2x}{dt^2} + x = \varepsilon f(x, y); \qquad |\varepsilon| \ll 1$$

We encounter weakly nonlinear oscillatory systems, each of which has a change in its amplitude (and hence in its energy), which is small during each period of the unperturbed oscillator. There is usually, but not always, a frequency correction. The following example is discussed in the text by Lin and Segel, page 326:

$$\frac{d^2x}{dt^2} + 2\varepsilon\frac{dx}{dt} + (1 + \varepsilon^2)x = 0 \tag{12.1}$$

with the solution

$$x(t) = x(0)\exp(-\varepsilon t)\cos(t + \phi) \tag{12.2}$$

In this example, the perturbed problem $\varepsilon \neq 0$ and the unperturbed problem have the same *oscillatory* time scale. Furthermore, the damping and oscillation are effectively decoupled. Of course, the damped oscillator is not periodic since the amplitude decreases in time.

We have great interest in limit cycles since they are stable periodic oscillations of finite amplitude, a phenomenon not present in linear systems. As we discussed in Chapter 11, if our dissipation term has the capacity to change sign as the system traverses its orbit in the phase plane, we have the potential to develop a limit cycle.

In our previous discussion of aperiodic systems, we relied primarily on qualitative arguments and postponed the development of a quantitative perturbation expansion until this chapter.

12.1 ORIENTATION AND INTRODUCTION OF OUR ASSUMPTIONS

A systematic asymptotic (long-time) perturbation technique has been developed over the past 50 years by Krylov and Bogoliubov and by Bogoliubov and Mitropolsky. We discuss the basic principles of the method, and you can always read the literature to obtain the details of the higher-order terms and the proofs, etc. In most cases, the lowest-order terms are the most interesting since they reveal the qualitative structure of the behavior of the system. There are three philosophically related methods.

12.1.1 The Elementary Method of Averaging (MOA)

This is the oldest of the three methods and is capable of correctly giving only the lowest-order corrections to the frequency and amplitude of the fundamental solution. Our unperturbed solution has both constant amplitude r and constant frequency ω_0.

The perturbation has the capacity to introduce slow time dependence in both the amplitude and the phase, and, thus, our perturbed solution might be written as

$$x(t) = r(\text{slow time dependence}) * \cos(\omega_0 t + \text{slow time dependence})$$

The MOA finds these slow time dependences. It is very easy to learn and straightforward to apply. We limit our discussion to this elementary method and refer to it as either the method of averaging, or MOA.

12.1.2 The Method of Krylov – Bogoliubov and Mitropolsky (KBM)

This method introduces a transformation of variables to the correct frequency or correct clock. It then seeks an expansion for the displacement in trigonometric functions that are periodic with period 2π. The amplitude and frequency are adjusted in each order of the perturbation so as to suppress secular terms. (See the texts by Krylov and Bogoliubov; Nayfeh; Minorsky; and Sanders and Verhulst in the Bibliography at the end of Chapter 15.)

12.1.3 The Method of Bogoliubov and Mitropolsky or the Method of Rapidly Rotating Phase (MRRP)

The MRRP seeks an expansion in polar coordinates for the amplitude or radius and the frequency. Thus, it is different from the KBM method that seeks a correct expression for the displacement and the frequency. The solutions can, with some significant algebraic manipulations, be transformed into one another. (See the texts cited in Section 12.1.2.)

12.1.4 The Basic Model

We always begin with a system that is a slightly perturbed SHO. The perturbation is written as $\varepsilon f(x, dx/dt)$. We organize our calculations so as to define an average radius that smoothly tracks the progress of the system on a time scale of the unperturbed motion. The average radius and average frequency change slowly since their derivatives are proportional to the smallness parameter. Thus, for example, if we have damped motion, we can picture our trajectory as a slowly winding or unwinding spiral that has superposed wiggles. See Figure 12.1.

The unsmooth or wiggly part represents the higher-order harmonics of the solution. They are usually defined in such a way that their average equals zero when integrated over a period of the unperturbed motion. Both the KBM and the MRRP are capable of finding the harmonics that are introduced by the perturbation. These terms can take the form of $\cos n\theta$ or $\sin n\theta$, $n \neq 1$, and are of importance in higher-order calculations and in treating resonance phenomena. (See the texts cited in Section 12.1.2.)

In this chapter, we seek the lowest-order approximation that is uniformly valid in time. Furthermore, the total effect of the perturbation must be such that during each time interval of length equal to the unperturbed period, the system can be viewed as a SHO. This condition on the perturbation is important. For example, consider a perturbation term containing a component that alters the effective spring constant and also has a negative damping

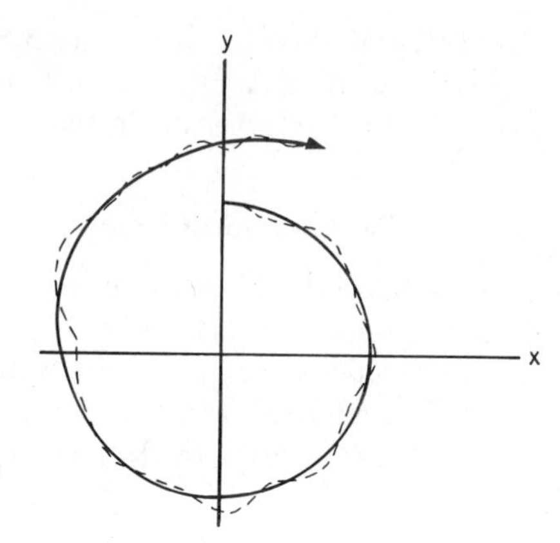

Figure 12.1 Keep in mind that we are only extracting the smooth part of the motion.

component:

$$\frac{d^2x}{dt^2} + x = \varepsilon\left(\frac{dx}{dt} + x^3\right); \qquad 0 < \varepsilon \ll 1 \tag{12.3}$$

The amplitude of the system increases during each cycle so that even though our parameter ε is small, the amplitude eventually becomes sufficiently large so that the nonlinear frequency correction term εx^3 is comparable to the linear term x. It is then no longer appropriate to consider our system as a slightly perturbed simple harmonic oscillator. (For a discussion of this point and related issues, see Chapter 14, Section 14.4.2, and the text by Kevorkian and Cole, pages 125–129.)

It is important to emphasize that, in most instances, the qualitative structure of the solutions is obtained in the lowest-order computation. As a word of caution, some systems require scaling of the variables in order to put the equation in the proper form. In these situations, a straightforward, i.e., careless, application of the method of averaging yields incorrect results. We discuss these points in Chapter 14.

12.2 THE METHOD OF AVERAGING (MOA)

We begin with our standard equation

$$\frac{d^2x}{dt^2} + x = \varepsilon f\left(x, \frac{dx}{dt}\right); \qquad |\varepsilon| \ll 1 \tag{12.4}$$

This yields a circle as the basic or unperturbed motion in the phase plane. In the MOA, we transform to polar coordinates and watch the changes in the radius and the frequency that result from the perturbation, $\varepsilon f(x, y)$. As discussed in Chapter 9, Section 9.3, there are two convenient sign conventions arising from the relationship between dx/dt and the variable y. We call them Convention I, in which we identify $dx/dt = y$, and Convention II, in which we identify $dx/dt = -y$. In both conventions, we say that the polar angle θ increases in time. Convention I yields clockwise motion and Convention II corresponds to counterclockwise motion.

The sign differences often play a mischievous role in that mistakes inadvertently enter the calculations. We exhibit the algebra explicitly, hoping thereby to show parallels in the arguments. The differences are only ones of sign.

12.2.1 Sign Convention I: *dx/dt = y*

$$x = r\cos\theta = r\cos(t+\phi) \tag{12.5a}$$

$$y = -r\sin\theta = -r\sin(t+\phi) \tag{12.5b}$$

$$\theta = t + \phi \tag{12.5c}$$

In general, both the amplitude r and the phase ϕ depend upon time. We want to construct our approximation procedure so as to retain the form given by Equations (12.5a) and (12.5b) since they maintain x and y on equal footing. We mean by this that we often want to characterize our second-order equation by a first-order vector equation in which the components are x and y. The relationship given by Equations (12.5a) and (12.5b) is a particularly simple and symmetric one.

We find a solution to Equation (12.4) by the method of variation of parameters. We recall from our discussion in Chapter 7 that the method begins by assuming the form of the solution of Equation (12.4) as given by Equations (12.5a) and (12.5b). We differentiate Equation (12.5a) and obtain

$$\frac{dx}{dt} = -r\sin(t+\phi) + \left[\frac{dr}{dt}\cos(t+\phi) - r\frac{d\phi}{dt}\sin(t+\phi)\right] \tag{12.6}$$

In order to have Equation (12.5b) with y identified as dx/dt, we impose a constraint condition:

$$\frac{dr}{dt}\cos(t+\phi) - r\frac{d\phi}{dt}\sin(t+\phi) = 0 \tag{12.7}$$

We then exhibit a solution as follows.

We differentiate Equation (12.6) and have

$$\begin{aligned}\frac{d^2x}{dt^2} = &-\frac{dr}{dt}\sin(t+\phi) - r\cos(t+\phi)\\ &- r\frac{d\phi}{dt}\cos(t+\phi)\end{aligned} \tag{12.8}$$

We substitute Equation (12.8) into Equation (12.4) and obtain

$$-\frac{dr}{dt}\sin(t+\phi) - r\frac{d\phi}{dt}\cos(t+\phi) = \varepsilon f(x, y) \tag{12.9}$$

We have two equations, Equations (12.7) and (12.9), for two unknowns, dr/dt and $d\phi/dt$, and we can solve and obtain

$$\frac{dr}{dt} = \dot{r} = -\varepsilon f(x, y)\sin(t+\phi) \tag{12.10a}$$

$$\frac{d\phi}{dt} = \dot{\phi} = \frac{-\varepsilon}{r} f(x, y)\cos(t+\phi) \tag{12.10b}$$

We pause to point out that we can obtain Equations (12.10a) and (12.10b) by simply writing Equation (12.4) in the form of two coupled first-order equations:

$$\frac{dx}{dt} = y \tag{12.11a}$$

$$\frac{dy}{dt} = -x + \varepsilon f(x, y) \tag{12.11b}$$

retaining Equations (12.5a) through (12.7).

We have immediately

$$x\dot{x} + y\dot{y} = r\dot{r} = \varepsilon f\left(x, \frac{dx}{dt}\right) y \tag{12.12a}$$

$$-x\dot{y} + y\dot{x} = r^2\dot{\theta} = r^2 - \varepsilon f\left(x, \frac{dx}{dt}\right) x \tag{12.12b}$$

Using Equations (12.5a) to (12.5c), we see that Equations (12.12a) and (12.12b) are equivalent to Equations (12.10a) and (12.10b), respectively. These equations, valid for a general perturbing term of the form $\varepsilon f(x, y)$, are due to Van der Pol (1922), but their systematic analysis is due to Krylov, Bogoliubov, and their co-workers. It is important to note that Equations (12.10a), (12.10b), (12.12a), and (12.12b) are exact since no approximations have been introduced.

12.2.2 Sign Convention II: $dx/dt = -y$

Introduce

$$x = r\cos\theta = r\cos(t+\phi) \tag{12.13a}$$

$$y = r\sin\theta = r\sin(t+\phi) \tag{12.13b}$$

$$\theta = t + \phi \tag{12.13c}$$

We follow the same procedures as in Section 12.2.1 by differentiating Equation (12.13a) and introducing the constraint equation

$$\frac{dr}{dt}\cos(t+\phi) - r\frac{d\phi}{dt}\sin(t+\phi) = 0 \tag{12.14}$$

This enables us to identify the variable y given by Equation (12.13b) with $-dx/dt$. We differentiate Equation (12.13a) again and obtain

$$\frac{d^2x}{dt^2} + x = -\frac{dr}{dt}\sin(t+\phi) - r\frac{d\phi}{dt}\cos(t+\phi)$$

or

$$-\frac{dr}{dt}\sin(t+\phi) - r\frac{d\phi}{dt}\cos(t+\phi) = \varepsilon f(x,-y) \tag{12.15}$$

We then solve Equations (12.14) and (12.15) and obtain

$$\frac{dr}{dt} = \dot{r} = -\varepsilon f(x,-y)\sin(t+\phi) \tag{12.16a}$$

$$\frac{d\phi}{dt} = \dot{\phi} = \frac{-\varepsilon}{r} f(x,-y)\cos(t+\phi) \tag{12.16b}$$

We can also obtain Equations (12.16a) and (12.16b) by writing Equation (12.4) as

$$\frac{dx}{dt} = -y \tag{12.17a}$$

$$\frac{dy}{dt} = x - \varepsilon f(x,-y) \tag{12.17b}$$

We then multiply Equation (12.17a) by x and Equation (12.17b) by y and obtain

$$x\dot{x} + y\dot{y} = r\dot{r} = -\varepsilon f(x,-y)\,y \tag{12.18a}$$

which upon division by r and the identification of $y = r\sin\theta$ given by Equation (12.13b) yields Equation (12.16a). We can then multiply Equation (12.17b) by x and Equation (12.17a) by y and subtract to obtain:

$$x\dot{y} - y\dot{x} = r^2\dot{\theta} = r^2 - \varepsilon f(x,-y)x \tag{12.18b}$$

If we divide by r^2 and use Equation (12.13c), we have Equation (12.16b).

Thus, we conclude that our two sign conventions, given by $dx/dt = +y$ and $dx/dt = -y$, yield equivalent equations. We will primarily use Convention II. We will, on occasion, discuss the results that are obtained using Convention I as well as using it in Example 3.2.

12.2.3 Basic Assumptions of the MOA

Now we come to our basic set of assumptions and the introduction of approximations. We use Sign Convention II and thus our basic relationships are given by Equations (12.13a) through (12.18b). We assume that our motion is close to being simple harmonic motion and that during a time interval equal to one period of the unperturbed motion, the radius r does not change appreciably. This is clear if we look, for example, at Equation (12.16a). The radial velocity $\dot{r}$ is proportional to the small parameter ε. The angular velocity $\dot{\theta}$ is measured in the phase plane by the change in angle that locates the point (x, y) on the trajectory. It consists of a rapidly varying part, corresponding to the unperturbed frequency $\omega_0 = 1$, and a slowly varying phase velocity, given by Equation (12.16b). The phase velocity $\dot{\phi}$ is proportional to ε and thus has a slow time variation.

We conclude that there are two time scales that characterize the motion. There is a fast scale of the angular motion $\dot{\theta} = 1$ that is always equal to the unperturbed frequency. We concentrate our attention on the slow scale of the radial motion and the phase change since it leads to cumulative effects while the rapidly varying terms are bounded for all time. Noting that the r.h.s. of Equations (12.16a) and (12.16b) are 2π-periodic functions with respect to the variable θ, we conclude that both dr/dt and $d\phi/dt$ are $O(\varepsilon)$ for times $t = O(1)$ and, consequently, change very little over intervals of length 2π. The method of averaging uses the 2π periodicity to bring about a separation of time scales. In a general problem, the first step involves finding a transformation of variables that puts the equation into a form suitable for application of the MOA. However, alas, it is not always clear that such a transformation exists or how to find it.

With these thoughts in mind, we determine the slow time dependence by integrating our equations of motion, given by Equations (12.16a) and (12.16b), over an interval of time equal to a period of the unperturbed motion, 2π. During the integration, we take the radius and the phase as constant since their variations only contribute in higher order.

$$\frac{d\bar{r}}{dt} = -\frac{\varepsilon}{2\pi}\int_0^{2\pi} f(\bar{r}\cos\theta, -\bar{r}\sin\theta)\sin\theta\, dt \tag{12.19a}$$

$$\frac{d\bar{\phi}}{dt} = -\frac{\varepsilon}{2\pi\bar{r}}\int_0^{2\pi} f(\bar{r}\cos\theta, -\bar{r}\sin\theta)\cos\theta\, dt \tag{12.19b}$$

We have divided our integrals by 2π. This defines the averaged variables. The overbar indicates that we are tracking the rate of change of the average radius and of the average phase. (Note that the radius and phase are taken as constants during the integrations; they are equal to their average values.)

We have from Equation (12.13c) that

$$\dot{\theta} = 1 + O(\varepsilon)$$

indicating that when the time goes from 0 to 2π, the angle θ also goes from 0 to 2π. We then rewrite Equations (12.19a) and (12.19b) as

$$\frac{d\bar{r}}{dt} = -\frac{\varepsilon}{2\pi}\int_0^{2\pi} f(\bar{r}\cos\theta, -\bar{r}\sin\theta)\sin\theta\, d\theta \tag{12.20a}$$

$$\frac{d\bar{\phi}}{dt} = -\frac{\varepsilon}{2\pi\bar{r}}\int_0^{2\pi} f(\bar{r}\cos\theta, -\bar{r}\sin\theta)\cos\theta\, d\theta \tag{12.20b}$$

Then our solution of Equation (12.4) is given by

$$x(t) = \bar{r}(t)\cos[t + \bar{\phi}(t)] = \bar{r}(t)\cos[\omega t + \bar{\phi}(0)]$$

where ω is the frequency of oscillation:

$$\omega = 1 + \dot{\bar{\phi}} \tag{12.21}$$

and $\bar{\phi}$ (0) is the initial phase.

These are the basic equations of the MOA. Often, the bar over the variables indicating the averaging process is left out since its omission seldom causes confusion.

12.3 EXAMPLES AND EXERCISES

In each of our examples, except Example 3.2, we use Sign Convention II in which we identify

$$\frac{dx}{dt} = -y$$

Example 3.1 Consider the Duffing oscillator:

$$\ddot{x} + x = -\varepsilon x^3; \qquad 0 < \varepsilon \ll 1$$
$$x(0) = a; \quad \dot{x}(0) = 0 \tag{12.22}$$

We have the equivalent set:

$$\dot{x} = -y \tag{12.23a}$$

$$\dot{y} = x + \varepsilon x^3 \tag{12.23b}$$

Our perturbation is a function of x alone, which means that we have a conservative system.

Let us see how the calculation of the change of the radius of the fundamental component of the oscillation proceeds. We have from Equations (12.18a), (12.19a), and (12.20a):

$$x\dot{x} + y\dot{y} = r\dot{r} = +\varepsilon x^3 y = +\varepsilon r^4 \cos^3\theta \sin\theta$$

$$\begin{aligned} \frac{d\bar{r}}{dt} &= +\frac{\varepsilon}{2\pi}\bar{r}^3 \int_0^{2\pi} \cos^3\theta \sin\theta \, dt \\ &= +\frac{\varepsilon}{2\pi}\bar{r}^3 \int_0^{2\pi} \cos^3\theta \sin\theta \, d\theta = 0 \end{aligned} \tag{12.24}$$

Thus, $\bar{r}$ is a constant equal to the initial amplitude a.

It is clear from this example that whenever our perturbation $\varepsilon f(x, dx/dt)$ is a polynomial in x, we have integrals of the form

$$\int_0^{2\pi} \cos^n\theta \sin\theta \, d\theta = 0 \Rightarrow \frac{d\bar{r}}{dt} = 0 \qquad \text{for any integer } n > 0 \tag{12.25}$$

Similarly, the equation for the change of the phase $\bar{\phi}$ follows from an application of Equations (12.18b), (12.19b), and (12.20b):

$$\begin{aligned} \frac{d\bar{\phi}}{dt} &= +\frac{\varepsilon}{2\pi}\bar{r}^2 \int_0^{2\pi} \cos^4\theta \, d\theta \\ &= +\frac{3\varepsilon}{8}\bar{r}^2 = +\frac{3\varepsilon}{8}a^2 \end{aligned} \tag{12.26a}$$

$$\bar{\phi} = +\frac{3\varepsilon}{8}a^2 t \tag{12.26b}$$

or using Equation (12.13c), we have

$$\theta = \left(1 + \frac{3\varepsilon}{8}a^2\right)t \tag{12.26c}$$

We previously obtained this result by the method of Poincaré and Lindstedt. (See Equation 10.35.)

Then our solution is written as

$$x = a\cos\left[\left(1 + \frac{3\varepsilon}{8}a^2\right)t\right] = a\cos\left[\omega t + \bar{\phi}(0)\right] \tag{12.27}$$

and the frequency of oscillation, ω, is

$$\omega = \left(1 + \frac{3\varepsilon}{8}a^2\right) \tag{12.28}$$

Our particular choice of initial conditions, given by Equation (12.22), is $\bar{\phi}(0) = 0$.

Our solution can be compared with Equation (10.36) in which we used the Poincaré–Lindstedt method to find both the fundamental and harmonic components of the motion. There is a trade-off. It is really very simple to use MOA and obtain a quick result. But, as has been mentioned, it misses the wiggles. ■

Example 3.2 We repeat our analysis of the Duffing oscillator as given by Equation (12.22) using Sign Convention I. (Note: This example just gives one more chance to review sign conventions. It surely is a bit of "overkill.") We begin by introducing

$$\frac{dx}{dt} = y \tag{12.29a}$$

$$\frac{dy}{dt} = -x - \varepsilon x^3 \tag{12.29b}$$

$$x = r\cos\theta \tag{12.30a}$$

$$y = -r\sin\theta \tag{12.30b}$$

We multiply Equation (12.29a) by x and Equation (12.29b) by y and add and obtain

$$x\dot{x} + y\dot{y} = r\dot{r} = -\varepsilon x^3 y = +\varepsilon r^4\cos^3\theta\sin\theta \tag{12.31}$$

Upon averaging, we get

$$\frac{d\bar{r}}{dt} = +\frac{\varepsilon}{2\pi}\bar{r}^3\int_0^{2\pi}\cos^3\theta\sin\theta\,dt$$

$$\frac{d\bar{r}}{dt} = +\frac{\varepsilon}{2\pi}\bar{r}^3\int_0^{2\pi}\cos^3\theta\sin\theta\,d\theta = 0 \tag{12.32}$$

We have the result that $\bar{r} = a =$ constant. We multiply Equation (12.29b) by $-x$ and Equation (12.29a) by y and add to obtain

$$-x\dot{y} + y\dot{x} = r^2\frac{d\theta}{dt} = r^2 + \varepsilon r^4\cos^4\theta. \tag{12.33}$$

Our equation for phase velocity becomes

$$\frac{d\phi}{dt} = +\varepsilon r^2\cos^4\theta \tag{12.34}$$

We then average and have

$$\frac{d\bar{\phi}}{dt} = +\frac{\varepsilon}{2\pi}\bar{r}^2\int_0^{2\pi}\cos^4\theta\,d\theta$$

$$= \frac{3\varepsilon}{8}\bar{r}^2 = +\frac{3\varepsilon}{8}a^2 \tag{12.35a}$$

$$\bar{\phi} = +\frac{3\varepsilon}{8}a^2 t \tag{12.35b}$$

We obtain the same frequency correction that we had in Equation (12.26b).

Recall that the angle θ is measured as increasing in a clockwise direction. That is to say, it is the negative of the angle θ that we refer to in Sign Convention II. ■

EXERCISE 3.1 Consider

$$\ddot{x} + x = -\varepsilon x^{2n+1}; \qquad 0 < \varepsilon \ll 1 \tag{12.36}$$

It is clear from our analysis of the Duffing oscillator that the radius is constant and that there is a frequency correction that comes from the integral

$$\frac{1}{2\pi}\int_0^{2\pi}\cos^{2n+2}\theta\,d\theta \tag{12.37}$$

Evaluate it and conclude that the corrected frequency is

$$\omega = 1 + \frac{\varepsilon\bar{r}^{2n}}{\pi}\frac{\Gamma\left(n+\frac{3}{2}\right)\Gamma\left(\frac{1}{2}\right)}{\Gamma(n+2)}$$

Simplify this to obtain

$$\omega = 1 + \frac{\varepsilon\bar{r}^{2n}(2n+1)!}{(n+1)!\,n!\,2^{2n+1}} \tag{12.38}$$

Observe that we have

$$n = 1: \qquad \omega = 1 + \frac{3\varepsilon}{8}\bar{r}^2 \tag{12.39}$$

$$n = 2: \qquad \omega = 1 + \frac{5\varepsilon}{16}\bar{r}^4 \quad ■ \tag{12.40}$$

Note that the perturbation in Equation (12.36) is an odd polynomial in x. This leads us to consider a general perturbation of the form $\varepsilon f(x)$, and upon expanding $f(x)$ in a Fourier series, we have

$$f(x) = \sum_{n=0}^{\infty} c_{2n+1} \cos(2n+1)\theta + \sum_{n=0}^{\infty} c_{2n} \cos(2n)\theta \qquad (12.41)$$

We note that only odd powers of $\cos\theta$ contribute to a frequency correction, the correction from the even powers vanishing upon integration over the interval $[0, 2\pi]$.

It is reasonable to ask: What are the effects of the even-power perturbations? We show in Chapter 14 that they lead to a change in the equilibrium point about which the oscillations take place, and contribute to the frequency corrections that are revealed through higher-order calculations.

Example 3.3 Consider the Van der Pol oscillator with an unperturbed frequency equal to 1.

$$\ddot{x} + x = +\varepsilon(1 - x^2)\dot{x} = -\varepsilon(1 - x^2)y; \qquad 0 < \varepsilon \ll 1 \qquad (12.42)$$

With $y = -dx/dt$, we identify

$$+\varepsilon(1 - x^2)\frac{dx}{dt} = -\varepsilon(1 - x^2)y$$

as the damping term. Recall that we have negative damping when the amplitude is growing and positive damping when the amplitude is decreasing. Thus, we see that if

$x^2 < 1$, we have negative damping;
$x^2 > 1$, we have positive damping.

Let us look at some diagrams in the phase plane. See Figures 12.2 and 12.3, noting the *sense* of angular velocity.

If our system begins with small amplitude, it grows since it is always in a region of negative damping, as shown by trajectory A. It eventually grows sufficiently to spend part of the time in a region of positive damping, i.e., where its amplitude is > 1. However, since it is still moving primarily in a region of negative damping, the amplitude continues to grow and approaches a closed curve in the phase plane called a limit cycle. The rate of approach to the curve is determined by the small parameter ε. As long as ε is sufficiently small, the limit cycle is approximately a circle of radius equal to 2. Thus, we find that

$$\overline{x^2} + \overline{y^2} = \overline{r^2} \to 4 \qquad \text{as} \qquad t \to \infty \qquad (12.43)$$

We can show that if $\bar{r} > 2$ or if $\bar{r} < 2$, the system approaches the limit cycle monotonically. We see that the limit cycle is a stable attractor for all orbits in the phase plane, and, hence, indicates stable periodic motion. Now that we

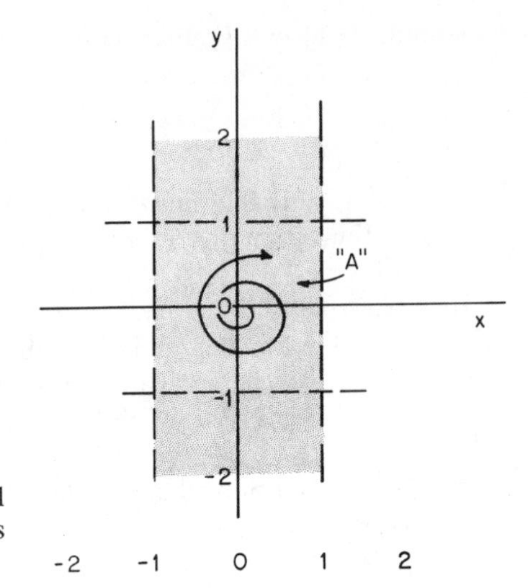

Figure 12.2 We start the Van der Pol oscillator with a small amplitude. It is initially in the region of slow growth.

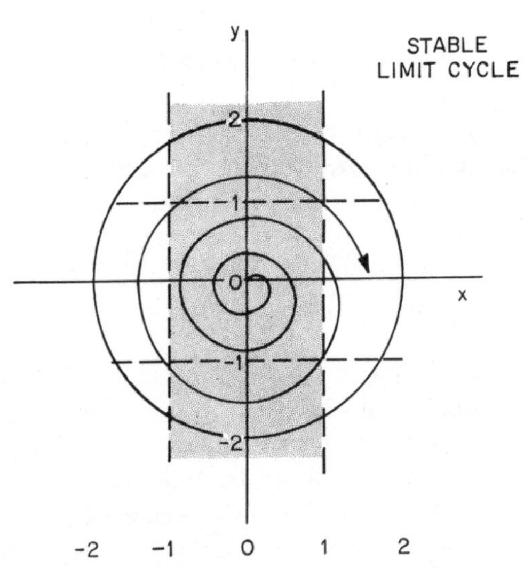

Figure 12.3 As the system draws energy from an energy reservoir, it enters a region of positive damping that slows its growth. Eventually, a balanced situation emerges in which the system spends some of its time in a region of positive damping and some of its time in a region of negative damping.

have some feeling for what is going to happen, let us proceed with our analysis. We write Equation (12.42) as a set of coupled first-order equations:

$$\dot{x} = -y$$
$$\dot{y} = +x + \varepsilon(1 - x^2)y$$

and obtain

$$x\dot{x} + y\dot{y} = r\dot{r} = \varepsilon(1 - x^2)y^2 \qquad (12.44a)$$
$$x\dot{y} - y\dot{x} = r^2\dot{\theta} = +r^2 + \varepsilon(1 - x^2)xy \qquad (12.44b)$$

We have

$$\dot{r} = \varepsilon(r \sin^2\theta - r^3 \sin^2\theta \cos^2\theta) \qquad (12.45a)$$
$$\dot{\phi} = \varepsilon(\sin\theta\cos\theta - r^2\cos^3\theta \sin\theta) \qquad (12.45b)$$

By integration, we can immediately find the averaged equations:

$$\dot{\bar{r}} = \frac{\varepsilon}{2\pi}\int_0^{2\pi}(\bar{r}\sin^2\theta - \bar{r}^3\sin^2\theta\cos^2\theta)\,d\theta \qquad (12.46a)$$
$$= \frac{\varepsilon\bar{r}}{2}\left(1 - \frac{\bar{r}^2}{4}\right) \qquad (12.46b)$$

and

$$\dot{\bar{\phi}} = \frac{\varepsilon}{2\pi}\int_0^{2\pi}(\sin\theta\cos\theta - \bar{r}^2\cos^3\theta\sin\theta)\,d\theta = 0 \qquad (12.47)$$

We can see that $\dot{\bar{\phi}} = 0$, indicating that at this level of approximation, there is no change in the frequency of oscillation. Since the initial phase is arbitrary, we choose it equal to zero.

We have $\dot{\bar{r}} = 0$ when $\bar{r} = 0$ and when $\bar{r} = r^* = 2$. An alternative to immediately averaging Equations (12.45a) and (12.45b) is to use trigonometric identities to recast the r.h.s. of Equations (12.45a) and (12.45b) in the forms

$$\dot{r} = \frac{\varepsilon r}{2}\left(1 - \frac{r^2}{4}\right) - \varepsilon\left(\frac{1}{2}r\cos 2\theta - \frac{1}{8}r^3\cos 4\theta\right) \qquad (12.48a)$$
$$\dot{\phi} = +\varepsilon\left[\frac{1}{2}\left(1 - \frac{r^2}{2}\right)\sin 2\theta - \frac{r^2}{8}\sin 4\theta\right] \qquad (12.48b)$$

We see that the terms on the r.h.s. of Equation (12.48a) are grouped such that the first has a nonzero average and yields Equation (12.46b). In other words, the radial velocity $\dot{r}$ has a smooth component and an oscillatory component that vanishes upon integration. Since there is no phase correction, Equation (12.48b) is a simple recasting of the terms on the r.h.s. of Equation (12.45b).

We are interested in finding the time dependence of the average radius $\bar{r}(t)$ and, therefore, we integrate Equation (12.46b) using standard techniques. We introduce the change of variables $z = \bar{r}^2$ and have

$$\frac{dz}{z(1 - z/4)} = \varepsilon\, dt$$

Integrating and introducing the initial value of the radius $\bar{r}(0)$, we have

$$\bar{r}(t) = \frac{2\bar{r}(0)}{\left\{\bar{r}(0)^2[1 - \exp(-\varepsilon t)] + 4\exp(-\varepsilon t)\right\}^{1/2}} \tag{12.49}$$

So we conclude that our system goes to a limit cycle of radius r^* as $t \to \infty$, independent of $\bar{r}(0)$. (Since our expansion is only valid for times $t = O(1/\varepsilon)$, it is not strictly correct to say that the radius approaches r^* as $t \to \infty$. However, it is possible to show, in autonomous systems, that higher-order terms do not change the qualitative character of the result. This is discussed in the text by Sanders and Verhulst.)

If the initial radius is < 2, the average radius increases. And if the initial radius > 2, it decreases until it reaches the limit-cycle value of $r = r^* = 2$. (See Figure 12.3.) The phase-plane plots given in Figures 12.4*a* and 12.4*b* are spirals that approach the limit cycle at a rate determined by the parameter ε. We develop a power-series expansion in terms of ε for the displacement and velocity. We choose $x(0) = 1$ and $\varepsilon = 0.3$ and we let the time run for 30 units, where one period of the unperturbed system is equal to 2π units. Note that the deviations that appear due to inclusion of the ε^2 terms.

It is possible to show that small motion in the neighborhood of the limit-cycle value returns the radius to the limit cycle. We can see this since if

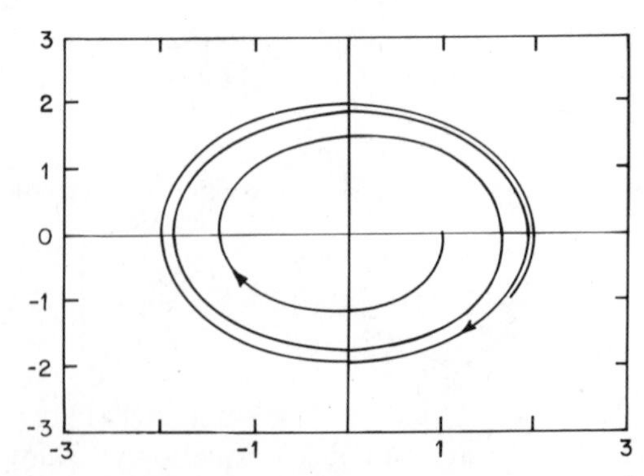

Figure 12.4a We have included terms proportional to ε^0 and ε^1.

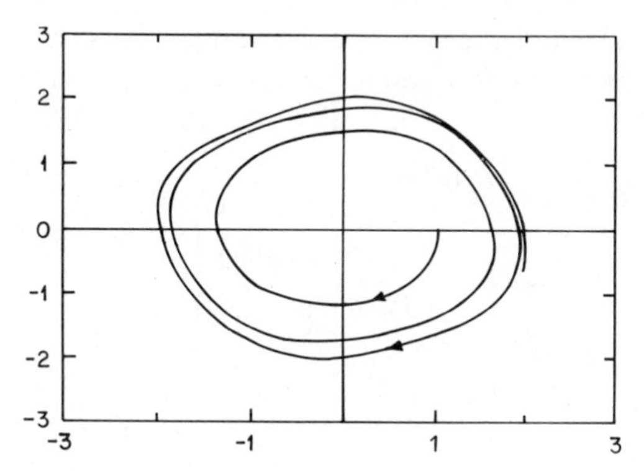

Figure 12.4b We have included terms proportional to ε^0, ε^1, and ε^2.

we introduce a small perturbation in the radius of the form

$$\bar{r} = r^* + u = 2 + u; \qquad |u| \ll 1$$

then Equation (12.46b) becomes

$$\frac{du}{dt} = \frac{\varepsilon}{2}(2 + u)\left[1 - \frac{1}{4}(4 + 4u + u^2)\right]$$

$$\frac{du}{dt} = -\varepsilon u + O(u^2)$$

So we conclude that small fluctuations around the limit cycle decay exponentially. We thus say that the limit cycle is a *stable attractor*.

Our equation for the displacement $x(t)$ becomes

$$x(t) = \bar{r}(t)\cos\left[t + \bar{\phi}(0)\right]$$

with a constant phase velocity and $\bar{r}(t)$ given by Equation (12.49). ■

EXERCISE 3.2 Observe that if we were to write the limit-cycle solution as $x(t) = 2\cos t$, it would not satisfy Equation (12.42). Show that it is a solution in the "average sense." This means that one should substitute the limit-cycle solution for $x(t)$ and dx/dt in Equations (12.44a) and (12.44b) and then average. ■

EXERCISE 3.3 Consider

$$\ddot{x} + x = +\varepsilon(a - bx^{2n})\frac{dx}{dt} = -\varepsilon(a - bx^{2n})y; \qquad a, b, \varepsilon > 0$$

Proceed as we did with the analysis of the Van der Pol oscillator and note that the phase velocity is constant and show that

$$\begin{aligned}\dot{\bar{r}} &= \frac{\varepsilon}{2\pi}\left(\int_0^{2\pi} a\bar{r}\sin^2\theta\,d\theta - \int_0^{2\pi} b\bar{r}^{2n+1}\cos^{2n}\theta\sin^2\theta\,d\theta\right)\\ &= \varepsilon\left[\frac{a\bar{r}}{2} - \frac{b\bar{r}^{2n+1}\Gamma\left(n+\frac{1}{2}\right)\Gamma\left(\frac{3}{2}\right)}{\pi(n+1)!}\right]\\ &= \left(\frac{\varepsilon\bar{r}}{2}\right)\left[a - \frac{b\bar{r}^{2n}(2n)!}{2^{2n}n!(n+1)!}\right]\end{aligned} \tag{12.50}$$

$$n=1: \qquad \dot{\bar{r}} = \frac{\varepsilon\bar{r}}{2}\left(a - \frac{b\bar{r}^2}{4}\right); \qquad r^* = 2\sqrt{\frac{a}{b}} \tag{12.51}$$

$$n=2: \qquad \dot{\bar{r}} = \frac{\varepsilon\bar{r}}{2}\left(a - \frac{b\bar{r}^4}{8}\right), \qquad r^* = 1.68\left(\frac{a}{b}\right)^{1/4} \tag{12.52}$$

$$n=3: \qquad \dot{\bar{r}} = \frac{\varepsilon\bar{r}}{2}\left(a - 5\frac{b\bar{r}^6}{64}\right); \qquad r^* = 1.53\left(\frac{a}{b}\right)^{1/6} \quad \blacksquare \tag{12.53}$$

EXERCISE 3.4 Consider

$$\ddot{x} + x = +\varepsilon\left(a - b|x|^{2n+1}\right)\frac{dx}{dt} = -\varepsilon\left(a - b|x|^{2n+1}\right)y; \qquad a, b, \varepsilon > 0 \tag{12.54}$$

This exercise is effectively the same as Exercise 3.3. The relevant integral is now

$$\frac{4}{2\pi}\int_0^{\pi/2}\cos^{2n+1}\theta\sin^2\theta\,d\theta = \frac{1}{\pi}\frac{n!\,\Gamma\left(\frac{3}{2}\right)}{\Gamma\left(n+\frac{5}{2}\right)} \tag{12.55}$$

Then

$$\dot{\bar{r}} = \varepsilon\bar{r}\left[\frac{a}{2} - \frac{b\bar{r}^{2n+1}n!\,\Gamma\left(\frac{3}{2}\right)}{\pi\Gamma\left(n+\frac{5}{2}\right)}\right] \tag{12.56}$$

Exhibit r^* for special cases:

$$n=0: \qquad r^* = \frac{3a\pi}{4b} = 2.36\left(\frac{a}{b}\right) \tag{12.57}$$

$$n=1: \qquad r^* = \left(\frac{15\pi a}{8b}\right)^{1/3} = 1.81\left(\frac{a}{b}\right)^{1/3} \tag{12.58}$$

Upon comparing these values of the limit-cycle radius r^* with those obtained in Exercise 3.3, we see that the radius gradually decreases as n increases. It approaches the limiting value $r^* = 1$ as $n \to \infty$ iff $a = b$. This makes sense if we think of our perturbation term as corresponding to a stiffer and stiffer velocity-dependent friction term. Then the sign change in this term that occurs as $|x|$ crosses 1 effectively confines the system to a very small neighborhood of $|r| = 1$. ■

Example 3.4 So far we have had either a change in the frequency or a change in the amplitude, but not both. This greatly simplifies the algebra. But we can consider, for example, a problem of the form:

$$\frac{d^2x}{dt^2} + x = \varepsilon(1 - x^2)\dot{x} - \varepsilon x^3; \qquad \varepsilon > 0 \tag{12.59}$$

We can think of this as a combination of a Van der Pol oscillator and Duffing oscillator. We then have a change in both the amplitude and the frequency:

$$\dot{\bar{r}} = \frac{\varepsilon\bar{r}}{2}\left(1 - \frac{\bar{r}^2}{4}\right)$$

and

$$\dot{\bar{\phi}} = +\frac{3\varepsilon}{8}\bar{r}^2$$

We cannot just simply solve the phase equation since the radius is time-dependent. But the radial equation can be solved as before, yielding Equation (12.49). We then have

$$\begin{aligned}\dot{\bar{\phi}} &= \frac{(3\varepsilon/8)\left[4\bar{r}(0)^2\right]}{\bar{r}(0)^2[1 - \exp(-\varepsilon t)] + 4\exp(-\varepsilon t)} \\ &= \frac{(3\varepsilon/8)\left[4\bar{r}(0)^2 \exp \varepsilon t\right]}{\bar{r}(0)^2(\exp \varepsilon t - 1) + 4}\end{aligned}$$

Then

$$\bar{\phi} = \tfrac{3}{2}\ln\left[r(\bar{0})^2(\exp \varepsilon t - 1) + 4\right] + c \tag{12.60}$$

If we write the initial phase as $\bar{\phi}(0)$, we have

$$\bar{\phi}(0) = 3\ln 2 + c \tag{12.61}$$

Thus, as long as either our radial or phase equation can be separately integrated, we can, in principle, solve the other equation. ■

EXERCISE 3.5 Consider the Rayleigh oscillator described by the equation:

$$\frac{d^2x}{dt^2} + x = +\varepsilon\left[\frac{dx}{dt} - \frac{(dx/dt)^3}{3}\right]$$

$$= -\varepsilon\left(y - \frac{y^3}{3}\right); \qquad 0 < \varepsilon \ll 1 \tag{12.62}$$

(This is effectively Equation (12.42), which we had for the Van der Pol oscillator. To see this, differentiate Equation (12.62) with respect to time and observe that it can be interpreted as a Van der Pol equation for the velocity.) Solve for r and obtain

$$\dot{\bar{r}} = \frac{\varepsilon \bar{r}}{2}\left(1 - \frac{\bar{r}^2}{4}\right)$$

which is the same result that we had for the Van der Pol oscillator. See Equation (12.46b).

Thus, the Rayleigh oscillator also describes a weakly nonlinear oscillatory system that has a limit cycle of average radius equal to 2. ■

EXERCISE 3.6 Consider

$$\ddot{x} + x = -\varepsilon\dot{x}^{2n+1} = +\varepsilon y^{2n+1}; \qquad 0 < \varepsilon \ll 1, \quad n > 0 \tag{12.63}$$

Show that the solution can be written as

$$x(t) = \bar{r}(t)\cos(t + \bar{\phi})$$

where the phase correction $\bar{\phi}$ is constant at this level of approximation and the amplitude $\bar{r}(t)$ decreases under the effect of the damping term. We want this equation to model a weakly damped harmonic oscillator. This introduces a condition on the magnitude of the initial value of the amplitude $\bar{r}(0)$. For example, we might require $\bar{r}(0) < 1$ so that we satisfy this condition for all values of n. In this case, we note that higher powers of n indicate weaker damping.

Solve for $\bar{r}(t)$ and obtain:

$$\bar{r}(t) = \frac{\bar{r}(0)}{\left[1 + \varepsilon\alpha 2n\bar{r}(0)^{2n}t\right]^{1/2n}} \tag{12.64}$$

where

$$\alpha = \frac{(2n+1)!}{(n+1)!\,n!\,2^{2n+1}} \tag{12.65}$$

Check this result for $n = 1$ and 2.

Observe that the coefficient α, except for a factor $\bar{r}(0)^{2n}$, is the same as the frequency-correction term that we obtained in Equation (12.38) of Exercise 3.1. Does this make sense?

Plot $\bar{r}(t)$ versus t with $\bar{r}(0) = 1$, $\varepsilon = 1/10$, for $n = 1, 2$, and 3. ■

We have restricted our discussion to autonomous systems in great part because they are easier to handle. One of their central characteristics is that the initial conditions play a passive role and generally do not affect the structure of the long-time behavior of the solution. This is not true for nonautonomous systems as is clear from Exercises 3.7 and 3.8. Although nonautonomous systems are of great interest, they are omitted since they require a level of treatment that is beyond what we have planned for the text.

EXERCISE 3.7 This exercise is based on an example on page 30 of the text by Sanders and Verhulst. Consider the nonautonomous system

$$\ddot{x} + x = -4\varepsilon(\sin^2 t)\dot{x}; \qquad |\varepsilon| \ll 1.$$

Pick two sets of initial conditions:

(a) $x(0) = A$, $\dot{x}(0) = 0$;
(b) $x(0) = 0$, $\dot{x}(0) = A$.

Find the dominant behavior for times $t < 1/\varepsilon$ with an error $O(\varepsilon)$ and show that in case (a) the coefficient in the exponential decay is $-3\varepsilon t/2$, while in case (b) it is $-\varepsilon t/2$. In each case choose $\varepsilon = 1/(100\pi)$ and plot $x_{\max}$ for $0 \le t \le 50\pi$. Introduce polar coordinates and observe that for the time interval under consideration the average phase can be taken as constant and equal to its initial value. Compare your results with those obtained in Exercise 1.2 of Chapter 11. Finally, noting that our equation is linear, we can use the method of elimination of the middle term discussed in Section 6.1.1 of Chapter 6 to convert our equation to a linear oscillator with frequency modulation. (It becomes a Mathieu equation.) ■

EXERCISE 3.8 Consider a modified Rayleigh oscillator,

$$\ddot{x} + x = \varepsilon\left[1 - (\cos^2 t)(\dot{x}^2)\right]\dot{x}; \qquad 0 < \varepsilon \ll 1.$$

Impose two sets of initial conditions:

(a) $x(0) = A$, $\dot{x}(0) = 0$;
(b) $x(0) = 0$, $\dot{x}(0) = A$.

Proceed as in Exercise 3.7 and show that for these boundary conditions, the average phase is $O(\varepsilon^2)$ for times $t = O(1)$ and thus constant at this level of approximation. Also establish that in case (a) one finds that $d\bar{r}/dt = 0$ when the average radius $\bar{r}^* = \sqrt{8}$, while in case (b) $d\bar{r}/dt = 0$ when the average radius $\bar{r}^* = \sqrt{8/5}$. Since the motion does not take place in the (x, y) plane, it is not correct to call the resulting curves "limit cycles." Furthermore, one cannot use stability theorems that were derived for autonomous systems. (One observes, for example, that the averaged equations contain the time t on an equal footing with the variables x and y. This leads to difficulties in determining the stability of the solutions.) If the system is allowed to run for times $t = O(1/\varepsilon)$ or $O(1/\varepsilon^2)$, the variation in the phase becomes important. Indeed, it is possible to show that the curve with average radius $\sqrt{8/5}$ grows and becomes equal to $\sqrt{8}$ when the time $t = O(1/\varepsilon^2)$. This takes some effort to establish analytically, but it is easy to verify using a computer to plot the trajectory and watching the time development. In addition, if one chooses any other initial phase, the condition $d\bar{r}/dt = 0$ leads to curves with average radii $\bar{r}^*$ between $\sqrt{8/5}$ and $\sqrt{8}$. For times $t = O(1/\varepsilon)$ all of these curves have radii that approach $\sqrt{8}$. So we see that nonautonomous systems have the capacity to exhibit a strong final-state dependence on the initial conditions and the time of observation. This exercise should serve as an additional warning that we have to be reasonably careful regarding the time validity of our approximations. (Readers desiring to pursue these questions at greater depth might want to refer to pages 27–32 and 79–80 of the text by Sanders and Verhulst.) ■

12.4 CONCLUSIONS AND CAUTIONS

At this level of approximation, the method of averaging is easy to apply and gives a good qualitative picture of the motion, with a minimum of algebraic complexity. It has a fundamental limitation in that it tracks only the smooth motion and neglects the harmonics, or wiggles. It also uses as the interval of integration one period of the unperturbed motion rather than employing a transformation to the frame of reference that views periodic motion as a fixed point. It then follows that it cannot track correctly, for all times, the changes in the radius of the trajectory in the phase plane as the orbit is traversed. Finally, we remark that by averaging our equations, we have really solved a different problem than the one with which we began. It is then necessary to show the relationship between our approximate solution to the averaged equations and the solution to the exact equations. This leads to a discussion of the error involved in our approximations and is discussed briefly in Chapter 15. (A detailed discussion is given in the text by Sanders and Verhulst.)

We have observed that we require that our perturbation terms have appropriately odd powers in order to yield a contribution upon averaging. By this we mean that our integral

$$\frac{d\bar{r}}{dt} = -\frac{\varepsilon}{2\pi}\int_0^{2\pi} f(\bar{r}\cos\theta, -\bar{r}\sin\theta)\sin\theta\, d\theta$$

is different from zero if $f(x, dx/dt)$ is an even function of $\cos\theta$ multiplied by an odd function of $\sin\theta$.

$$\frac{d\bar{\phi}}{dt} = -\frac{\varepsilon}{2\pi\bar{r}}\int_0^{2\pi} f(\bar{r}\cos\theta, -\bar{r}\sin\theta)\cos\theta\, d\theta$$

where we need $f(x, dx/dt)$ to be an odd function of $\cos\theta$ multiplied by an even function of $\sin\theta$ in order for this average to be nonzero.

These results are characteristic of the first or lowest-order approximations. Thus, for example, we found that we required oddness to obtain a limit cycle. We also found that only odd-power conservative perturbations contributed to changes in the frequency of the fundamental component of the oscillation. Unfortunately, as we will discuss in Chapter 14, this conclusion is not always maintained when a higher-order computation is made. This is part of the price we pay for such a simple and straightforward calculational procedure. In Chapter 13, we introduce the method of multiple time scales. It is significantly more complicated than the MOA, but like the KBM method and the MRRP, it is a systematic procedure that is capable of being extended to higher-order computations.

CHAPTER 13

THE METHOD OF MULTIPLE TIME SCALES (MMTS)

This method is probably the most universally applicable of all of the techniques that we discuss. It can be applied to linear and nonlinear problems, singular differential equations that occur in boundary-layer theory, parametric oscillation problems such as the Mathieu equation, etc. It is straightforward to apply and it is clear at each step how to calculate higher-order terms. Perhaps the only difficulty is in trying to understand why it works rather than how it works. It has been developed in a number of equivalent forms; the most popular ones are known as two-timing and the MMTS. (See the text by Kevorkian and Cole for a detailed exposition of the method of two-timing. This text also contains the proof by Morrison of the equivalence of the method of averaging and the multiple time methods, through second order, for a broad class of problems.) We use the MMTS as developed by Nayfeh and also use his notation. As a note of caution, it has been shown that if the MMTS is applied to a problem that does not possess well-separated time scales, it may give an incorrect result. (See the text by Sanders and Verhulst.)

However, as with most asymptotic methods, it is advisable (useful) to compare your solution to that obtained by a different method to make certain that you are on the right track. This is one of our motivations for studying both the MMTS and the averaging methods. In addition, since these are the two most universally used techniques, a command of both gives you access to a greater variety of the literature.

13.1 REVIEW

1. Simple harmonic motion (SHM) is characterized as periodic motion with constant amplitude and constant frequency. This means that after each

period, the amplitude returns to its initial value. The frequency is independent of the amplitude. So we conclude that two identical simple harmonic oscillators that have differing initial conditions still have the same period of oscillation.

2. If we consider SHM, we have a natural time scale or clock that describes the timing of the motion. If we look at the system with a strobe light that is adjusted to this natural time scale so that it flashes once each period, then the oscillator appears as a stationary point. We find that when the period of the oscillator is perturbed, we must readjust our stroboscopic device in order to maintain our view of the system as a fixed point. The correction of the frequency of our flashing light is an ongoing process as we attempt to keep the system so it appears as a fixed point for longer and longer times.
3. If we perturb our oscillator by a cubic term in the displacement, we observe that we have to suppress secular generating terms. These arise from our perturbation expansion in terms of the unperturbed solution and the unperturbed frequency. Thus, we artificially generate terms that effectively force the system at its natural frequency. This situation can be ameliorated by introducing a new time scale whenever our perturbation has an odd power in the displacement.
4. If we perturb our oscillator with a cubic term in the velocity, a traditional perturbation expansion is nonuniform in time. This comes from the coupling of the small parameter in the expansion and the time of observation. We have to disentangle the slow time variation of the fundamental component of the amplitude from the rapid oscillations of its harmonics. If we fail to do this, the periodic underpinning of the motion is lost. The method of averaging accomplishes this separation of time scales and reorganizes our expansion to make it uniformly valid in time.
5. One of our primary observations was that our traditional perturbation schemes really fail already for linear perturbations either in the displacement or velocity. Thus, the nonlinearity of our problems is not the central cause of the difficulty. We did not see it in linear problems because we can solve those problems exactly and are not obliged to use perturbation theory.

13.2 THE CONCEPT OF TIME SCALES: THE MMTS

We now introduce the concept of independent time scales. This means that the physical processes described by the equation take place on different and independent time scales. So we can pick the scale to seek a solution. We then obtain a solution that is correct on that time scale, but incomplete if viewed from another time scale. It is a little bit like looking through a microscope with a particular objective lens. You pick the lens and then you see what there is to see. If you pick a different lens, you see something apparently independent.

We are seeking a continuous description, but often we do not really transform that continuity into our actual observation. For example, consider an analog watch with a second hand, minute hand, and hour hand. You do not have a 1/10 second hand, or day hand, etc. Assume you record observations continuously using the second hand. Alternatively, you can consider the hour as the basic unit of time and make continuous hourly observations; you will have an overlap with the person who uses the second-hand observations, but your picture of the processes that are taking place might be quite different.

This is all natural to us, but it is not straightforward to put into a mathematical framework because the time scales are not independent, but only appear to be.

Our basic assumption is that the time t is viewed as being constructed from a succession of independent time scales given by the relationship

$$t_n = \varepsilon^n t; \qquad 0 < \varepsilon \ll 1; \qquad n = 0, 1, 2, \ldots$$

You think of the time scales as readings measured in different units; then you have

$$t_0 = \text{seconds}$$
$$t_1 = \text{minutes}$$
$$t_2 = \text{hours}$$

and our small parameter is then $\varepsilon = 1/60$.

The introduction of new time scales is not a new expansion, and t, εt, and $\varepsilon^2 t$ measure the same time on different scales. (This analogy is introduced in the book by Nayfeh entitled *Introduction to Perturbation Techniques*.)

So you can think of the process taking place as viewed by a fast clock. Then there are things that appear to be constants but yet really vary in time, especially because of "blurred vision," since the solution is approximate. So when you solve differential equations associated with a given time scale, you can only conclude that they do not vary on the time scale that you are using for your observation.

We do this all of the time. We often look at flowers, trees, etc., and say that they appear not to be growing; yet if we use the proper time scale, our conclusions would change. This is also the situation when we look at a chemical process that takes place in several steps. We choose to view the steps as independent in scale. Finally, when we have a complicated system that is approaching equilibrium, we often find it convenient to wait until all of the rapid transients have died off and we are left with only the slowest term. We then develop a description of the process in the so-called relaxation-time approximation or regime. In many instances, our fundamental approximation is to assume that our true time t is decomposed into two independent time scales, t_0 and t_1. Why do we expect this to work? Let us say that we are observing an oscillatory system that has a very slowly decreasing amplitude and we view the system at intervals associated with its natural frequency. For example, say that our system oscillates with a period of 1 second and the amplitude decreases about 1% per day. Suppose that we maintain the system as an oscillator for many days, restricting our measurements to a few minutes each morning. Then, while we are making our measurements, we can think of the amplitude as constant. On the other hand, we can neglect the oscillatory motion when computing the slow decrease in our amplitude. This separation was accomplished in the MOA by the averaging process. In the MMTS, it is accomplished by the introduction of the time scales t_0, t_1, t_2, etc.

Our interest remains fixed upon systems that are close to the SHO. Thus, we have to replace our total time derivatives by the partial derivatives, each associated with one of our independent time scales. (Remember the watch and think of the scales as seconds, minutes, hours, etc.)

We then seek a solution to our problems in an expansion both in powers of the small parameter ε and in the time scales. We seek expansions that are uniform in time and thus we suppress secular terms in each order of the expansion.

Now let us see how this works with some simple examples.

Example 2.1 Consider

$$\frac{d^2x}{dt^2} + 2\varepsilon\frac{dx}{dt} + (1+\varepsilon^2)x = 0 \tag{13.1}$$

We previously encountered this equation as Equation (12.1). As we mentioned, the time scale of the decay is well separated from the time scale of the oscillations. Thus, in this situation, the MMTS gives the exact result. However, since the verification involves a second-order calculation, it is necessary to first study the material discussed in Chapter 14 before coming to this conclusion. ■

Example 2.2 Consider

$$\frac{dx}{dt} = -\varepsilon(1+\cos t)x(t); \qquad x(0) = 1 \tag{13.2a}$$

The solution is found immediately to be

$$x(t) = \exp(-\varepsilon t)\exp(-\varepsilon \sin t) \tag{13.2b}$$

We encountered a similar equation in Chapter 11 in the form of Equation (11.7c) and plotted the result in Figure 11.1. Observe that there are two independent time scales. One time scale is associated with the smoothly varying component, $\exp(-\varepsilon t)$, and the other with the oscillations about the decaying exponential. If one solves this equation using the MMTS, the assumed separation is in fact realized and one obtains Equation (13.2b) as the result. This is an interesting exercise that can be done after studying the material covered in Example 2.3. ■

Example 2.3 Consider the Duffing oscillator:

$$\ddot{x} + x = -\varepsilon x^3; \qquad 0 < \varepsilon \ll 1, \quad x(0) = B, \quad \dot{x}(0) = 0 \tag{13.3}$$

We seek a solution that differs from the exact solution by $O(\varepsilon)$ for times $t < 1/\varepsilon$. There is some confusion possible regarding the time validity of the approximation. This arises from the interconnection between the length of time over which we observe the system and our desire to specify an error using the "O" symbol. We devote Chapter 15 to a detailed exposition of the error analysis associated with both the MOA and the MMTS.

We seek only the lowest-order corrections, and to this end, we write our differential operator in terms of the fast time t_0 and the first slow time t_1:

$$\frac{d}{dt} = \frac{dt_0}{dt}\frac{\partial}{\partial t_0} + \frac{dt_1}{dt}\frac{\partial}{\partial t_1} + \text{h.o.t.} \tag{13.4a}$$

$$= 1\frac{\partial}{\partial t_0} + \varepsilon\frac{\partial}{\partial t_1} + \text{h.o.t.} \tag{13.4b}$$

We introduce the symbolic operators D_i, $i = 0, 1, 2, \ldots,$ defined as

$$D_i \equiv \frac{\partial}{\partial t_i} \tag{13.4c}$$

Thus

$$D_0 = \frac{\partial}{\partial t_0}; \qquad D_1 = \frac{\partial}{\partial t_1}; \qquad D_0 D_1 = \frac{\partial^2}{\partial t_0\,\partial t_1}; \qquad \text{etc.} \tag{13.4d}$$

Equation (13.4a) implies that times t_0 and t_1 are treated as independent variables. We neglect time t_2 and realize that when the length of time of our observation is as long as $1/\varepsilon$, this time scale introduces corrections that are of $O(\varepsilon)$. Thus, we restrict the time of observation to times t that are less than $1/\varepsilon$. ■

The second-order differential operator is given by

$$\frac{d^2}{dt^2} = D_0^2 + 2\varepsilon D_0 D_1 + \text{h.o.t.} \tag{13.5}$$

and consists of two parts:

(1) The first part is a component that follows the rapid motion: D_0^2. Since it is independent of the small parameter in the problem, we identify it with the unperturbed aspect of the system, i.e., the SHO.
(2) The second part is a component that follows the slow motion: $2\varepsilon D_0 D_1$. It depends on the first power of the small parameter ε and thus we identify it with the slow variation in the amplitude and the frequency, both of which are constant for the SHO and vary on a slow time scale t_1.

Since we are treating the various time scales as independent, we develop an expansion for the displacement in terms of each scale as well as in the small parameter ε:

$$x(t) = x_0(t_0, t_1; \varepsilon) + \varepsilon x_1(t_0, t_1; \varepsilon) + O(\varepsilon^2) \tag{13.6}$$

We are seeking a solution that gives the displacement $x(t)$ with an error $O(\varepsilon)$ for times $t < 1/\varepsilon$. We substitute this proposed solution in Equation (13.3) and compare terms order by order in ε:

$$D_0^2 x_0 + x_0 + \varepsilon\left(D_0^2 x_1 + x_1 + 2D_0 D_1 x_0\right) = -\varepsilon x_0^3 + O(\varepsilon^2) \tag{13.7}$$

Our equation to order ε^0 is

$$D_0^2 x_0 + x_0 = 0$$

with the solution

$$x_0(t) = a(t_1)\cos\left[t_0 + \phi(t_1)\right] \tag{13.8}$$

We have a partial differential equation, so the amplitude and frequency can depend upon the higher-order time scales, which in our case is t_1. This is a central aspect of the MMTS. It pushes slower time dependency onto the higher-order terms in the expansion. It says that if you restrict your view to only the fast time scale, you are unable to observe the variation in the amplitude and frequency. However, if you use a clock that makes observations or measurements on the time scale of the slower motion, the variation is revealed. Remember that we are only going to include the slow time t_1 and limit our time of observation to $t < 1/\varepsilon$.

We have to keep this in mind as we solve for the time behavior of the apparent constants $a(t_1)$ and $\phi(t_1)$. In order to find the coefficients, we have to set up an equation for the first-order correction to the displacement x_1 and impose the condition that it be free of secular terms. Keep in mind that we are free to choose the slow time t_1 and to adjust its rate to eliminate secular terms in the amplitude and frequency.

We do not solve for this correction, but only use the secularity condition. This will be clearer if you write the equation for x_1:

$$\begin{aligned} D_0^2 x_1 + x_1 &= -x_0^3 - 2D_0 D_1 x_0 \\ &= -a^3\cos^3(t_0+\phi) + 2(D_1 a)\sin(t_0+\phi) \\ &\quad + 2(aD_1\phi)\cos(t_0+\phi) \\ &= \left[-\tfrac{3}{4}a^3 + 2(aD_1\phi)\right]\cos(t_0+\phi) \\ &\quad -\tfrac{1}{4}a^3\cos[3(t_0+\phi)] + 2(D_1 a)\sin(t_0+\phi) \end{aligned} \tag{13.9}$$

Notice that our partial differential operator on the l.h.s. is the same as we had for the zero-order equation. (That is to say, it is the differential operator of the SHO.) This pattern is maintained in higher-order calculations, i.e., the l.h.s. operator is unchanged and the r.h.s. contains known lower-order terms.

We are going to encounter the same secular-term problem as we had in our previous methods of analysis unless we choose

$$D_1 a = 0 \qquad \text{and} \qquad D_1\phi = \tfrac{3}{8}a^2 \tag{13.10a}$$

leading to

$$a = a(t_2) \tag{13.10b}$$

and

$$\phi = \tfrac{3}{8}a^2 t_1 + \tilde{\phi}(t_2) \tag{13.10c}$$

where $a(t_2)$ and $\tilde{\phi}(t_2)$ are constants on the time scale t_1 but can vary on the longer time scales.

We justify our suppression of the secular terms by the argument that the correction to the displacement x_1 must remain small compared to the main term x_0 for all times. Our free parameter in the MMTS is the slow time-scale dependence of the amplitude and the phase or frequency. Alternatively, we can say that we are using our freedom to choose the slow time scale t_1 and adjust its rate so as to eliminate secular terms.

We have seen this secularity condition in our previous investigation of the Duffing oscillator. (It is true, that now we encounter it as a partial differential equation, but this does not change our analysis in a substantial way.)

We stop here and do not solve for the displacement correction term x_1 since we are already neglecting time-correction terms of order t_2. Remember that when

$$t \sim \frac{1}{\varepsilon}: \qquad t_1 = O(1); \quad t_2 = \varepsilon^2 t = O(\varepsilon)$$

so that the displacement x_1 and the time scale t_2 are both of $O(\varepsilon)$. So you must either include both of them or stop with the determination of the dependence of the amplitude a and the phase ϕ on the slow time t_1.

Our solution is then written as

$$x(t) = a\cos\left[\left(1 + \varepsilon\frac{3}{8}a^2\right)t\right] + O(\varepsilon) \qquad \text{for } t < 1/\varepsilon \ \blacksquare \tag{13.11}$$

Thus, as with the method of averaging, we find the frequency and amplitude corrections to the fundamental and leave to higher-order calculations the determination of the structure of the harmonics. If you think about this a bit, you see that the higher-order calculations become more and more complicated as new scales are introduced and become interwoven. Indeed, the second-order calculations are reasonably complex. The third- and fourth-order calculations are messy enough, so that you will have difficulty finding a textbook that does even one sample problem. It is important to emphasize that if you have to perform this type of calculation, it is good advice to do it by both the MMTS and the MOA.

Example 2.4 Consider the Van der Pol oscillator:

$$\ddot{x} + x = \varepsilon(1 - x^2)\dot{x}; \qquad 0 < \varepsilon \ll 1, \quad x(0) = a \tag{13.12}$$

We introduce the expansions given in Equations (13.4a) to (13.4d) and (13.6) for the differential operators and the displacement. We have the SHO as our lowest-order equation, as required by our underlying assumptions:

$$D_0^2 x_0 + x_0 = 0 \tag{13.13a}$$

The solution is

$$x_0 = a(t_1)\cos[t_0 + \phi(t_1)] \tag{13.13b}$$

Then the first-order equation becomes

$$\begin{aligned} D_0^2 x_1 + x_1 &= [1 - a^2\cos^2(t_0 + \phi)][-a\sin(t_0 + \phi)] \\ &\quad + 2(D_1 a)\sin(t_0 + \phi) + 2(aD_1\phi)\cos(t_0 + \phi) \\ &= [-a(1 - \tfrac{1}{4}a^2) + 2(D_1 a)]\sin(t_0 + \phi) \\ &\quad + \tfrac{1}{4}a^3\sin[3(t_0 + \phi)] + 2(aD_1\phi)\cos(t_0 + \phi) \end{aligned} \tag{13.14}$$

We encounter secular generating terms

$$[-a(1 - \tfrac{1}{4}a^2) + 2(D_1 a)]\sin(t_0 + \phi) \tag{13.15a}$$

and

$$2(aD_1\phi)\cos(t_0 + \phi) \tag{13.15b}$$

We suppress these terms by imposing the condition that our equations be free of them in each order of the expansion.

Thus, we require

$$D_1\phi = 0 \Rightarrow \phi = \phi(t_2) \tag{13.16a}$$

$$2(D_1a) = a\left(1 - \tfrac{1}{4}a^2\right) \tag{13.16b}$$

We solve for the amplitude by standard calculus techniques and obtain

$$a(t_1) = \frac{2}{\left[1 + \left\{\left[4/a(0)^2\right] - 1\right\}\exp(-t_1)\right]^{1/2}} \tag{13.17}$$

Recall that we discussed the Van der Pol oscillator at some length in Example 3.3 of Chapter 12. (Equation (13.17) is equivalent to Equation (12.49).) The radius of the limit cycle is, in this order of approximation, independent of the small parameter ε. (There is a dependence that is revealed through the higher-order terms.)

So we see that we obtain the same equations and the same resolution as we did by the averaging method. ■

13.3 THE EQUIVALENCE OF THE MOA AND MMTS

Let us consider the general problem

$$\frac{d^2x}{dt^2} + x = \varepsilon f\left(x, \frac{dx}{dt}\right) = \varepsilon f(x, -y)$$

$$y = \frac{-dx}{dt}; \qquad |\varepsilon| \ll 1 \tag{13.18}$$

in which we allow the perturbational term $\varepsilon f(x, -y)$ to be a nonanalytic function such as $|x|$ or $|y|$. We will show that we obtain precisely the same conditions using the MMTS for the time variation of the amplitude and the frequency as we did by the MOA, establishing thereby the equivalence of the two methods in the lowest order of the perturbation expansion. Let us outline the procedure that we will follow.

Introduce the multiple time scales and the expansion of the differential operator and the displacement given by Equations (13.4a) to (13.4d), (13.5), and (13.6). The lowest-order solution is given by Equation (13.8). We write the equation for the next order term

$$D_0^2 x_1 + x_1 = -2 D_0 D_1 x_0 + f(x_0, -y_0) \tag{13.19}$$

with

$$x_0 = a \cos(t_0 + \phi) \tag{13.20a}$$

$$y_0 = a \sin(t_0 + \phi) \tag{13.20b}$$

and we wish to identify those components of $f(x_0, -y_0)$ that are proportional to $\cos(t_0 + \phi)$ and $\sin(t_0 + \phi)$. We group these terms together with those generated from the differential operator and set them equal to zero. This leads to conditions on the amplitude and frequency that suppress secular terms. This is precisely what we did in our analysis of the Duffing oscillator and the Van der Pol oscillator. However, if $f(x, -y)$ is either nonanalytic or a sufficiently complicated function, we identify the secular generating terms by expanding $f(x_0, -y_0)$ in a Fourier series:

$$\begin{aligned} f(x_0, -y_0) = & \sum_{n=0}^{\infty} a_n \cos[n(t_0 + \phi)] \\ & + \sum_{n=1}^{\infty} b_n \sin[n(t_0 + \phi)] \end{aligned} \tag{13.21}$$

We are only interested in the coefficients a_1 and b_1 since only the terms proportional to $\cos t_0$ and $\sin t_0$ are secular generating terms. We have

$$a_1 = \frac{1}{\pi} \int_0^{2\pi} f(x_0, -y_0) \cos(t_0 + \phi) \, dt_0 \tag{13.22a}$$

$$b_1 = \frac{1}{\pi} \int_0^{2\pi} f(x_0, -y_0) \sin(t_0 + \phi) \, dt_0 \tag{13.22b}$$

We have to suppress secular generating terms on the r.h.s. of Equation (13.19). If we use our solution x_0 and introduce

$$\theta = t_0 + \phi$$

our condition for suppression of secular generating terms becomes

$$D_1 a = -\frac{1}{2\pi} \int_0^{2\pi} f(a \cos\theta, -a \sin\theta) \sin\theta \, d\theta \tag{13.23a}$$

$$a D_1 \phi = -\frac{1}{2\pi} \int_0^{2\pi} f(a \cos\theta, -a \sin\theta) \cos\theta \, d\theta \tag{13.23b}$$

These are equivalent to the equations that we obtained by the method of averaging for the time variation of the amplitude and frequency (Equations (12.19a) and (12.19b)). This establishes the equivalence of the two methods in this order of approximation. It is also clear that Equations (13.23a) and (13.23b) lead to Equations (13.10a), (13.16a), and (13.16b) as special cases. Let us see how these equations apply when $f(x, y)$ is a nonanalytic function.

Example 3.1

$$\ddot{x} + x = \varepsilon f\left(x, \frac{dx}{dt}\right) = \varepsilon f(x, -y)$$

$$y = \frac{-dx}{dt}; \qquad 0 < \varepsilon \ll 1 \tag{13.24a}$$

with

$$f\left(x, \frac{dx}{dt}\right) = (1 - |x|)\frac{dx}{dt} = -(1 - |x|)y \tag{13.24b}$$

This is a Van der Pol type of equation with a weakened restoring force. After we introduce the perturbation expansion, we have to solve Equations (13.23a) and (13.23b). If we take cognizance of the absolute value signs and evaluate the integrals, we obtain

$$\begin{aligned} D_1 a &= \frac{a}{2\pi}\int_0^{2\pi} \sin^2\theta\, d\theta \\ &\quad - \frac{a^2}{2\pi}\int_0^{2\pi} |\cos\theta| \sin^2\theta\, d\theta \\ &= \frac{a}{2} - \frac{2}{3\pi}a^2 \end{aligned} \tag{13.25}$$

We also have

$$D_1\phi = \frac{1}{2\pi}\int_0^{2\pi} \cos\theta \sin\theta\, d\theta - \frac{a}{2\pi}\int_0^{2\pi} |\cos\theta| \sin\theta \cos\theta\, d\theta = 0 \tag{13.26}$$

We obtain a limit cycle whose radius r^* is given by the condition $D_1 a = 0$. Then

$$r^* = \frac{3\pi}{4} \approx 2.36$$

The radius is slightly larger than 2, as expected. This is the same result that we obtained in Exercise 3.4, Chapter 12, Equation (12.57). ■

13.4 COMPLEX NOTATION

The use of the complex representation for the cosine and sine functions greatly simplifies the algebra. We write

$$\cos x = \frac{1}{2}[\exp ix + \exp(-ix)]$$

and

$$\sin x = \frac{1}{2i}[\exp ix - \exp(-ix)]$$

Let us consider, for example, the Duffing oscillator as given by Equation (13.3). We proceed as before and our lowest-order solution to Equation (13.7) then takes the form

$$x_0(t) = A \exp it_0 + A^* \exp(-it_0) \tag{13.27}$$

We indicate the complex conjugate of the amplitude A as A^*. The amplitude A is complex and can be expressed as $A = (a/2)\exp i\phi$, where both a and ϕ are real. The factor $(\frac{1}{2})$ is convenient since it makes Equations (13.8) and (13.27) equivalent. We write

$$[A \exp it_0 + A^* \exp(-it_0)]^3 = A^3 \exp 3it_0 + 3A^2A^* \exp it_0 + \text{c.c.}$$

where c.c. indicates the complex conjugate. Thus, we obtain the equivalent of Equation (13.9) in the form

$$\begin{aligned} D_0^2x_1 + x_1 &= -x_0^3 - 2D_0D_1x_0 \\ &= \left(-3A^2A^* - 2iD_1A\right)\exp it_0 \\ &\quad - A^3 \exp 3it_0 + \text{c.c.} \end{aligned} \tag{13.28}$$

It is clear that we have to suppress the terms that multiply $\exp(\pm it_0)$ since they are equivalent to the sine and cosine terms that we have discussed in previous treatments.

Now write our complex amplitude A in terms of a and ϕ. Remembering the factor $(\frac{1}{2})$, we express the r.h.s. of Equation (13.28) as

$$\begin{aligned} &\left[-\tfrac{3}{8}a^3 - iD_1a + (aD_1\phi)\right]\exp(it_0 + i\phi) \\ &\quad - \tfrac{1}{8}a^3 \exp(3it_0 + 3i\phi) + \text{c.c.} \end{aligned} \tag{13.29}$$

Then our secularity condition requires that we set the real and imaginary coefficients of $\exp[\pm i(t_0 + \phi)]$ equal to zero. Thus, we write

$$-i(D_1a) = 0 \qquad \text{or} \qquad a = a(t_2) \tag{13.30}$$

indicating that the amplitude a is constant on the time scale t_1. Also, we have

$$-\tfrac{3}{8}a^3 + a(D_1\phi) = 0$$

This leads to an equation for the phase given by

$$\phi = \tfrac{3}{8}a^2 t_1 + \tilde{\phi}(t_2) \tag{13.31}$$

So we observe that the technique that employs complex notation involves the identification and suppression of terms of the form $\exp(\pm it_0)$. This notation is often preferred since it eliminates the need to use lots of trigonometric identities. We use it in all of our future calculations.

In solving problems using complex notation, we have to find the terms that multiply $\exp(\pm it_0)$. This is easily done if we recall the expansion of

$$(a+b)^N = \sum_{m=0}^{N} \binom{N}{m} a^{N-m} b^m; \qquad N \text{ an integer} \tag{13.32}$$

where

$$\binom{N}{m} = \frac{N!}{(N-m)!m!}$$

In our problems, we have

$$\begin{aligned}&[A \exp it_0 + A^* \exp(-it_0)]^N \\ &\quad = \sum_{m=0}^{N} \binom{N}{m} A^{N-m} A^{*m} \exp[i(N-2m)t_0]\end{aligned} \tag{13.33}$$

If N is odd ($N = 2n+1$), we have a term of the form $\exp(\pm it_0)$. This occurs when $m = n$ and $m = n+1$. We identify only the term $m = n$ since the other term just gives the complex conjugate. Thus, we have for the coefficient of $\exp it_0$

$$\frac{(2n+1)!}{(n+1)!n!} A^{n+1} A^{*n} \tag{13.34a}$$

Remember that $A = \frac{1}{2}a \exp i\phi$. Thus, the coefficient given in Equation (13.34a) can also be written

$$a^{2n+1} \frac{(2n+1)!}{2^{2n+1}(n+1)!n!} \exp i\phi \tag{13.34b}$$

13.5 EXERCISES AND EXAMPLES

Problems for Solution

EXERCISE 5.1 Solve the Van der Pol oscillator as given by Equation (13.12) using complex notation. One has to find the terms that multiply exp it_0. One expands the term

$$\left(1 - x_0^2\right)\dot{x}_0 = \left\{1 - \left[A^2 \exp 2it_0 + 2AA^* + A^{*2}\exp(-2it_0)\right]\right\}$$
$$\cdot\left[iA \exp it_0 - iA^* \exp(-it_0)\right]$$

There are two factors that yield terms proportional to exp it_0.
The rest of the analysis is straightforward. ■

Example 5.1

$$\ddot{x} + x = -\varepsilon x^{2n+1}; \qquad x(0) = B, \quad \dot{x}(0) = 0 \tag{13.35}$$

We want to find the lowest-order correction to the frequency of oscillation. Using Equation (13.34a), our secular equation is

$$0 = -2i(D_1 A) - \frac{(2n+1)!}{(n+1)!n!}A^{n+1}A^{*n}$$

Then write our complex amplitude A in terms of its real amplitude a and the phase ϕ. Observe that the coefficient i occurs only in front of the derivative term, indicating that there is no correction to the amplitude. This follows since the system is conservative. Using Equation (13.34b), we write our phase equation as

$$\left[(-2i)\left(\frac{ia}{2}\right)D_1\phi\right]\exp i\phi = \frac{(2n+1)!a^{2n+1}}{(n+1)!n!2^{2n+1}}\exp i\phi$$

Thus, one finds

$$\phi(t_1) = \left[a^{2n}\frac{(2n+1)!}{2^{2n+1}(n+1)!n!}\right]t_1 + \tilde{\phi}(t_2)$$

yielding a frequency

$$\omega = 1 + \varepsilon a^{2n}\frac{(2n+1)!}{2^{2n+1}(n+1)!n!}$$

This is the same result that we obtained for Exercise 3.1 in Chapter 12. ■

EXERCISE 5.2 Consider

$$\ddot{x} + x = -\varepsilon \dot{x}^{2n+1}; \qquad n > 0 \tag{13.36}$$

Observe that this problem is analyzed in the same way as Example 5.1. Show that one obtains a decaying amplitude and use Equation (13.34b) to write

$$D_1 a = -a^{2n+1} \frac{(2n+1)!}{2^{2n+1}(n+1)!n!} \tag{13.37}$$

Integrate and obtain the same result as we had for Exercise 3.6 of Chapter 12, namely,

$$a(t_1) = \frac{a(0)}{\left[1 + \alpha 2na(0)^{2n} t_1\right]^{1/2n}} + \tilde{a}(t_2) \tag{13.38a}$$

with

$$\alpha = \frac{(2n+1)!}{2^{2n+1}(n+1)!n!} \qquad ■ \tag{13.38b}$$

EXERCISE 5.3 (This exercise is derived from a problem in the text by Smith entitled "Precession of Mercury.") Consider a quadratic oscillator in the form

$$\ddot{x} + x = C + \varepsilon x^2; \qquad x(0) = B, \quad \dot{x}(0) = 0, \quad 0 < \varepsilon \ll 1$$

where the constant C is $O(1)$. Find the first-order solution incorporating the first-order correction to the frequency of oscillation. Recall that our basic tenet is that the unperturbed problem is a simple harmonic oscillator. This means that our first step is to introduce a shift in the origin of the form $u = x - C$ in order to write our equation as $\ddot{u} + u = \varepsilon(u + C)^2$. It is then simple to use the MMTS to find the solution and frequency correction to lowest order. Show that $x = C + (B - C)\cos[(1 - \varepsilon C)t]$. Repeat the analysis using the MOA. ■

EXERCISE 5.4 Refer to Equation (13.9). Impose the secularity condition and solve for x_1. We can include x_1 in our expression for $x(t)$ as given by Equation (13.11) and thereby obtain an approximate solution for the Duffing oscillator that is in error by as little as $O(\varepsilon^2)$ for times $t = O(1)$. The error analysis is the subject of Chapter 15. ■

Comment on Exercises and Examples

(a) If you have an equation of the form

$$\ddot{x} + x = \varepsilon(1 - |x|)\dot{x} \tag{13.39}$$

the complex notation can be confusing and it is better to use the sine and cosine representations.

(b) If you have an equation of the form of the Van der Pol oscillator, then it is easy to find the coefficient of the $\exp it_0$ terms. However, if you have a problem of the form

$$\ddot{x} + x = \varepsilon(1 - x^2)\dot{x}^3 \tag{13.40}$$

you observe that there are three terms that multiply $\exp it_0$. Thus, it often requires some organization to solve problems with higher-order damping terms.

CHAPTER 14

HIGHER-ORDER CALCULATIONS

It is essential to be able to develop higher-order perturbation expansions and use them to reveal aspects of weakly nonlinear oscillatory systems that are not present in the lowest-order results. Since this is our primary goal rather than to develop higher-order expansions for their own sake, we limit our discussion to the MMTS. The MOA can be developed in such a manner as to obtain, for each of the problems that we discuss, the same results as are found by the MMTS. The texts in the Bibliography at the end of Chapter 15 contain detailed discussions of both of these techniques. One might want to read the relevant sections of the book entitled *Perturbation Methods* by Nayfeh since he takes great pains to solve problems by a variety of techniques and to demonstrate that the results are the same or equivalent.

Organization of the Chapter. We begin by developing calculations through the second order for conservative systems, and in doing so, examine the role of the initial conditions in the expansion procedure. We then turn our attention to the quadratic oscillator in preparation for a discussion of scaling. It will become clear that the analysis of these systems involves a lot of algebra and one has to be careful to organize the calculations with some care.

14.1 SECOND-ORDER CALCULATIONS FOR CONSERVATIVE SYSTEMS

In this section, we develop the MMTS for conservative systems through the second order in the small parameter ε. We find the frequency corrections that enable us to follow the system for times $t = O(1/\varepsilon)$ with an error of $O(\varepsilon^2)$.

The algebra becomes messy and we find that it is best to use complex rather than trigonometric notation.

We begin by developing the second-order calculation for the Duffing oscillator.

$$\ddot{x} + x = -\varepsilon x^3; \qquad 0 < \varepsilon \ll 1 \tag{14.1a}$$

$$x(0) = B; \qquad \dot{x}(0) = 0 \tag{14.1b}$$

Note that we have picked special initial conditions in which the initial velocity is equal to zero. Since the system is bounded and conservative, its trajectory in the phase plane is a closed orbit. Consequently, the system precisely retraces its path after every interval equal to its period. Thus, we are free to choose as the origin of our time coordinate the point where the velocity is equal to zero. This choice of the initial conditions leads to a simplification of the algebra.

The calculation involves the following parts:

(a) Expand the displacement through $O(\varepsilon^2)$:

$$x = x_0 + \varepsilon x_1 + \varepsilon^2 x_2 + O(\varepsilon^3)$$

(b) Develop the differential operator through powers of ε^2 and introduce the second slow time $t_2 = \varepsilon^2 t$. Write

$$\frac{d}{dt} = D_0 + \varepsilon D_1 + \varepsilon^2 D_2 + \text{h.o.t.}$$

Then we have

$$\frac{d^2}{dt^2} = D_0^2 + 2\varepsilon D_0 D_1 + \varepsilon^2 (2 D_0 D_2 + D_1^2) + \text{h.o.t.} \tag{14.2}$$

Expand our perturbing term εx^3 through powers of ε^2. This involves finding both x_0 and x_1. Then we have

$$\varepsilon x^3 = \varepsilon x_0^3 + 3\varepsilon^2 x_0^2 x_1 + O(\varepsilon^3)$$

(c) Impose the secularity condition on the differential equations for x_1 and x_2. We will see that this leads to some ambiguities and we will show how these are resolved.

We have for the ε^0 term:

$$D_0^2 x_0 + x_0 = 0$$

with the solution

$$x_0 = A(t_1, t_2) \exp it_0 + \text{c.c.} \tag{14.3}$$

This is a partial differential equation, so we note that the complex amplitude A and, consequently, the amplitude a and the phase ϕ can vary on higher-order time scales. We indicate this possible time dependence as $A(t_1, t_2)$, or

$$A = \tfrac{1}{2}a(t_1, t_2)\exp[i\phi(t_1, t_2)]$$

Observe that to $O(\varepsilon)$, the initial conditions are given by Equation (14.1b), with $B = a$ and $\phi(0) = 0$.

We now develop the ε equation:

$$\begin{aligned} D_0^2 x_1 + x_1 & \\ &= -2D_0 D_1 x_0 - x_0^3 \\ &= -2iD_1 A \exp it_0 + 2iD_1 A^* \exp(-it_0) \\ &\quad -\left[A^3 \exp 3it_0 + 3A^2 A^* \exp it_0 + 3AA^{*2}\exp(-it_0) + A^{*3}\exp(-3it_0)\right] \end{aligned} \tag{14.4}$$

and collect coefficients of $\exp it_0$ and $\exp(-it_0)$ since these lead to secular terms. Recalling that the secularity conditions are not independent but are related by complex conjugation, we just consider the conditions on the term containing $\exp it_0$. Thus, we require

$$\left(-2iD_1 A - 3A^2 A^*\right)\exp it_0 = 0$$

or

$$D_1 A = \frac{3i}{2} A^2 A^* \tag{14.5a}$$

There is an equivalent condition on A^*:

$$\left(+2iD_1 A^* - 3AA^{*2}\right)\exp(-it_0) = 0$$

or

$$D_1 A^* = \frac{-3i}{2} AA^{*2} \tag{14.5b}$$

Notice that Equation (14.5b) is the complex conjugate of Equation (14.5a) and does not give any new condition on the amplitude A. We set the real and imaginary parts of Equation (14.5a) separately equal to zero and obtain conditions on the amplitude a and the phase ϕ:

$$(D_1 a)\exp i\phi + ai(D_1\phi)\exp i\phi = \frac{3i}{8}a^3 \exp i\phi$$

Thus, we have

$$D_1 a = 0 \Rightarrow a = a(t_2)$$

and

$$D_1 \phi = \frac{3}{8}a^2 \Rightarrow \phi = \frac{3}{8}a^2 t_1 + \tilde{\phi}(t_2) \tag{14.6}$$

The amplitude a is constant to this order and can vary only on a slower time scale, i.e., t_2. Similarly, the constant of integration in the phase equation can vary on higher time scales. We indicate this by writing them as $a(t_2)$ and $\tilde{\phi}(t_2)$. Our equation for x_1 now becomes

$$D_0^2 x_1 + x_1 = -A^3 \exp 3it_0 - A^{*3} \exp(-3it_0) \tag{14.7}$$

It can be solved by the trial solution

$$x_1 = E \exp 3it_0 + E^* \exp(-3it_0)$$

and we find $E = +A^3/8$ and $E^* = +A^{*3}/8$. Thus, we write

$$x_1 = \frac{A^3}{8} \exp 3it_0 + C \exp it_0 + \text{c.c.} \tag{14.8a}$$

where

$$C = \tfrac{1}{2} c \exp i\psi \tag{14.8b}$$

We have included a contribution from the homogeneous equation with the amplitude c and the phase ψ both real. Our equation for $x(t)$ to this order is

$$x(t) = A \exp it_0 + \varepsilon\left(C \exp it_0 + \frac{A^3}{8} \exp 3it_0\right) + \text{c.c.} \tag{14.9}$$

We check that we have satisfied our initial conditions given by Equation (14.1b). The condition that $\dot{x}(0) = 0$ leads to $\psi(0) = 0$, since we already had $\phi(0) = 0$. Our condition on the initial amplitude is

$$x(0) = B = a + \varepsilon c + \frac{\varepsilon a^3}{32} + O(\varepsilon^2) \tag{14.10}$$

We are now ready to compute the contribution of $O(\varepsilon^2)$:

$$D_0^2 x_2 + x_2 = -2D_0 D_2 x_0 - D_1^2 x_0 - 2D_0 D_1 x_1 - 3x_0^2 x_1 \tag{14.11a}$$

We exhibit each of the terms separately:

$$\text{(a)}\quad -2D_0D_2x_0 = -2i(D_2A)\exp it_0 + \text{c.c.} \tag{14.11b}$$

$$\text{(b)}\quad -D_1^2x_0 = -(D_1^2A)\exp it_0 + \text{c.c.} \tag{14.11c}$$

$$\text{(c)}\quad -2D_0D_1x_1$$

$$= -2\left[\frac{3i}{8}(D_1A^3)\exp 3it_0 + i(D_1C)\exp it_0 + \text{c.c.}\right] \tag{14.11d}$$

$$\text{(d)}\quad -3x_0^2x_1 = -3\left[A^2\exp 2it_0 + 2AA^* + A^{*2}\exp(-2it_0)\right]$$

$$\times\left[\frac{A^3}{8}\exp 3it_0 + \frac{A^{*3}}{8}\exp(-3it_0) + C\exp it_0 + C^*\exp(-it_0)\right] \tag{14.11e}$$

We are interested only in suppressing the secular generating terms, and thus we group the coefficients of $\exp it_0$ and set them equal to zero. (We can check our algebra by separately making certain that the coefficients of $\exp(-it_0)$ lead to the same secularity condition.)

$$\left(-2iD_2A - D_1^2A - 2iD_1C - 3A^2C^* - 6AA^*C - \tfrac{3}{8}A^3A^{*2}\right)\exp it_0 = 0 \tag{14.12}$$

At this stage in the analysis, we make a specific choice for the coefficient C of the homogeneous contribution to x_1. We are free to make any choice that enables us to satisfy the suppression of the secular generating terms. We choose $C = 0$. This means that we are choosing only the inhomogeneous solution for x_1 and that we will not use C to help us satisfy the initial condition given by Equations (14.1b) and (14.10). We will find that this leads to a particular correction to the frequency of oscillation. After we have performed this calculation, we will make a different choice of the coefficient C and we will find that this leads to different coefficients for some of the terms in the expression for the frequency of oscillation. If it is correct that we can make any choice for C that enables us to satisfy the initial conditions, how is it possible to obtain two different but correct frequencies? This is indeed puzzling, but it is explained in the next section. It illuminates many aspects of the calculation and the satisfaction of the initial conditions.

14.1.1 First Choice for the Homogeneous Solution: $C = 0$

Our secularity condition as given by Equation (14.12) becomes

$$-2iD_2A - D_1^2A - \tfrac{3}{8}A^3A^{*2} = 0 \tag{14.13}$$

In order to evaluate $D_1^2 A$, we use Equations (14.5a) and (14.5b):

$$\begin{aligned} D_1^2 A &= D_1\left(\frac{3i}{2}A^2A^*\right) = \frac{3i}{2}(2AA^*D_1A) + \frac{3i}{2}(A^2D_1A^*) \\ &= \frac{3i}{2}2A\left(\frac{3i}{2}A^2A^*\right)A^* + \frac{3i}{2}A^2\left(-\frac{3i}{2}A^{*2}A\right) \\ &= \frac{-9}{4}A^3A^{*2} \end{aligned} \tag{14.14}$$

Thus, our secularity condition given by Equation (14.13) becomes

$$2iD_2A = +\tfrac{9}{4}A^3A^{*2} - \tfrac{3}{8}A^3A^{*2} = \tfrac{15}{8}A^3A^{*2} \tag{14.15}$$

We equate the coefficients of $\exp i\phi$ and obtain

$$iD_2a - aD_2\phi = \frac{15}{256}a^5$$

$$D_2a = 0 \Rightarrow a = a(t_3) \tag{14.16a}$$

$$\tilde{\phi} = \frac{-15}{256}a^4t_2 + \tilde{\tilde{\phi}}(t_3) \tag{14.16b}$$

where we indicate higher-order time scales by t_3. Thus, the frequency of oscillation is

$$\omega(C=0) = \left[1 + \frac{3}{8}\varepsilon a^2 - \frac{15}{256}\varepsilon^2a^4 + O(\varepsilon^3)\right] \tag{14.17}$$

With the choice of the homogeneous solution $C = 0$, our displacement $x(t)$ given by Equation (14.9) is written

$$\begin{aligned} x(t) &= a\cos\left[\left(1 + \frac{3}{8}\varepsilon a^2 - \frac{15}{256}\varepsilon^2a^4\right)t\right] \\ &\quad + \frac{\varepsilon a^3}{32}\cos\left[3\left(1 + \frac{3}{8}\varepsilon a^2\right)t\right] + O(\varepsilon^2) \qquad \text{for} \quad t = O(1/\varepsilon) \end{aligned} \tag{14.18}$$

14.1.2 Second Choice for the Homogeneous Solution: $C \neq 0$

We return to Equation (14.12) and choose

$$C = -\tfrac{1}{8}A^2A^* \tag{14.19}$$

This fixes the phase of C to be the same as that of A since we have

$$C = \tfrac{1}{2}c \exp i\psi; \qquad A = \tfrac{1}{2}a \exp i\phi$$
$$c = \tfrac{-1}{32}a^3; \qquad \psi = \phi \tag{14.20}$$

We also see by referring to Equation (14.10) that this particular choice of a homogeneous solution leads to a satisfaction of the initial condition $x(0)$ entirely by the zero-order solution given in terms of the amplitude a.

$$x(0) = B = a + O(\varepsilon^2) \tag{14.21}$$

We now investigate what effect our choice of C, as given by Equation (14.19), has on the secularity condition given by Equation (14.12). First, we compute

$$-2iD_1C = -2i\left(-\tfrac{1}{8}\right)D_1(A^2A^*) \tag{14.22a}$$

We use the secularity conditions given by Equations (14.5a) and (14.5b) and obtain

$$\begin{aligned} -2iD_1C &= +\left(\frac{i}{4}\right)(2A)\left(\frac{3i}{2}\right)(A^2A^*)A^* \\ &\quad +\left(\frac{i}{4}\right)A^2\left(\frac{-3i}{2}\right)(A^{*2}A) \\ &= \tfrac{-3}{8}(A^3A^{*2}) \end{aligned} \tag{14.22b}$$

This result, together with Equations (14.14) and (14.19), enables us to write Equation (14.12) as

$$-2iD_2A + \tfrac{9}{4}A^3A^{*2} - \tfrac{3}{8}A^3A^{*2}$$
$$-3A^2\left(-\tfrac{1}{8}\right)A^{*2}A - 6AA^*\left(-\tfrac{1}{8}\right)(A^2A^*) - \tfrac{3}{8}A^3A^{*2} = 0$$

or

$$2iD_2A = \tfrac{21}{8}A^3A^{*2} \tag{14.23}$$

As before, we equate real and imaginary parts in order to obtain the amplitude a and the phase ϕ. Again, we conclude that the amplitude is a constant. Our equation for the phase takes the form

$$-aD_2\phi = \frac{21}{256}a^5$$

or

$$\tilde{\phi} = -\frac{21}{256}a^4t_2 + \tilde{\tilde{\phi}}(t_3) \tag{14.24}$$

and the frequency of oscillation, ω, becomes

$$\omega(C \neq 0) = 1 + \frac{3}{8}\varepsilon a^2 - \frac{21}{256}\varepsilon^2 a^4 + O(\varepsilon^3) \tag{14.25}$$

We observe that the ε^2 correction as given by Equation (14.25) differs from that given by Equation (14.17). This discrepancy is the primary subject of Section 14.2.

The displacement $x(t)$ is given by

$$x(t) = A \exp it_0 + \varepsilon\left(\frac{-1}{8}A^2A^* \exp it_0 + \frac{A^3}{8}\exp 3it_0\right) + \text{c.c.} \tag{14.26a}$$

Alternatively, we can write $x(t)$ as

$$x(t) = a\cos\left[\left(1 + \frac{3}{8}\varepsilon a^2 - \frac{21}{256}\varepsilon^2 a^4\right)t\right] + \varepsilon\left(\frac{a^3}{32}\right)\left\{\cos\left[3\left(1 + \frac{3}{8}\varepsilon a^2\right)t\right] - \cos\left[\left(1 + \frac{3}{8}\varepsilon a^2\right)t\right]\right\} + O(\varepsilon^2) \quad \text{for} \quad t = O(1/\varepsilon) \tag{14.26b}$$

Note that $x(t)$ given by Equation (14.18) differs from $x(t)$ given by Equation (14.26b). This is explained in Section 14.2.

14.2 THE ROLE OF THE INITIAL CONDITIONS IN THE EXPANSION PROCEDURE

One of the characteristics of simple harmonic motion is that the frequency of oscillation is independent of the amplitude. By contrast, a conservative nonlinear oscillator has a frequency of oscillation that depends upon the amplitude, and this leads to some confusion regarding the meaning of various terms in the perturbation expansion. In this section, we try to explain the origin of the confusion as well as the source of its resolution. We develop our discussion through an analysis of the Duffing oscillator as given by Equations (14.1a) and (14.1b). The total energy of the oscillator is

$$E = \tfrac{1}{2}\dot{x}^2 + \tfrac{1}{2}x^2 + \tfrac{1}{4}\varepsilon x^4 \tag{14.27a}$$

We pick initial conditions for our Duffing oscillator such that at time $t = 0$, the displacement $x(0)$ is a maximum and equal to the amplitude of oscillation B. The velocity is equal to zero when $x = B$. This condition occurs at the so-called "turning points" of the motion. Our system executes periodic motion bounded by these turning points. We chose this condition or, equivalently, the starting point for our time variable t since it leads to particularly simple equations of motion. Then we have the relationship between the total energy E and the maximum amplitude x_{max} given by

$$E = \tfrac{1}{2}x_{max}^2 + \tfrac{1}{4}\varepsilon x_{max}^4 \tag{14.27b}$$

The question to ask at this point is: What is the relationship between x_{max} and the quantities denoted by the variable a in the equations for the frequency, Equations (14.17) and (14.25)? Let us first look at Equation (14.17). It arose from a choice of the C term that corresponded to neglecting the homogeneous solution in x_1 and choosing to satisfy the initial condition, given by Equation (14.10), in the form

$$x(0) = B = a + \varepsilon\frac{a^3}{32} + O(\varepsilon^2) = x_{max} \tag{14.10}$$

We obtained an expression for the frequency given by Equation (14.17) in terms of the variable a:

$$\omega(C = 0) = \left[1 + \frac{3}{8}\varepsilon a^2 - \frac{15}{256}\varepsilon^2 a^4 + O(\varepsilon^3)\right] \tag{14.17}$$

We indicate the initial displacement as x_{max} since the velocity is initially equal to zero. However, we notice that $x_{max} \neq a$! This means that when we refer to the quantity denoted by a in the equation for frequency, we are not referring to the maximum displacement $x = B$. Alternatively, we have chosen as the origin of our perturbation expansion the point corresponding to the maximum displacement when the small parameter $\varepsilon = 0$. It is a legitimate choice, but possibly confusing.

Before we proceed further, let us see what the situation is like for our other choice of the quantity C. In that case, the initial condition is satisfied in the form given by Equation (14.21):

$$x(0) = B = \hat{a} + O(\varepsilon^2) = x_{max} \tag{14.21}$$

and the frequency of oscillation given by

$$\omega(C \neq 0) = 1 + \frac{3}{8}\varepsilon\hat{a}^2 - \frac{21}{256}\varepsilon^2\hat{a}^4 + O(\varepsilon^3) \tag{14.25}$$

We distinguish the two a's by calling this one $\hat{a}$.

Now we see that if we are to compare Equations (14.17) and (14.25), we should note that we have two different a's. To see how they are related, we write both of them in terms of the maximum displacement B. To this order, we have

$$B = \hat{a} = a + \frac{\varepsilon a^3}{32} \qquad \text{or} \qquad a = \hat{a} - \frac{\varepsilon \hat{a}^3}{32} \tag{14.28}$$

Thus, we can write Equation (14.17) in terms of $\hat{a}$ and obtain:

$$\omega(C = 0) = 1 + \frac{3}{8}\varepsilon\left(\hat{a} - \frac{\varepsilon \hat{a}^3}{32}\right)^2 - \frac{15}{256}\varepsilon^2\hat{a}^4 + O(\varepsilon^3)$$

$$\omega = 1 + \frac{3}{8}\varepsilon\hat{a}^2 - \left(\frac{3}{8}\right)\left(\frac{1}{16}\right)\varepsilon^2\hat{a}^4 - \frac{15}{256}\varepsilon^2\hat{a}^4 + O(\varepsilon^3) \tag{14.29}$$

Equation (14.29) is the same as Equation (14.25). This result sheds some light on the interplay between the homogeneous equation and the initial conditions. It is unfortunate that this aspect of the solution does not appear in a lowest-order calculation, since it reveals part of the structure of the perturbation expansion. We can now see why we have two different expressions for the displacement $x(t)$ as given by Equations (14.18) and (14.26b).

To recapitulate, we see that in the case of the frequency as given by Equation (14.17), we chose to develop a perturbation expansion in terms of the unperturbed radius a. We did not alter this choice when we turned on the perturbation. By contrast, in the case of the frequency given by Equation (14.25), we chose to express it in terms of the maximum displacement $B = \hat{a}$. The choice of the expansion parameter affects the coefficients of the terms in the expansion. This should serve as a warning when comparing results obtained by different methods or results given in different sources. For example, in the book *Perturbation Methods* by Nayfeh, page 176, he obtains the result of Equation (14.17). This agrees, as he points out, with the traditional choice used in the method of averaging. Furthermore, note that since our system is bounded and conservative, we have closed trajectories and a freedom to choose our initial phase $\phi = 0$. Equivalently, we have the energy as a "calibration point" for our system. If we discuss dissipative systems, greater care must be exercised when comparing different expressions. In order to reinforce some of these ideas, we do the "finite-amplitude pendulum" problem as Example 4.2.

We will see, once again, that the interplay amongst the initial conditions, the amplitude, and the homogeneous solution is not a simple one.

Example 2.1 We illustrate in this simple example how one has to be careful regarding how the maximum displacement changes when a perturbation is introduced. Let us begin with the simple harmonic oscillator as our unperturbed system with energy $E = 8$. Let the maximum displacement x_{max} be 4.

(Neglect units throughout.) Then

$$V(x) = \tfrac{1}{2}x^2$$

and

$$E = \frac{1}{2}\left(\frac{dx}{dt}\right)^2 + \frac{1}{2}x^2$$

This system is perturbed so that the potential becomes

$$U(x) = \frac{1}{2}x^2 + \frac{\varepsilon}{4}x^4; \qquad \varepsilon = 0.02$$

(a) Assume that the energy of the perturbed system is $E_1 = 8$. We seek the maximum displacement $x_{1,\max}$ correct to two significant figures. Then

$$E_1 = \tfrac{1}{2}(x_{1,\max})^2 + \tfrac{1}{4}\varepsilon(x_{1,\max})^4$$

or, to the lowest order,

$$x_{1,\max} \approx \sqrt{2E_1}\left(1 - \frac{\varepsilon E_1}{2}\right)$$

Given that $\varepsilon = 0.02$, we have $x_{1,\max} \approx 3.7$.

(b) Consider the same form of the perturbation, but now maintain the maximum displacement at $x_{2,\max} = 4$. What is the total energy E_2 correct to two significant figures? We simply substitute the value of $x_{2,\max}$ in the equation for the energy and obtain

$$E_2 = 9.3$$

(c) With these results established, plot $V(x)$ versus x and $U(x)$ versus x on the same set of axes. Observe the turning points and E, E_1, E_2, $x_{\max}$, $x_{1,\max}$, and $x_{2,\max}$.

(d) The periods of oscillation are $T(E)$, $T_1(E_1)$, and $T_2(E_2)$ and are associated with each of the systems with the same index. Which system has the greatest period? Which has the smallest? This example is interesting because it reveals a possible confusion associated with specifying the nature of the perturbation. In particular, it is essential to make clear what is fixed and what is being varied. ■

Example 2.2 We will illustrate how the concept of "choice of origin" of a perturbation expansion can influence the form of the terms of the expansion. Consider Figure 14.1 in which we have points A and B separated by a distance ε that is small compared with each of their distances to another point P.

(Point A is closer than B is to point P.) The quantities x, y, and ε are vectors so that a distance formula will include $\cos(x, \varepsilon)$ or $\cos(y, \varepsilon)$. To simplify the discussion we consider the special case in which the points are collinear. The observer at point A considers his or her distance as the

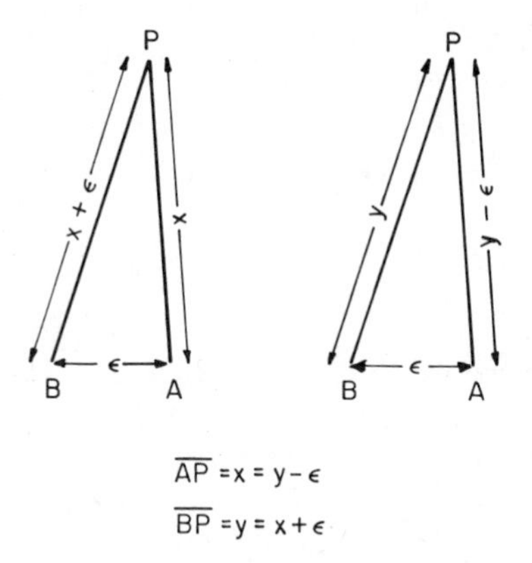

Figure 14.1 Here we find a perfectly simple example of why one has to try and clearly specify the nature of what is meant by the unperturbed system. One obtains a different form of the perturbation depending on whether one uses x or y as the basis of the unperturbed system.

unperturbed one and thus computes the quantity

$$\frac{1}{x} - \frac{1}{x+\varepsilon} = +\frac{\varepsilon}{x^2} - \frac{\varepsilon^2}{x^3} + O(\varepsilon^3)$$

as a measure of the difference in distance of points A and B from point P. On the other hand, the observer at point B chooses his or her distance as the unperturbed one and finds the difference of the distances to be

$$\frac{1}{y-\varepsilon} - \frac{1}{y} = +\frac{\varepsilon}{y^2} + \frac{\varepsilon^2}{y^3} + O(\varepsilon^3)$$

These expressions are the same to the lowest order and differ in the second order. The difference corresponds to a different choice for the origin of the perturbation expansion. This is quite similar to the situation we observed for the difference in the forms of the frequency for the Duffing oscillator. ■

14.3 A QUADRATIC OSCILLATOR

In all of our previous examples, the effect of the perturbation term led to a correction in the lowest-order calculation to either the frequency or amplitude of the unperturbed system. We will see that the quadratic oscillator requires for its analysis a second-order expansion.

Consider the "quadratic" oscillator given by the equation

$$\ddot{x} + x = -\varepsilon x^2; \qquad x(0) = a - \frac{\varepsilon a^2}{2}$$
$$\dot{x}(0) = 0; \qquad 0 < \varepsilon \ll 1 \tag{14.30}$$

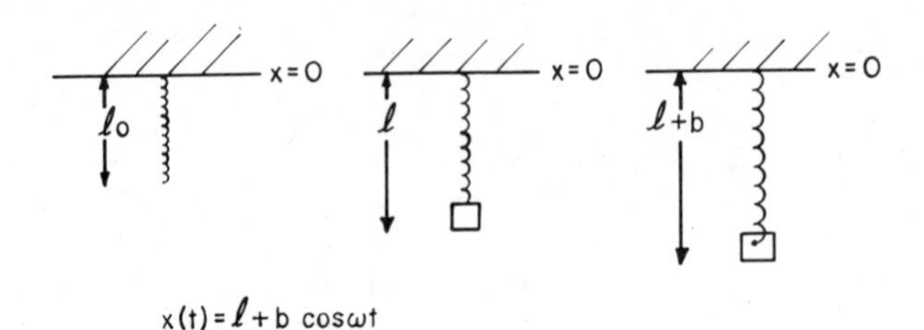

Figure 14.2 The oscillations of the mass m are restricted to the region $l - b \leq x \leq l + b$.

We want to find the lowest-order correction to the frequency of oscillation. In addition, we want to know whether there is, to the same order in ε, a correction to the amplitude. We use the MMTS and anticipate that we have to develop an expansion through the second order in ε.

We introduce a brief digression in Example 3.1.

Example 3.1 We illustrate the simplifications that arise when one chooses an appropriate origin of coordinates. Consider a mass m and a massless spring of unstretched length l_0 and spring constant k. See Figure 14.2.

When the mass is attached, the spring attains a length l such that $mg = k(l - l_0)$. It is then pulled down a small distance b and released. It executes simple harmonic motion described by the equation:

$$x(t) = l + b\cos\omega t; \qquad \omega = \sqrt{\frac{k}{m}}$$

For a linear system, it is often convenient to choose the origin of the coordinate system by introducing $y = x - l$. Then $y(t) = b\cos\omega t$. The gravitational field brings about a change in the zero point of the oscillation of the system. We find a similar result for the quadratic oscillator described by Equation (14.30), and in anticipation of this result, we chose the special initial conditions. ■

Return to the analysis of Equation (14.30).

(a) First, we construct a power-series expansion for the displacement:

$$x(t) = x_0(t_0, t_1, t_2; \varepsilon) + \varepsilon x_1(t_0, t_1, t_2; \varepsilon) + \varepsilon^2 x_2(t_0, t_1, t_2; \varepsilon) + O(\varepsilon^3) \tag{14.31a}$$

(b) Now we have to develop the second-order expansion of the differential operator:

$$\frac{d^2}{dt^2} = D_0^2 + 2\varepsilon D_0 D_1 + \varepsilon^2(2D_0D_2 + D_1^2) + \text{h.o.t.} \tag{14.31b}$$

(c) We also need

$$\varepsilon x^2 = \varepsilon x_0^2 + \varepsilon^2 2x_0x_1 + O(\varepsilon^3) \tag{14.31c}$$

(d) We expand the differential equation in powers of ε and have

$$O(\varepsilon^0): \qquad D_0^2x_0 + x_0 = 0$$

The solution is

$$x_0 = A(t_1, t_2)\exp it_0 + \text{c.c.} \tag{14.32a}$$

$$A = \tfrac{1}{2}a\exp i\phi; \qquad \phi(0) = 0 \tag{14.32b}$$

(e) We now find the solution to $O(\varepsilon^1)$:

$$D_0^2x_1 + x_1 = -\left[A^2\exp 2it_0 + 2AA^* + A^{*2}\exp(-2it_0)\right] - 2iD_1A\exp it_0 + 2iD_1A^*\exp(-it_0) \tag{14.33}$$

In order to suppress secular terms, it is necessary to set $D_1A = 0$ and, hence, neither the amplitude nor the phase has t_1 dependence. Thus, we write

$$A(t_2); \qquad \text{or} \quad a(t_2) \quad \text{and} \quad \phi(t_2)$$

We solve Equation (14.33) by choosing an appropriate trial solution and find

$$x_1 = -2AA^* + \left(\frac{A^2}{3}\right)\exp 2it_0 + \left(\frac{A^{*2}}{3}\right)\exp(-2it_0) - \left(\frac{1}{3}\right)AA^*\exp it_0 - \frac{1}{3}A^*A\exp(-it_0) \tag{14.34}$$

Notice that we have chosen the homogeneous solution

$$-C\exp it_0 + \text{c.c.}$$

$$C = \frac{1}{2}c\exp i\psi; \qquad \psi = 0; \qquad c = \frac{a^2}{6} \tag{14.35}$$

Our choice of the form of the homogeneous solution frees us from having to determine the time dependence of another phase. The term $-2AA^*$ contributes $-\varepsilon a^2/2$ to the initial conditions. We have chosen our initial conditions to emphasize that the perturbation leads to a shift in the "zero-point" or equilibrium position of the oscillation and that the oscillations are not symmetrical with respect to $x = 0$. The asymmetry follows because the perturbation does not change sign when $x \to$

$-x$. This is obvious if we write the lowest-order approximation to the perturbation term as

$$-\varepsilon x_0^2 = -\varepsilon(a\cos t_0)^2 = -\varepsilon\left(\frac{a^2}{2}\right) - \varepsilon\left(\frac{a^2}{2}\right)\cos 2t_0 \quad (14.36)$$

(f) We now construct the expansion to $O(\varepsilon^2)$:

$$D_0^2 x_2 + x_2 = -2D_0 D_2 x_0 - D_1^2 x_0 - 2D_0 D_1 x_1 - 2x_0 x_1 \quad (14.37)$$

We observe that

$$D_1^2 x_0 = 0 = D_0 D_1 x_1 \quad (14.38)$$

since A does not depend on the time scale t_1.

Expanding and gathering the terms that multiply $\exp it_0$, we have

$$\left[-2iD_2 A - 2\left(-2A^2A^* + \frac{A^2A^*}{3}\right)\right]\exp it_0 = 0 \quad (14.39a)$$

or

$$(iD_2 a - aD_2\phi)\exp i\phi = \frac{10}{3}A^2A^* = \frac{5}{12}a^3 \exp i\phi \quad (14.39b)$$

in order that secular terms be suppressed. Then

$$D_2 a = 0 \Rightarrow a = a(t_3) \quad (14.40a)$$

and

$$\tilde{\phi}(t_2) = \frac{-5}{12}a^2 t_2 + \tilde{\tilde{\phi}}(t_3) \quad (14.40b)$$

We are interested in the frequency of oscillation. It is

$$\omega = 1 - \frac{5}{12}\varepsilon^2 a^2 + O(\varepsilon^3) \quad (14.41)$$

Referring to Equation (14.34), we also have the solution:

$$\begin{aligned} x(t) &= a\cos\left[\left(1 - \frac{5}{12}\varepsilon^2 a^2\right)t\right] \\ &\quad + \varepsilon\left(-\tfrac{1}{2}a^2 + \tfrac{1}{6}a^2\cos 2t - \tfrac{1}{6}a^2\cos t\right) \\ &\quad + O(\varepsilon^2) \qquad \text{for } t = O\left(\frac{1}{\varepsilon}\right) \end{aligned} \quad (14.42)$$

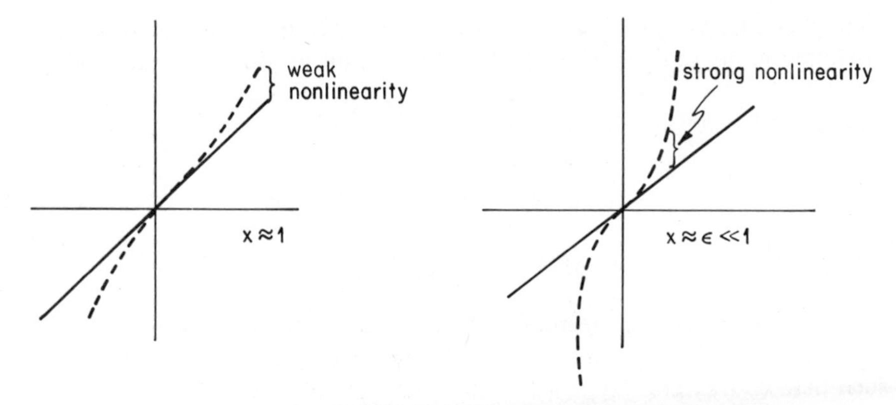

Figure 14.3 Our perturbed system is considered to deviate slightly from a simple harmonic oscillator. If the amplitude is large, then we speak of a weak nonlinearity. On the other hand, we can have a strong nonlinearity if the amplitude is sufficiently small.

EXERCISE 3.1 Consider Equation (14.30) and impose the initial conditions $x(0) = a$; $\dot{x}(0) = 0$. Repeat the calculations and obtain the equation for $x(t)$ analogous to Equation (14.42). Convince yourself that there is no substantial change in the algebra. ■

14.4 SCALING

In all of our examples so far, ε has been the small parameter in the problem. This meant that our amplitude of oscillation could be sizable. For example, we found that the radius of the limit cycle of the Van der Pol oscillator was $r^* = 2$. Also, our only condition on the magnitude of the amplitude of oscillation of the Duffing oscillator was that

$$\varepsilon a^3 \ll a \qquad \text{and} \qquad 0 < \varepsilon \ll 1$$

Thus, even though x can be appreciable, the nonlinear terms are small. See Figure 14.3.

We now consider problems in which it is the amplitude itself that serves as the small parameter. Thus, although the nonlinearity may be strong, we are in a region of small amplitude. See Figure 14.3.

This requires the development of some new strategies. We are given that the amplitude is small, so in the absence of a given small parameter, we introduce an artificial one. Namely, we write

$$x \to \varepsilon x$$

where $x = O(1)$. We then develop our expansion in powers of ε, and at the end of the calculation, we set $\varepsilon = 1$.

In effect, the parameter ε is a bookkeeping device that helps us to properly order the terms in the expansion. It requires a bit of practice to get it right.

Example 4.1 Consider the oscillations of a system described by the equation

$$\ddot{x} + \frac{x}{1 + x^2} = 0; \qquad |x| \ll 1$$
$$x(0) = a; \qquad \dot{x}(0) = 0 \tag{14.43}$$

Find the lowest order correction to the frequency of oscillation. Since we are given that our displacement is small, we introduce the artificial parameter ε such that $x \to \varepsilon x$ and expand the denominator to obtain

$$\varepsilon\ddot{x} + \varepsilon x\left[1 - (\varepsilon x)^2\right] = 0 + \text{h.o.t.} \tag{14.44}$$

or

$$\ddot{x} + x = +\varepsilon^2 x^3 \tag{14.45}$$

We observe that Equation (14.45) is of the same form as the Duffing oscillator given by Equation (14.1), with ε replaced by $-\varepsilon^2$. Thus, the frequency is written as

$$\omega = 1 - \tfrac{3}{8}(\varepsilon a)^2 \tag{14.46a}$$

Since ε was introduced only for convenience, we set it equal to 1 and our frequency is correctly given as

$$\omega = 1 - \tfrac{3}{8}a^2 \qquad \blacksquare \tag{14.46b}$$

EXERCISE 4.1 A particle executes small-amplitude oscillations in a potential well described by

$$V(x) = A\cosh x$$

The equation of motion is

$$\ddot{x} = \frac{-dV}{dx} \tag{14.47}$$

The particle is released from rest at time $t = 0$, such that $x(0) = a \ll 1$.

(a) Plot $V(x)$ versus x for $|x| < 1$.
(b) Find the dominant terms in the equation of motion so that you will be able to solve for the lowest-order frequency correction.

(c) In order to obtain the frequency correction, it is helpful to "fix" Equation (14.47) so that it is in the same form as the Duffing oscillator, Equation (14.1). Rescale the time so that the unperturbed frequency is equal to 1. Then since the displacement is small, scale the amplitude so as to introduce a small parameter. The result is

$$x(t) = a \cos \omega t \tag{14.48a}$$

with

$$\omega = \left(1 + \frac{a^2}{16}\right)\sqrt{A} \qquad \blacksquare \tag{14.48b}$$

14.4.1 The Pendulum

Example 4.2 This is a detailed example involving many algebraic manipulations. Consider the oscillations of a pendulum described by the equation

$$\ddot{x} + \sin x = 0; \qquad |x| \ll 1 \tag{14.49a}$$

$$x(0) = B; \quad \dot{x}(0) = 0 \tag{14.49b}$$

We want to find the frequency of oscillation and the displacement for finite amplitudes of oscillation. Specifically, we have in mind to develop a second-order perturbation expansion. Thus, we expand $\sin x$ and find

$$\sin x = x - \frac{x^3}{3!} + \frac{x^5}{5!} + O(x^7) \tag{14.50}$$

If $|x| \ll 1$, the neglected terms are small as compared with those that are included. This provides the basis for a perturbation expansion. We scale the amplitude by introducing a small parameter ε. We then construct a power-series expansion in this parameter and equate the corresponding terms in the equation. After we have proceeded with the expansion to the desired degree of accuracy, we set the parameter equal to 1. If we have done things correctly, we will have a consistent expansion in terms of the amplitude of oscillation. Let us see how it works.

(a) First, we scale the displacement by $\sqrt{\varepsilon}$. We see that this leads to an expansion in powers of ε:

$$x \to \sqrt{\varepsilon}\, x \tag{14.51}$$

Then Equation (14.49a) becomes

$$\sqrt{\varepsilon}\, \ddot{x} + \sin\sqrt{\varepsilon}\, x = 0 \tag{14.52}$$

We expand $\sin\sqrt{\varepsilon}\,x$ and obtain

$$\sin\sqrt{\varepsilon}\,x = \sqrt{\varepsilon}\left(x - \frac{\varepsilon x^3}{3!} + \varepsilon^2 \frac{x^5}{5!} + \text{h.o.t.}\right)$$

We see that after dividing through by a factor $\sqrt{\varepsilon}$, Equation (14.49a) has become

$$\ddot{x} + x - \frac{\varepsilon x^3}{3!} + \frac{\varepsilon^2 x^5}{5!} = 0 \tag{14.53}$$

where we have neglected h.o.t. in the expansion. We now introduce expansions of the displacement and the differential operator:

$$x = x_0 + \varepsilon x_1 + \varepsilon^2 x_2 + O(\varepsilon^3) \tag{14.54a}$$

$$\frac{d^2}{dt^2} = D_0^2 + 2\varepsilon D_0 D_1 + \varepsilon^2\left(D_1^2 + 2D_0 D_2\right) + O(\varepsilon^3) \tag{14.54b}$$

The expansion of nonlinear terms in Equation (14.53) becomes

$$-\frac{\varepsilon x_0^3}{3!} + \varepsilon^2\left(\frac{x_0^5}{5!} - \frac{3x_0^2 x_1}{3!}\right) + O(\varepsilon^3) \tag{14.55}$$

(b) The solution to Equation (14.53) to $O(\varepsilon^0)$ is

$$x_0 = A \exp it_0 + \text{c.c.} \tag{14.56a}$$

$$A = \tfrac{1}{2}\hat{a} \exp i\phi; \qquad \phi(0) = 0 \tag{14.56b}$$

$$x(0) = \hat{a}; \qquad \dot{x}(0) = 0 \tag{14.56c}$$

We have introduced the notation $\hat{a}$ to indicate that our amplitude $\hat{a}$ is equal to the maximum displacement B. This notation is helpful later in the analysis.

(c) We now construct the solution to Equation (14.53) to $O(\varepsilon)$

$$\begin{aligned} D_0^2 x_1 + x_1 &= +\tfrac{1}{6}x_0^3 - 2D_0 D_1 x_0 \\ &= \tfrac{1}{6}\left(A^3 \exp 3it_0 + 3A^2 A^* \exp it_0 + \text{c.c.}\right) \\ &\quad + \left(-2iD_1 A \exp it_0 + \text{c.c.}\right) \end{aligned} \tag{14.57}$$

We need to suppress terms multiplying $\exp it_0$. Thus, our secularity conditions becomes

$$D_1 A = \frac{-i}{4} A^2 A^* \tag{14.58a}$$

and

$$D_1 A^* = \frac{+i}{4} A A^{*2} \tag{14.58b}$$

Hence,

$$\phi(t_1) = \tfrac{-1}{16}\hat{a}^2 t_1 + \tilde{\phi}(t_2, t_3) \tag{14.59}$$

We then assume a suitable trial solution for $x_1(t)$ and find

$$x_1(t) = -\left(\frac{A^3}{48}\exp 3it_0 + \text{c.c.}\right) + \left(\frac{A^2A^*}{48}\exp it_0 + \text{c.c.}\right) \tag{14.60}$$

We have chosen our homogeneous solution so that we have

$$x_1(0) = \dot{x}_1(0) = 0$$

Note that we have chosen to satisfy our initial conditions by the lowest-order solution x_0. We review this point after our calculation is completed.

Observe how our choice of power of the expansion parameter in Equation (14.51) enables us to get the appropriate terms to enter the expansion in the correct order. Clearly, this requires some practice. We now calculate terms of $O(\varepsilon^2)$:

$$D_0^2 x_2 + x_2 = \frac{1}{2}x_0^2 x_1 - \frac{x_0^5}{120} - 2D_0 D_1 x_1 - D_1^2 x_0 - 2D_0 D_2 x_0 \tag{14.61}$$

In order to satisfy our secularity condition, we identify the coefficients of $\exp it_0$ in each term on the r.h.s. of Equation (14.61). After using the secularity conditions expressed by Equations (14.58a) and (14.58b) and some related algebra, we have

$$\frac{+1}{2}x_0^2 x_1 \to \frac{1}{48}A^3 A^{*2}$$

$$\frac{-x_0^5}{120} \to \frac{-1}{12}A^3 A^{*2}$$

$$-2D_0 D_1 x_1 \to \frac{-1}{96}A^3 A^{*2}$$

$$-D_1^2 x_0 \to \frac{+1}{16}A^3 A^{*2}$$

Then our secularity condition becomes

$$2D_0D_2(A\exp it_0) = -\frac{1}{96}A^3A^{*2}\exp it_0 \tag{14.62}$$

We observe that the amplitude is constant and that our equation for the phase becomes

$$-\hat{a}D_2\phi = -\frac{1}{96}\frac{\hat{a}^5}{32} = -\frac{1}{3072}\hat{a}^5 \tag{14.63a}$$

or

$$\tilde{\phi}(t_2) = +\frac{1}{3072}\hat{a}^4t_2 + \tilde{\tilde{\phi}}(t_3) \tag{14.63b}$$

Our expression for the frequency is

$$\omega = \left[1 - \frac{1}{16}\hat{a}^2 + \frac{1}{3072}\hat{a}^4 + O(\hat{a}^6)\right] \tag{14.64}$$

where we have put the parameter ε equal to 1. It is important to observe that we have satisfied our initial conditions given by Equation (14.49b) by the zero-order solution. Therefore, we have

$$x(0) = B = \hat{a} + O(\hat{a}^5) = x_{\max} \tag{14.65}$$

Our expression for the displacement is

$$\begin{aligned} x(t) = \hat{a}\cos&\left\{\left[1 - \frac{1}{16}\hat{a}^2 + \frac{1}{3072}\hat{a}^4 + O(\hat{a}^6)\right]t\right\} \\ &+ \frac{\hat{a}^3}{192}\left[\cos\left\{\left[1 - \frac{1}{16}\hat{a}^2 + O(\hat{a}^4)\right]t\right\}\right. \\ &\left. -\cos\left\{3\left[1 - \frac{1}{16}\hat{a}^2 + O(\hat{a}^4)\right]t\right\}\right] + O(\hat{a}^5) \\ &\text{for } t = O(1/\hat{a}^2) \end{aligned} \tag{14.66}$$

We are now ready to repeat the calculation of $x(t)$, neglecting the homogeneous solution. The first change comes at Equation (14.60), which becomes

$$x_1(t) = -\left(\frac{A^3}{48}\exp 3it_0 + \text{c.c.}\right) \tag{14.67}$$

The initial condition, given by Equation (14.49b), is then

$$x(0) = B = a - \frac{a^3}{192} + O(a^5) \tag{14.68}$$

We distinguish our amplitude a from the amplitude $\hat{a}$, since the former is not equal to the maximum displacement B.

When we consider our secularity condition, we note that the term $-D_0 D_1 x_1$ does not contribute, since x_1 lacks a term multiplying $\exp it_0$. Collecting terms, we have

$$+\frac{1}{2}x_0^2 x_1 \to -\frac{1}{96}A^3 A^{*2}$$

$$-\frac{x_0^5}{120} \to -\frac{1}{12}A^3 A^{*2}$$

$$-D_1^2 x_0 \to +\frac{1}{16}A^3 A^{*2}$$

Thus, our secularity condition becomes

$$2D_0 D_2 (A \exp it_0) = \frac{-1}{32}A^3 A^{*2} \exp it_0 \tag{14.69}$$

The amplitude of oscillation is constant and our equation for the phase becomes

$$-aD_2\phi = -\frac{1}{32}\frac{1}{32}a^5$$

or

$$\tilde{\phi}(t_2) = +\frac{1}{1024}a^4 t_2 + \tilde{\tilde{\phi}}(t_3) \tag{14.70}$$

Our frequency becomes

$$\omega = 1 - \frac{1}{16}a^2 + \frac{1}{1024}a^4 + O(a^6) \tag{14.71}$$

and our displacement $x(t)$ is given by

$$x(t) = a\cos\left\{\left[1 - \frac{1}{16}a^2 + \frac{1}{1024}a^4 + O(a^6)\right]t\right\}$$
$$-\frac{a^3}{192}\cos\left\{3\left[1 - \frac{1}{16}a^2 + O(a^4)\right]t\right\} + O(a^5) \qquad \text{for } t = O(1/a^2) \tag{14.72}$$

where we have put our parameter ε equal to 1. We have a situation that parallels the results for the Duffing oscillator. We have two apparently different expressions for the frequency and the displacement given by Equations (14.64), (14.66), (14.71), and (14.72). We now know to look at the initial conditions. We have Equation (14.65):

$$x(0) = B = \hat{a} + O(\hat{a}^5)$$

and Equation (14.68):

$$x(0) = B = a - \frac{a^3}{192} + O(a^5)$$

We can relate the two a's by comparing them to the maximum displacement B:

$$B = \hat{a} = a - \frac{a^3}{192} \qquad \text{or, equivalently,} \qquad a = \hat{a} + \frac{\hat{a}^3}{192} \tag{14.73}$$

to this order.

Thus, for example, we can replace each a occurring in Equations (14.71) and (14.72) by using its relationship to $\hat{a}$ given by Equation (14.73). These equations then become Equations (14.64) and (14.66), respectively. (We find a detailed derivation of the frequency and displacement for the pendulum problem on pages 62–66 in the excellent text by Bogoliubov and Mitropolsky. They neglect the homogeneous solution and hence do not express their final results in terms of the maximum displacement B.) This example should clearly point out that you have to understand the meaning of symbols you use and their relationships to the ones used by others. It is not a good practice to simply compare formulas given in different texts. This sounds like gratuitous advice, but forewarned is forearmed. ■

14.4.2 Nonlinear Damping

Consider the nonlinear damped oscillator discussed in the text by Kevorkian and Cole, pages 127–129:

$$\begin{aligned} &\ddot{x} + x = -\varepsilon\alpha\left(\frac{dx}{dt}\right)^3 - 3\nu\varepsilon^2\frac{dx}{dt}; \qquad \varepsilon, \alpha, \nu > 0, \quad \alpha, \nu = O(1) \\ &x(0) = a; \qquad \dot{x}(0) = 0 \end{aligned} \tag{14.74}$$

Without solving, we conclude that if $\alpha = 0$, we have exponential decay:

$$x(t) \sim \exp\left(-\tfrac{3}{2}\nu\varepsilon^2 t\right) \cdot (\text{oscillatory component})$$

Similarly, if $\nu = 0$, we know that the amplitude decays like $1/\sqrt{t}$. (We solved this problem in Chapter 13 as Exercise 5.2.)

Let us write Equation (14.74) as a set of coupled first-order equations:

$$\dot{x} = -y \tag{14.75a}$$

$$\dot{y} = x - 3\nu\varepsilon^2 y - \varepsilon\alpha y^3 \tag{14.75b}$$

We want to think of our nonlinear term as a perturbation of a linear system. If we focus on the linear components, we observe that the trace of our matrix equals $-3\nu\varepsilon^2$ (indicating positive damping) and the determinant is 1. This is fine except that the coefficient of the linear damping term is $O(\varepsilon^2)$ and the coefficient of the nonlinear damping term is $O(\varepsilon)$. Thus, we have the following question: How can we "scale," or adjust, our amplitudes so as to develop a perturbation expansion in which the linear aspects dominate for all sufficiently small amplitudes? In order to frame an answer, we consider an initial condition in which the amplitude is sufficiently great that the cubic damping term dominates. Thus, if we have the inequality

$$\alpha\varepsilon\left(\frac{dx}{dt}\right)^3 > 3\nu\varepsilon^2\frac{dx}{dt}; \qquad t \approx 0 \tag{14.76}$$

the system starts to decay like $1/\sqrt{t}$. However, after some time, the inequality surely fails. This happens after

$$\alpha\varepsilon x_{\max}^3 = 3\nu\varepsilon^2 x_{\max}$$

or

$$x_{\max} = \sqrt{\varepsilon}\sqrt{\frac{3\nu}{\alpha}} \tag{14.77}$$

since at this point the two terms are comparable. Then for all times after this, the exponential-decay term dominates. So we see that there is an exchange of dominance in the time evolution of the decay when the amplitude is $O(\sqrt{\nu\varepsilon/\alpha})$. This means that the natural scaling given by the small parameter ε is insufficient to characterize the motion of the system for all times. If we want to develop an expansion for the amplitude that will be uniform in time, we must scale the amplitude a like $O(\sqrt{\varepsilon})$. This is done by introducing

$$\hat{x} = \frac{x}{\sqrt{\varepsilon}}; \qquad \hat{y} = \frac{y}{\sqrt{\varepsilon}} \tag{14.78}$$

Then Equation (14.74) becomes

$$\frac{d^2\hat{x}}{dt^2} + \hat{x} = -\varepsilon^2\left[\alpha\left(\frac{d\hat{x}}{dt}\right)^3 + 3\nu\frac{d\hat{x}}{dt}\right] \tag{14.79a}$$

where the initial conditions are now

$$\hat{x}(0) = \frac{a}{\sqrt{\varepsilon}}; \qquad \frac{d\hat{x}(0)}{dt} = 0 \tag{14.79b}$$

(Alternatively, we can return to our set of coupled first-order equations, Equations (14.75a) and (14.75b), which now take the form

$$\frac{d\hat{x}}{dt} = -\hat{y} \tag{14.80a}$$

$$\frac{d\hat{y}}{dt} = \hat{x} - \varepsilon^2(3\nu\hat{y} + \alpha\hat{y}^3) \tag{14.80b}$$

and observe that both damping terms enter in the same order of the calculation.) We now have a simple problem in which we have a lightly damped oscillator. We introduce the same expansion procedures that we developed previously, with ε^2 as our expansion parameter:

$$\hat{x} = \hat{x}_0 + \varepsilon^2\hat{x}_1 + \text{h.o.t.}$$

$$\frac{d^2}{dt^2} = D_0^2 + 2\varepsilon^2 D_0 D_1 + \text{h.o.t.}$$

and obtain

$$\hat{x}_0 = A \exp it_0 + \text{c.c.} \tag{14.81a}$$

$$A = \tfrac{1}{2}\hat{r} \exp i\phi; \qquad \phi(0) = 0; \qquad \hat{r} = \frac{a}{\sqrt{\varepsilon}} \tag{14.81b}$$

We then solve to $O(\varepsilon^2)$:

$$\begin{aligned} D_0^2\hat{x}_1 + \hat{x}_1 = &-2iD_1 A \exp it_0 + 2iD_1 A^* \exp(-it_0) \\ &- [3\nu iA \exp it_0 - 3\nu iA^* \exp(-it_0)] \\ &- \alpha[iA \exp it_0 - iA^* \exp(-it_0)]^3 \end{aligned} \tag{14.82}$$

The condition for suppression of secular terms leads to the requirement

$$2D_1 A + 3\nu A + 3\alpha A^2 A^* = 0 \tag{14.83}$$

There is no adjustment to the phase in this order. The equation for the amplitude takes the form

$$D_1\hat{r} = -\tfrac{3}{2}\nu\hat{r} - \tfrac{3}{8}\alpha\hat{r}^3$$

or

$$\frac{d\hat{r}}{\hat{r}(1 + \alpha\hat{r}^2/4\nu)} = -\frac{3}{2}\nu\, dt_1 \tag{14.84}$$

Note the standard integral

$$\int_{z(0)}^{z} \frac{dz}{z(1 + bz^2)} = \frac{1}{2} \ln \frac{z^2[1 + bz(0)^2]}{(1 + bz^2)z(0)^2} \tag{14.85}$$

We integrate Equation (14.84) and rearrange terms to obtain

$$\hat{r}(t) = \frac{2\sqrt{\nu/\alpha}}{\{[1 + (4\nu\varepsilon/\alpha a^2)](\exp 3\nu t_1) - 1\}^{1/2}} \tag{14.86a}$$

Finally, we express this result in terms of the true displacement $x(t)$

$$x(t) = \frac{2\sqrt{\varepsilon\nu/\alpha}\, \cos(t + \phi)}{\{[1 + (4\nu\varepsilon/\alpha a^2)](\exp 3\nu t_1) - 1\}^{1/2}} \tag{14.86b}$$

Let us see if our result makes sense. It satisfies the initial condition $x(0) = a$ if we choose $\phi(0) = 0$. The solution satisfies the initial condition on the velocity with an error $O(\varepsilon)$. For sufficiently short times, we can expand the exponential in Equation (14.86b) and obtain

$$\exp(3\nu\varepsilon^2 t) \sim 1 + 3\nu\varepsilon^2 t \tag{14.87}$$

or, upon rearranging, we have

$$x(t) = \frac{a\cos(t + \phi)}{(1 + 3\varepsilon\alpha a^2 t/4)^{1/2}} \tag{14.88}$$

This is precisely what we would have obtained if we made the coefficient of the linear damping term equal zero. Thus, as we anticipated, or "arranged our expansion," the primary damping initially comes from the cubic term. For sufficiently long times t, we also have

$$x(t) \sim \sqrt{\varepsilon}\, \exp(-3\nu\varepsilon^2 t/2)\cos(t + \phi) \tag{14.89}$$

indicating that the cubic damping term has become unimportant and the system is entirely under the influence of the linear terms.

EXERCISE 4.2 Consider Equations (14.75a) and (14.75b). Use the MOA to obtain $r(t)$. In addition, show that if the initial amplitude is large, the cubic damping term dominates or that the amplitude initiates its decay like $1/\sqrt{t}$. ■

14.4.3 An Oscillator with Quadratic and Cubic Terms

Example 4.3 Consider an oscillator problem in which there are quadratic and cubic terms:

$$\ddot{z} + z = -\alpha z^2 - \beta z^3; \qquad \alpha, \beta = O(1), \quad |z| \ll 1 \tag{14.90a}$$

$$z(0) = a - \frac{\alpha a^2}{2}; \qquad \dot{z}(0) = 0 \tag{14.90b}$$

(This problem is discussed in Nayfeh's text, *Introduction to Perturbation Methods*.)

We think of our system as a simple harmonic oscillator that is perturbed by nonlinear terms. Thus, we require the initial amplitude to be sufficiently small so that the quadratic and cubic terms are small compared with the linear ones. The system is conservative and the initial velocity is zero; thus, the nonlinear terms are small for all time.

Since there are no small parameters, we have to scale the amplitude of the oscillation so that the linear terms dominate. This is necessary if we are to satisfy our requirement that the unperturbed problem be a SHO. This is accomplished by requiring that $|z| \ll 1$. In recognition of this, we introduce

$$z = \varepsilon x \tag{14.91}$$

Then Equation (14.90) becomes

$$\ddot{x} + x = -\varepsilon \alpha x^2 - \varepsilon^2 \beta x^3 \tag{14.92}$$

We note that if the amplitude of our oscillation is sufficiently small, there is a regime in which the quadratic term dominates the cubic term. We also know from our experience with a "quadratic" oscillator that there is no secularity condition in the lowest order. Since the cubic term leads to a secularity condition in the lowest order, we have to be certain that we have scaled our variables accordingly. In other words, we want the quadratic terms to enter the computation one step earlier than the cubic ones. We expand the amplitude and the differential operator and obtain, to order ε,

$$D_0^2 x_0 + x_0 = 0 \tag{14.93}$$

with the solution

$$x_0 = A \exp it_0 + \text{c.c.}; \qquad A = \tfrac{1}{2} r \exp i\phi \tag{14.94a}$$

$$r(0) = \frac{a}{\varepsilon}; \qquad \phi(0) = 0 \tag{14.94b}$$

(Note that since $a = O(\varepsilon)$, $r(0)$ is of order 1.)

We now solve to order ε:

$$D_0^2 x_1 + x_1 = -2iD_1 A \exp it_0 + 2iD_1 A^* \exp(-it_0) \\ - \alpha\left[A \exp it_0 + A^* \exp(-it_0)\right]^2 \tag{14.95}$$

Note that we have the same equation as we had previously for the quadratic oscillator. (See Equation 14.33.) This is not surprising, since at this point in the expansion, the cubic terms have not appeared. Our secularity condition becomes

$$D_1 A = 0 \tag{14.96}$$

so that $A = A(t_2)$. It is straightforward to solve for x_1:

$$x_1 = \alpha\left[-2AA^* + \frac{A^2}{3}\exp 2it_0 + \frac{A^{*2}}{3}\exp(-2it_0) \\ - C \exp it_0 - C^* \exp(-it_0)\right] \tag{14.97a}$$

$$C = \frac{1}{2} c \exp i\psi; \qquad c = \frac{r^2}{6}; \qquad \psi(0) = \phi(0) \tag{14.97b}$$

A particular choice for the constant C might be $C = AA^*$ leading to $\psi = 0$. [Keep in mind that in order to satisfy the initial conditions with the zero-order approximation we only require that $C = A^3/3$ and $\psi(0) = \phi(0)$.] Observe that the term $-2\alpha AA^*$ contributes to the satisfaction of the initial conditions given by Equation (14.90b).

We are now ready to find the term of $O(\varepsilon^2)$. Our equation is

$$D_0^2 x_2 + x_2 = -2D_0 D_2 x_0 - D_1^2 x_0 - 2\alpha x_0 x_1 - \beta x_0^3 \tag{14.98}$$

We expand x_0 and x_1 and collect the coefficients of $\exp it_0$ and require

$$-2iD_2 A - 2\alpha^2\left(-2A^2A^* + \frac{A^2A^*}{3}\right) - 3\beta A^2 A^* = 0 \tag{14.99}$$

The amplitude to this order is constant, and the phase equation is

$$rD_2\phi = \frac{-5}{12}\alpha^2 r^3 + \frac{3}{8}\beta r^3 \tag{14.100a}$$

or

$$\phi = \frac{-5}{12}\alpha^2 r^2 t_2 + \frac{3}{8}\beta r^2 t_2 \tag{14.100b}$$

or, finally, setting $\varepsilon = 1$, $z = x$, and $r = a$, we have as our frequency

$$\omega = 1 - \frac{5}{12}\alpha^2 a^2 + \frac{3}{8}\beta a^2 + O(a^4) \tag{14.101}$$

Thus, we see that both the quadratic and the cubic terms contribute correction terms to the frequency of oscillation. If we set the coefficients $\beta = 0$ and $\alpha = \varepsilon$, we obtain Equation (14.41), and if we set $\alpha = 0$, we obtain the frequency correction associated with the Duffing oscillator. (This is given, for example, by the lowest-order corrections in Equation (14.25).) If we review the computation, we observe that contributions from the quadratic and cubic terms are disconnected. Thus, it is expected that the frequency correction be a sum of the correction terms from the quadratic and the Duffing oscillator. (This is the same result as is given in Nayfeh's book, *Introduction to Perturbation Techniques*, page 167.)

This is an important example since it is done incorrectly in a variety of textbooks. Why? Often, one forgets to scale the variables. Furthermore, there are old-fashioned methods, for example, "The Method of Equivalent Linearization," or "The Method of Harmonic Balance," that are totally restricted to lowest-order calculations and, therefore, cannot be adapted to problems of this type. (See Section 14.4.4.) We also note in this regard that in order to study this problem, it is necessary to develop the more elaborate method of averaging or use the method of multiple time scales. ■

14.4.4 Comment on the Method of Harmonic Balance

The method of harmonic balance, or equivalent linearization, was introduced in the text by Krylov and Bogoliubov. It is also discussed in the classic books by Bogoliubov and Mitropolsky and by Minorsky as well as the modern book by Jordan and Smith. It is shown to be equivalent to the elementary method of averaging, i.e., the MOA, and is widely used because it is very simple to learn and to apply. However, we have not given an exposition of it because, although it is useful, it gives unreliable results. In particular, it fails to give the correct quantitative result when the underlying problem requires a scaling of the variables. This happens because there is a mixing of terms in the orders of the expansion. We illustrate this by discussing Problem 25 that is given on page 122 of the text by Jordan and Smith.

Example 4.4 Consider

$$\ddot{x} = x^2 - x^3 \tag{14.102}$$

We locate the equilibrium points at $x = 0$ and $x = 1$, and study the small-amplitude oscillations about $x = 1$. In particular, we seek the lowest-order correction to the frequency of oscillation. To do this, we introduce $x = 1 + u$,

$|u| \ll 1$, and we have

$$\ddot{u} + u = -2u^2 - u^3 \tag{14.103a}$$

We choose initial conditions

$$u(0) = a - a^2; \qquad \dot{u}(0) = 0 \tag{14.103b}$$

Then this problem is identical to Example 4.3 with $\alpha = 2$ and $\beta = 1$. The frequency is then given by Equation 14.101:

$$\begin{aligned} \omega &\approx 1 - \frac{(4)(5)}{12}a^2 + \frac{3}{8}a^2 \\ &\approx 1 - \frac{31}{24}a^2 \end{aligned} \tag{14.104}$$

Let us now briefly outline the method of Harmonic balance. It assumes a solution of Equation (14.103a) in the form

$$u = c + a\cos\omega t \tag{14.105}$$

One then substitutes this form of the solution into Equation (14.103a) and equates the coefficients of the constant terms and also those terms containing $\cos\omega t$. This yields the small shift in the zero point of oscillation. It also gives us the lowest-order correction to the frequency associated with the cancellation of the secular generating terms in the expansion. Upon substitution of the trial solution, we have

$$\begin{aligned} -a\omega^2\cos\omega t + a\cos\omega t + c &= -\left(c^2 + 2ac\cos\omega t + a^2\cos^2\omega t\right) \\ &= -\left(c^3 + 3c^2a\cos\omega t + 3ca^2\cos^2\omega t + a^3\cos^3\omega t\right) \end{aligned} \tag{14.106}$$

We first write the trigonometric identities

$$\begin{aligned} \cos^2\omega t &= \tfrac{1}{2}(1 + \cos 2\omega t) \\ \cos^3\omega t &= \tfrac{1}{4}(3\cos\omega t + \cos 3\omega t) \end{aligned}$$

Equating constant terms in Equation (14.106) we have

$$c + 2c^2 + a^2 + c^3 + \tfrac{3}{2}ca^2 = 0 \tag{14.107}$$

Since the oscillations are small, we have to the lowest order $c = -a^2$. Equating the coefficients of $\cos\omega t$, we find

$$-a\omega^2 + a + 4ac + 3c^2a + \tfrac{3}{4}a^3 = 0 \tag{14.108}$$

After substituting $c = -a^2$ and retaining the lowest-order terms, we have

$$\omega^2 = 1 - \frac{13}{4}a^2 \quad \text{or} \quad \omega \approx 1 - \frac{13}{8}a^2 \tag{14.109}$$

Thus, we have a result that is in disagreement with Equation (14.104). Reviewing Equation (14.99), we see that there are three terms that are included in the suppression of secular terms in this problem. The term $2\alpha^2 A^2 A^*/3$ that comes from the interplay of x_1 and x_0 is missed by the method of harmonic balance. This is not surprising since the method is limited to first-order calculations. Although the numerical discrepancy between Equations (14.104) and (14.109) is small, for small amplitudes, the method of harmonic balance has deficiencies in principle. Thus, it should be used with caution. ■

14.5 LIMIT CYCLES ARISING FROM "QUADRATIC TERMS"

We found in our discussion of the first-order calculations that the interplay of linear and cubic velocity terms was capable of leading to limit-cycle behavior. For example, both the Rayleigh oscillator and the Van der Pol oscillator have this characteristic. This result is so striking that if you are a bit careless, you will come to the conclusion that one requires linear and cubic terms in order to obtain limit cycles. We now show how it is possible to obtain a limit cycle with a system of equations that contain only linear and quadratic velocity terms. We consider two related examples.

Example 5.1 We want to show how a limit cycle arises from a set of coupled first-order equations that contain linear and quadratic terms. (This example is based on the problem discussed in the book by Andronov et al. on pages 264–265. They attribute the example to Bautin.)

$$\frac{dx}{dt} = y \tag{14.110a}$$

$$\frac{dy}{dt} = -x + \rho y + \beta xy + \alpha y^2 + \delta x^2 \tag{14.110b}$$

$$\beta, \alpha, \delta = O(1)$$

We see that there are no small parameters multiplying the nonlinear terms, so we have to consider variables x and y themselves to be small. Thus, we write

$$x = \varepsilon x_1(t_0, t_1, t_2; \varepsilon) + \varepsilon^2 x_2(t_0, t_1, t_2; \varepsilon) + \varepsilon^3 x_3(t_0, t_1, t_2; \varepsilon) + \text{h.o.t.} \tag{14.111a}$$

$$y = \varepsilon y_1(t_0, t_1, t_2; \varepsilon) + \varepsilon^2 y_2(t_0, t_1, t_2; \varepsilon) + \varepsilon^3 y_3(t_0, t_1, t_2; \varepsilon) + \text{h.o.t.} \tag{14.111b}$$

We want the problem to be a simple harmonic oscillator as its lowest approximation, so we have to assume a scaling of the amplitude of the oscillation. If the damping-term coefficient ρ scales like ε, we know from previous calculations that we have an increasing spiral and that in the lowest order of the perturbation expansion, the quadratic terms do not suppress it. (This becomes clear when we solve for x_2 and y_2 and observe that they have no terms multiplying $\exp it_0$.)

With this observation, we assume that the coefficient ρ scales like ε^2. We write $\rho = \varepsilon^2\sigma$. Then the damping term does not enter into the first-order calculation. (In the MMTS, one often has to guess the correct scaling and then proceed with the calculation and see if the result makes sense.) We now develop the expansion through $O(\varepsilon^3)$.

Our differential operator is

$$\frac{d^2}{dt^2} = D_0^2 + \varepsilon 2D_0D_1 + \varepsilon^2(2D_0D_2 + D_1^2) + \text{h.o.t.} \qquad (14.112)$$

and we have, to $O(\varepsilon)$,

$$D_0^2x_1 + x_1 = 0 \qquad (14.113)$$

and

$$x_1 = A(t_1, t_2)\exp it_0 + \text{c.c.} \qquad (14.114a)$$

$$y_1 = D_0x_1 = iA(t_1, t_2)\exp it_0 + \text{c.c.} \qquad (14.114b)$$

We now solve to $O(\varepsilon^2)$:

$$D_0^2x_2 + x_2 = -2D_0D_1x_1 + \beta x_1y_1 + \alpha y_1^2 + \delta x_1^2 \qquad (14.115)$$

We see that the terms x_1y_1, x_1^2, and y_1^2 have no terms that go like $\exp it_0$. Thus, we have

$$D_1x_1 = 0 \quad \text{or} \quad A = A(t_2) \qquad (14.116)$$

Then, substituting our known solutions x_1 and y_1, given by Equations (14.114a) and (14.114b), we have

$$\begin{aligned} D_0^2x_2 &+ x_2 \\ &= \beta[A\exp it_0 + A^*\exp(-it_0)][iA\exp it_0 - iA^*\exp(-it_0)] \\ &\quad + \alpha[-A^2\exp 2it_0 + 2AA^* - A^{*2}\exp(-2it_0)] \\ &\quad + \delta[A^2\exp 2it_0 + 2AA^* + A^{*2}\exp(-2it_0)] \end{aligned} \qquad (14.117)$$

We can write Equation (14.117) as

$$D_0^2 x_2 + x_2 = \left(i\beta A^2 - \alpha A^2 + \delta A^2\right)\exp 2it_0 + \left(-i\beta A^{*2} - \alpha A^{*2} + \delta A^{*2}\right)\exp(-2it_0) + 2AA^*(\alpha + \delta) \tag{14.118}$$

We then assume a solution of the form

$$x_2 = \left[B(t_1, t_2)\exp 2it_0 + \text{c.c.}\right] + 2AA^*(\alpha + \delta) \tag{14.119}$$

We then have

$$B = -\tfrac{1}{3}(i\beta - \alpha + \delta)A^2 \tag{14.120}$$

or

$$x_2 = 2AA^*(\alpha + \delta) - \tfrac{1}{3}A^2(i\beta - \alpha + \delta)\exp 2it_0 - \tfrac{1}{3}A^{*2}(-i\beta - \alpha + \delta)\exp(-2it_0) \tag{14.121a}$$

$$y_2 = -\tfrac{1}{3}(2i)A^2(i\beta - \alpha + \delta)\exp 2it_0 + \tfrac{1}{3}(2i)A^{*2}(-i\beta - \alpha + \delta)\exp(-2it_0) \tag{14.121b}$$

We choose not to include the homogeneous solution.

We now solve to $O(\varepsilon^3)$:

$$D_0^2 x_3 + x_3 = -2D_0D_1x_2 - 2D_0D_2x_1 - D_1^2x_1 + \sigma y_1 + \beta(x_1y_2 + x_2y_1) + \alpha(2y_1y_2) + \delta(2x_1x_2) \tag{14.122}$$

We multiply out all the terms and collect those that are coefficients of $\exp it_0$. We have

$$\begin{aligned}
\sigma y_1 &\to \sigma iA \\
\beta(x_1y_2 + x_2y_1) &\to \beta\left[A^*A^2\left(-\tfrac{2}{3}i\right)(i\beta - \alpha + \delta) + 2AA^*(\alpha + \delta)(iA) + \tfrac{1}{3}(iA^2A^*)(i\beta - \alpha + \delta)\right] \\
2\alpha y_1y_2 &\to 2\alpha\left(-\tfrac{2}{3}\right)A^2A^*(i\beta - \alpha + \delta) \\
2\delta x_1x_2 &\to 2\delta\left[2A^2A^*(\alpha + \delta) - \tfrac{1}{3}A^2A^*(i\beta - \alpha + \delta)\right]
\end{aligned} \tag{14.123}$$

We note that $D_0D_1x_2$ has no terms of the form $\exp it_0$, so we need not include them in our search for secular generating terms. Also, we have $D_1 x_1 = 0$, since A does not have t_1 dependence. We write amplitude A as $A = \frac{1}{2} a \exp i\phi$. So

$$D_2 A = \frac{1}{2}(D_2 a) \exp i\phi + i\left(\frac{a}{2}\right)(D_2\phi) \exp i\phi \tag{14.124}$$

We are now ready to collect those terms related to growth of our amplitude a. These contribute to generating a possible limit cycle. We equate to zero the coefficient of $\exp it_0$ on the r.h.s. of Equation (14.122). Thus, using the relations given by Equation (14.123), we have

$$\begin{aligned} iD_2 a = \Bigg[& i\sigma\frac{a}{2} - \frac{2i}{3}\frac{a^3}{8}\beta(-\alpha + \delta) \\ & + 2i\frac{a^3}{8}\beta(\alpha + \delta) + \frac{i}{3}\frac{a^3}{8}\beta(-\alpha + \delta) \\ & - 2\alpha\left(\frac{2}{3}\right)(i\beta)\left(\frac{a^3}{8}\right) - 2\delta\left(\frac{1}{3}\right)(i\beta)\left(\frac{a^3}{8}\right)\Bigg] \end{aligned} \tag{14.125}$$

We factor out $(i/3)\beta(a^3/8)$. We then have

$$\begin{aligned} iD_2 a &= i\left[\frac{a\sigma}{2} + \frac{1}{3}\frac{\beta a^3}{8}(+2\alpha - 2\delta + 6\alpha + 6\delta - \alpha + \delta - 4\alpha - 2\delta)\right] \\ &= i\left\{\frac{a\sigma}{2} - \frac{\beta a^3}{24}[3(\alpha + \delta)]\right\} \end{aligned} \tag{14.126a}$$

Thus,

$$D_2 a = a\left[\frac{\sigma}{2} + \frac{\beta a^2}{8}(\alpha + \delta)\right] \tag{14.126b}$$

We have a limit cycle whose radius a^* is

$$a^* = \sqrt{\frac{4\sigma}{|\beta(\alpha + \delta)|}} \tag{14.127}$$

if

$$\beta(\alpha + \delta) < 0; \qquad \sigma > 0 \tag{14.128}$$

Recall that our displacement x is $O(\varepsilon)$, so the true radius of the limit cycle, r^* is given by

$$r^* = \sqrt{\frac{4\rho}{|\beta(\alpha + \delta)|}}$$

and is thus proportional to the square root of ρ.

So we see that if we have an appropriate combination of signs, we can obtain a limit cycle from a system of polynomial equations of degree 2. ■

EXERCISE 5.1 This exercise is closely related to Example 5.1. Consider a set of coupled equations of the form

$$\frac{dz_1}{dt} = \sigma z_1 + z_2 + \alpha z_1^2; \qquad \alpha = O(1) \tag{14.129a}$$

$$\frac{dz_2}{dt} = -z_1 + A z_1^2; \qquad A = O(1) \tag{14.129b}$$

Show that if $\sigma = O(\varepsilon^2) > 0$ and $\alpha A < 0$, we obtain a limit cycle with radius

$$a^* = \sqrt{\frac{2\sigma}{|A\alpha|}}$$

■

14.6 CONCLUSIONS

This has been a complicated chapter. We have studied the role of the homogeneous solutions in our equations as well as the role of the initial conditions in specifying the expansion parameters. The problems become algebraically complicated when we have both quadratic and cubic terms, but it is possible to develop a systematic procedure to study them. In the absence of a small parameter, we have to scale the amplitude so that the system remains a perturbed simple harmonic oscillator. It is clear that you have to be careful when you have both quadratic and cubic terms because they can introduce delicate cancellations that dramatically change the structure of the solutions. Finally, one is now equipped to analyze many of the nonlinear models appearing in the scientific literature.

CHAPTER 15

ERROR ANALYSIS

In Chapters 12 to 14, we used the MOA and the MMTS to generate solutions in the form of expansions in a small parameter to a variety of problems. A central characteristic of these approximation schemes is to alter fundamentally the equation we are attempting to solve. In the MOA, we introduce a set of averaged equations. In the MMTS, we introduce artificially separated time scales. Each method employs a systematic approximation scheme to solve the altered equations. We obtained frequency corrections for conservative systems so that our approximate solution maintained a closed orbit in the phase plane for longer and longer times. This also led to the need to determine higher-order harmonic components of the basic solutions. We have, for example, carried out such calculations in Chapter 14 for the Duffing oscillator. We have also considered nonconservative systems that led to limit cycles. In such cases, we introduced the concept of stability due to Poincaré. Namely, if we have a stable limit cycle, then once an approximate solution comes sufficiently close, it remains nearby. If you think of the limit cycle as surrounded by an annulus, as shown in Figure 11.2, then once inside the annulus, the orbit remains inside. Finally, in this regard, we attach an error to the neglected terms in the expansion. By many standards, the error analysis we have given is not entirely satisfactory. We are now ready to give a more detailed analysis and thereby to establish how well our approximate solutions follow the exact solutions and for how long.

15.1 OUTLINE OF OUR APPROACH

The basic equation we discuss is

$$\ddot{x} + x = \varepsilon f(x); \qquad x(0) = a, \quad \dot{x}(0) = 0 \tag{15.1}$$

where $f(x)$ is a polynomial in x. We assume that we have found a first-order approximate solution $z^1(t)$ by either the MMTS or MOA and we are now interested in bounding the difference between the approximate and the exact solutions. To accomplish this, we introduce a remainder function $R(t)$ that is a measure of their difference:

$$R \equiv x - z^1 \tag{15.2}$$

We then introduce a differential equation for $R(t)$ and convert it to an integral equation of the form:

$$\begin{aligned} R(t) = {} & \text{integral of } [\text{a known function multiplied by } \sin(t - t')] \\ & + \text{integral of } [\ R(t') \text{ multiplied by } \sin(t - t')] \end{aligned}$$

(The sine function is chosen because it is a solution to the homogeneous equation.)

In determining a bound on $R(t)$, we encounter integrals of the form $I(A)$, $J(A)$, and $K(A)$, given by Equations (15.6a), (15.9a), and (15.12a), respectively. At that point, we introduce the Lipshitz condition, given by Equation (15.3), and Gronwall's lemma, given by Equations (15.4) and (15.5), in order to establish error bounds. We find that Gronwall's lemma is a basic tool to establish error bounds on our approximate solutions for short times $t = O(1)$ and for longer times $t = O(1/\varepsilon)$. We limit our discussion to conservative systems, since the technique is almost the same for dissipative systems and there is a significant increase in the algebraic complexity of the arguments. We have based our discussion on the methods outlined in the text by Smith, and one can refer to it for a more rigorous and detailed mathematical treatment. In addition, the text by Sanders and Verhulst has a detailed discussion of averaging techniques, including an error analysis for long times $t = O(1/\varepsilon^2)$. In many ways, the treatments given in these references complement one another.

15.2 BASIC TOOLS FOR THE ANALYSIS OF THE ERROR

We first collect a few results as preparation for a discussion of the error analysis.

(i) *The Lipshitz Condition*. This is a condition that is used in establishing uniqueness of a solution to a differential equation. We use it in order to establish a simple bound. It is discussed in books on differential equations and in the text by Sanders and Verhulst.

Definition: If there exists a constant L such that for every (x, t) and (z, t) in a domain D,

$$|f(x,t) - f(z,t)| \leq L|x - z| \tag{15.3}$$

we say that the function f satisfies a Lipshitz condition with respect to x and z in the domain D. The constant L is called the Lipshitz constant. It can depend on the domain D, but not on the variables x and z. The mean value theorem says that if $\partial f/\partial x$ is uniformly bounded in D, f satisfies a Lipshitz condition in D. In this chapter, we limit our discussion to functions $f(x)$ that are polynomials in their argument and these arguments are bounded. Thus, they satisfy the mean value theorem and, consequently, the Lipshitz Condition.

Example 2.1 Consider the equation

$$\frac{dy}{dx} = 2\sqrt{y}; \qquad y(0) = 0$$

The function $2\sqrt{y}$ does not satisfy the Lipshitz condition since near $y = 0$, $\sqrt{y} > y$.

Also note that the equation has two solutions,

$$y \equiv 0 \text{ and } y = x^2. \qquad \blacksquare$$

(ii) Recall Gronwall' lemma, which we discussed in Chapter 6, Section 6.4, Equations (6.69) to (6.77). We use it in the following form.

Given the inequality

$$|R| \leq |E(A)| + \varepsilon L \int_0^t |R(t')| \, dt' \tag{15.4}$$

with $E(A)$ a constant and L a Lipshitz constant, we have

$$|R| \leq |E(A)| \exp \varepsilon L t \tag{15.5}$$

(iii) We encounter integrals of the form

$$I(A) = \varepsilon^2 \int_0^t \cos[(1 + \varepsilon A)t'] \cos t' \, dt' \tag{15.6a}$$

$$= \varepsilon^2 \left(\frac{\sin[(2 + \varepsilon A)t]}{2(2 + \varepsilon A)} + \frac{\sin \varepsilon A t}{2(\varepsilon A)} \right) \tag{15.6b}$$

We are given that A is a constant of $O(1)$. If time $t = O(1)$, then

$$\frac{\varepsilon^2 \sin \varepsilon A t}{\varepsilon A} = O(\varepsilon^2); \qquad \text{so}$$

$$I(A) = O(\varepsilon^2) \qquad \text{for } t = O(1) \tag{15.7}$$

(The other term remains $O(\varepsilon^2)$ for all t.)
If time $t = O(1/\varepsilon)$ or $\varepsilon t = O(1)$, then

$$\frac{\varepsilon^2 \sin \varepsilon A t}{\varepsilon A} = O(\varepsilon) \qquad \text{and}$$

$$I(A) = O(\varepsilon) \qquad \text{for } t = O\left(\frac{1}{\varepsilon}\right) \tag{15.8}$$

Thus, we see that the "order" of the integral changes when time t grows from $t = O(1)$ to $t = O(1/\varepsilon)$. (Keep in mind that we mean that the error is as little as $O(\varepsilon^2)$ for $t = O(1)$ and is not worse than $O(\varepsilon)$ for $t = O(1/\varepsilon)$.)

Consider

$$J(A) = \varepsilon^2 \int_0^t \sin[(1 + \varepsilon A)t'] \sin t' \, dt' \tag{15.9a}$$

$$= \varepsilon^2 \left[\frac{\sin \varepsilon A t}{2(\varepsilon A)} - \frac{\sin[(2 + \varepsilon A)t]}{2(2 + \varepsilon A)} \right] \tag{15.9b}$$

The error analysis we have given for $I(A)$ holds for $J(A)$ since they only differ by a minus sign between the two terms.

Thus,

$$J(A) = O(\varepsilon^2) \qquad \text{for } t = O(1) \tag{15.10}$$

$$J(A) = O(\varepsilon) \qquad \text{for } t = O\left(\frac{1}{\varepsilon}\right) \tag{15.11}$$

Next we consider

$$K(A) = \varepsilon^2 \int_0^t \cos[(1 + \varepsilon A)t'] \sin t' \, dt' \tag{15.12a}$$

$$= \varepsilon^2 \left(\frac{1 - \cos[(2 + \varepsilon A)t]}{2(2 + \varepsilon A)} - \frac{1 - \cos \varepsilon A t}{2(\varepsilon A)} \right) \tag{15.12b}$$

By techniques similar to those we used in discussing $I(A)$, we can establish that

$$K(A) = O(\varepsilon^2) \qquad \text{for } t = O(1) \tag{15.13}$$

and

$$K(A) = O(\varepsilon) \qquad \text{for } t = O\left(\frac{1}{\varepsilon}\right) \tag{15.14}$$

15.3 ERROR ANALYSIS FOR THE ALTERED SIMPLE HARMONIC OSCILLATOR

We illustrate our method of analysis by considering the altered simple harmonic oscillator given by

$$\ddot{x} + x = -\varepsilon x; \qquad x(0) = a, \quad \dot{x}(0) = 0 \tag{15.15}$$

It is a particularly simple system, yet it is typical of any conservative system that has a frequency correction in lowest-order perturbation theory. In addition, we have already analyzed it by a variety of techniques and we are familiar with its solution.

Let us say that we have obtained an approximate solution to Equation (15.15) in the form

$$z^1(t) = a \cos\left[\left(1 + \frac{\varepsilon}{2}\right)t\right] \tag{15.16}$$

$$z^1(0) = a, \qquad \dot{z}^1(0) = 0$$

This is not an exact solution to Equation (15.15), so we write its "equation" as

$$\ddot{z}^1 + z^1 = \varepsilon f(z^1) + \rho(\varepsilon, t); \qquad \text{with } f(z^1) = -z^1 \tag{15.17}$$

The quantity $\rho(\varepsilon, t)$ is called the residue and it is a measure of how much our approximate solution $z^1(t)$ differs from the exact solution $x(t)$. If we substitute $z^1(t)$ given by Equation (15.16) into Equation (15.17), we find

$$\rho(\varepsilon, t) = -\frac{\varepsilon^2 a}{4}\cos\left[\left(1+\frac{\varepsilon}{2}\right)t\right] \tag{15.18}$$

We now introduce our remainder function R, given by Equation (15.2), $R \equiv x - z^1$, that measures the difference between our exact solution and our approximate solution. We then find the differential equation satisfied by R. We just substitute $R + z^1$ for $x(t)$ in Equation (15.15) and use Equations (15.16) and (15.18):

$$\ddot{R} + R = \varepsilon\left[f(R + z^1) - f(z^1)\right] - \rho(\varepsilon, t) \tag{15.19}$$

Notice $R(0) = \dot{R}(0) = 0$ since the approximate solution $z^1(t)$ satisfies the exact boundary conditions.

We have

$$\ddot{R} + R = -\varepsilon R + \frac{\varepsilon^2 a}{4}\cos\left[\left(1+\frac{\varepsilon}{2}\right)t\right] \tag{15.20}$$

We then convert Equation (15.20) to an integral equation with the formal solution

$$\begin{aligned} R(t) = {} & \varepsilon^2\frac{a}{4}\int_0^t \cos\left[\left(1+\frac{\varepsilon}{2}\right)t'\right]\sin(t-t')\,dt' \\ & - \varepsilon\int_0^t R(t')\sin(t-t')\,dt' \end{aligned} \tag{15.21}$$

The first term yields integrals of the form of $I(A)$ and $K(A)$ given by Equations (15.6a) and (15.12a), respectively.

$$\begin{aligned} E(a) &\equiv \int_0^t \frac{\varepsilon^2 a}{4}\cos\left[\left(1+\frac{\varepsilon}{2}\right)t'\right]\sin(t-t')\,dt' \\ &= \sin t\left\{\int_0^t \frac{\varepsilon^2 a}{4}\cos\left[\left(1+\frac{\varepsilon}{2}\right)t'\right]\cos t'\,dt'\right\} \\ &\quad - \cos t\left\{\int_0^t \frac{\varepsilon^2 a}{4}\cos\left[\left(1+\frac{\varepsilon}{2}\right)t'\right]\sin t'\,dt'\right\} \end{aligned}$$

Referring to Equations (15.6b) and (15.12b), we conclude that

$$E(a) = O(\varepsilon^2) \qquad \text{for } t = O(1) \tag{15.22}$$

$$E(a) = O(\varepsilon) \qquad \text{for } t = O(1/\varepsilon) \tag{15.23}$$

We now return to Equation (15.21) and take absolute values:

$$|R| \leq |E(a)| + \varepsilon \int_0^t |R(t')|\, dt' \tag{15.24}$$

We can use Gronwall's lemma, given by Equations (15.4) and (15.5), to write

$$|R| \leq |E(a)| \exp \varepsilon t \tag{15.25}$$

Thus, we clearly see that the bound on our remainder term for times $0 < t \leq 1/\varepsilon$ is dominated by the bound on $E(a)$. We find that this result is maintained in our analysis of nonlinear oscillatory systems. We conclude that

$$|x - z^1| = O(\varepsilon^2) \qquad \text{for } t = O(1) \tag{15.26a}$$

$$|x - z^1| = O(\varepsilon) \qquad \text{for } t = O\left(\frac{1}{\varepsilon}\right) \tag{15.26b}$$

Note: Since we know that the exact solution to our altered simple harmonic oscillator is

$$x(t) = a\cos(\sqrt{1 + \varepsilon}\, t)$$

we can verify that our error bound is correct. All that is necessary is to expand the square root in the argument of the cosine function so as to write $x(t)$ as

$$x(t) = a\cos\left\{\left[1 + \frac{\varepsilon}{2} + O(\varepsilon^2)\right]t\right\}$$

We then have

$$|x - z^1| = \left\{a\cos\left[\left(1 + \frac{\varepsilon}{2}\right)t\right]\right\}\left\{\cos\left[\left(O(\varepsilon^2)\right)t\right] - 1\right\}$$
$$- a\left[\sin\left(1 + \frac{\varepsilon}{2}\right)t\right]\sin\left[O(\varepsilon^2)t\right] \tag{15.27}$$

This yields the inequality given by Equations (15.26a) and (15.26b).

15.4 THE DUFFING OSCILLATOR

We are now ready to establish error bounds for our approximate solutions to the Duffing oscillator. We will see that we have already developed all the necessary material and that we can establish bounds for any weakly nonlinear oscillatory system. Our equation is

$$\ddot{x} + x = -\varepsilon x^3; \qquad x(0) = a, \quad \dot{x}(0) = 0 \tag{15.28}$$

We begin by deriving a bound for $t = O(1)$. We are interested only in the lowest-order approximation, so we can refer to Equation (10.37) and write

$$z^1 = a\cos\left[\left(1 + \frac{\varepsilon 3a^2}{8}\right)t\right] + \frac{\varepsilon a^3}{32}(\cos 3t - \cos t) + O(\varepsilon^2) \quad \text{for } t = O(1) \tag{15.29}$$

Note that $z^1(0) = a$; $\dot{z}^1(0) = 0$ satisfy the exact boundary conditions.

Notice that we use the unperturbed frequency $\omega_0 = 1$ in the argument of the correction term since it has an ε as a multiplier. In other words,

$$\varepsilon\left|\cos\left[\left(1 + \frac{\varepsilon 3a^2}{8}\right)t\right] - \cos t\right| = O(\varepsilon^2) \qquad \text{for } t = O(1) \tag{15.30a}$$

and

$$\varepsilon\left|\cos\left[\left(1 + \frac{\varepsilon 3a^2}{8}\right)t\right] - \cos t\right| = O(\varepsilon) \qquad \text{for } t = O\left(\frac{1}{\varepsilon}\right) \tag{15.30b}$$

Following the procedures that we used in our analysis of the altered simple harmonic oscillator, we introduce the equation

$$\ddot{z}^1 + z^1 = -\varepsilon(z^1)^3 + \rho(\varepsilon, t) \tag{15.31}$$

We substitute our trial solution given by Equation (15.29) into Equation (15.31) and obtain our residue function $\rho(\varepsilon, t)$:

$$\rho(\varepsilon, t) = -\left(\varepsilon^2\frac{9a^5}{64}\right)\cos t + \left(\varepsilon^2\frac{3a^5}{32}\right)(\cos^2 t)(\cos 3t - \cos t) \tag{15.32}$$

We now introduce our remainder function $R(t) \equiv x - z^1$ and Equation (15.28) becomes

$$\begin{aligned} \ddot{R} + R &= -\rho(\varepsilon, t) - \left[\varepsilon x^3 - \varepsilon(z^1)^3\right] \\ R(0) &= \dot{R}(0) = 0 \end{aligned} \tag{15.33}$$

We convert Equation (15.33) to an integral equation and have

$$R(t) = -\int_0^t \rho(\varepsilon, t') \sin(t - t')\, dt'$$

$$- \varepsilon \int_0^t \left\{ x^3(t') - \left[z^1(t') \right]^3 \right\} \sin(t - t')\, dt' \qquad (15.34)$$

Initially, we want to bound the first term for times $t = O(1)$. If we refer to integrals $I(A)$, $J(A)$, and $K(A)$ given by Equations (15.6a) through (15.14), we conclude, using Equation (15.32), that

$$|E(a)| = \left| \int_0^t \rho(\varepsilon, t') \sin(t - t')\, dt' \right|$$

$$= O(\varepsilon^2) \qquad \text{for } t = O(1) \qquad (15.35)$$

In obtaining this result, we are careful to note that

$$\varepsilon^2 \int_0^t \cos t' \sin(t - t')\, dt' \qquad (15.36)$$

yields a bound $O(\varepsilon^2)$ for $t = O(1)$.

Our perturbation $f(x) = x^3$ satisfies a Lipshitz condition; therefore, we obtain, from Equation (15.34),

$$|R(t)| \le |E(a)| + \varepsilon L \int_0^t |R(t')|\, dt' \qquad (15.37)$$

where L is the Lipshitz constant. We can now use Gronwall's lemma given by Equation (15.5) to obtain

$$|R(t)| \le |E(a)| \exp \varepsilon L t \qquad \text{for } t = O(1) \qquad (15.38)$$

Now say that we want to have the bound be valid for times as long as $t = O(1/\varepsilon)$. It is necessary to refer to Equations (15.30a), (15.30b), and (15.36). In particular, the integral in Equation (15.36) now has a "poorer" bound:

$$\varepsilon^2 \int_0^t \cos t' \sin(t - t')\, dt' = O(\varepsilon) \qquad \text{for } t = O(1/\varepsilon) \qquad (15.39)$$

The other integrals are evaluated by the same procedures and our bound for the remainder $R(t)$ becomes

$$|R(t)| \le |E(a)| \exp \varepsilon L t = O(\varepsilon) \qquad \text{for } t = O\left(\frac{1}{\varepsilon}\right) \qquad (15.40)$$

With these observations regarding error bounds on our solutions, we return to Equation (15.29) and write our approximate solution to the Duffing oscillator, given by Equation (15.28), as

$$x(t) = a\cos\left[\left(1 + \frac{\varepsilon 3a^2}{8}\right)t\right] + \frac{\varepsilon a^3}{32}(\cos 3t - \cos t) + O(\varepsilon^2) \qquad \text{for } t = O(1) \tag{15.41}$$

Also we have

$$x(t) = a\cos\left[\left(1 + \frac{\varepsilon 3a^2}{8}\right)t\right] + O(\varepsilon) \qquad \text{for } t = O\left(\frac{1}{\varepsilon}\right) \tag{15.42}$$

We are required to omit the harmonic-correction terms in our approximate solution if we are going to use it for times as long as $t = O(1/\varepsilon)$. Why? We know from our studies in Chapter 14, leading to Equation (14.26b), that there is a frequency-correction term of the form

$$-\varepsilon^2 \frac{21a^4}{256}$$

When the time t is $O(1/\varepsilon)$, this introduces a frequency correction of $O(\varepsilon)$. Thus, referring to Equation (15.30b), we observe that the frequency correction and the harmonic correction are both $O(\varepsilon)$ for $t = O(1/\varepsilon)$ and one must include or omit both.

15.5 CONCLUSIONS

For times as long as $t = O(1/\varepsilon)$, we have been able to obtain bounds for the difference between the exact solution and our approximate solutions for the altered simple harmonic and Duffing oscillator equations. In our derivation, we used a remainder function $R(t)$ that satisfied $R(0) = \dot{R}(0) = 0$. If we had used an approximate solution that had a small error in satisfying the initial conditions, this would have been easily incorporated into a term of similar structure to $\rho(\varepsilon, t)$. We observed that our residue function $\rho(\varepsilon, t)$ had to be as small as $O(\varepsilon^2)$ for $t = O(1)$ in order to produce our bound for $t = O(1/\varepsilon)$. (This point is discussed at length in Chapter 4 of the text by Smith).

It is clear that we can follow the same procedure for any conservative perturbation that is a nonlinear polynomial. We mean that if by some method, such as the MMTS, one obtains an approximate solution with a residue ρ that is $O(\varepsilon^2)$ for times $t = O(1)$ or is $O(\varepsilon)$ for times $t = O(1/\varepsilon)$, we can use the Lipshitz condition and Gronwall's lemma to obtain a bound like that given by

Equation (15.38) or (15.40). By our methods, we are unable to obtain a bound that is valid for times as long as $t = O(1/\varepsilon^2)$. Techniques to obtain these bounds are given in the text by Sanders and Verhulst.

In addition, it is possible with a little more effort to obtain similar bounds for dissipative systems that have perturbations that are polynomials in the displacement and the velocity. This is done in the books by Smith and by Sanders and Verhulst. One obtains results that are equivalent to those given by Equations (15.38) and (15.40).

We make the observation that one has to be careful in interpreting the meaning of the phrase "arbitrarily long time" or "arbitrarily small ε." In all of our analyses, we are able to consider perturbations that are small, as characterized by the parameter ε, for times up to but no longer than $O(1/\varepsilon)$ reflecting the requirement that the product εt be $O(1)$. Thus, there is a coupling between the permitted time validity of our approximation and the size of the perturbation.

In conclusion, one can safely assume that both the MMTS and the MOA generate a systematic perturbation expansion that leads to error terms that can be bounded for times $t = O(1/\varepsilon)$ for weakly nonlinear oscillatory problems of the form

$$\ddot{x} + x = \varepsilon f\left(x, \frac{dx}{dt}\right); \qquad |\varepsilon| \ll 1$$

and where the perturbing term $\varepsilon f(x, dx/dt)$ satisfies the Lipshitz condition.

BIBLIOGRAPHY FOR CHAPTERS 7 TO 15

1. Lotka, A. J., *Elements of Physical Biology*, Williams and Wilkins, Baltimore, 1924. Reprinted as *Elements of Mathematical Biology*, Dover, New York, 1956. This book has an excellent introduction to the modeling techniques that are used to understand exponential and resource-limited growth and their application to chemical reactions, epidemics, etc. The text also has the development of the Lotka part of the Volterra–Lotka equations. There are lots of applications and data analysis. Since this is a reprint of the book originally written in 1924, you can find some interesting comments and articles of historical interest.
2. D'Ancona, U., *Struggle for Existence*, Brill, Leiden, 1954. This book outlines in complete detail the Volterra theory and models of predator–prey interaction, competition between species, relationship to questions and techniques of statistical mechanics circa 1920, and other questions. The book is extremely readable.
3. Gause, G. F., *Struggle for Existence*, Williams and Wilkins, Baltimore, 1934. Reprinted by Hafner, New York, 1969. This text complements the one by D'Ancona. It contains a careful exposition of the Volterra theory and methods that

were used to check its validity by experiments. Data are included. It is interesting and easy reading, conveying the spirit of inquiry prevalent at the time.

4. Krylov, N. M., and N. N. Bogoliubov, *Introduction to Nonlinear Mechanics*, Princeton University Press, Princeton, 1947. N. N. Bogoliubov, and Y. A. Mitropolsky, *Asymptotic Methods in the Theory of Nonlinear Oscillations*, Gordon and Breach, New York, 1961. These are the classic texts that provide a systematic and thorough discussion of the method of averaging and its application to nonlinear oscillatory systems. The earlier text is based strongly on intuitive arguments that find their mathematical justification in the later work.
5. Minorsky, N., *Nonlinear Oscillations*, Van Nostrand, New York, 1962. Reprinted by Krieger, Huntington, New York, 1974. We find here a careful introduction to the subject with particular emphasis on the exposition of the methods used by Russian mathematicians before their material was available in English. It is excellent to read in conjunction with the books by Bogoliubov and co-workers.
6. Andronov, A. A., and S. E. Khaiken, *Theory of Oscillators*, Princeton University Press, Princeton, 1949. A. A. Andronov, A. A. Vitt, and S. E. Khaiken, *Theory of Oscillators*, Addison-Wesley, Reading, Massachusetts, 1966. Reprinted by Dover, New York, 1987. These texts develop in detail the methods of Poincaré and others in matters related to the theory of oscillators. There are a great number of applications to both mechanical and electrical devices. They give a fine background to anyone wishing to watch the interplay of relatively sophisticated mathematics with applications to engineering problems.
7. Nayfeh, A. H., *Introduction to Perturbation Techniques*, Wiley, New York, 1981. A. H. Nayfeh, *Perturbation Methods*, Wiley, New York, 1973. A. H. Nayfeh, *Problems in Perturbation*, Wiley, New York, 1985. A. H. Nayfeh, and D. T. Mook, *Nonlinear Oscillations*, Wiley, New York, 1979. Nayfeh has written three books on this subject and coauthored one with Mook. They are all very reader-oriented and *Introduction to Perturbation Techniques* is the easiest to read.

   Problems in Perturbation is primarily a solutions book to the latter text. It should be emphasized that these two books provide a discussion containing all the steps in the calculations of a broad spectrum of material. This is especially relevant to anyone wishing to develop a feel for the techniques used in the study of oscillatory systems. Particular attention is given to the method of multiple time scales.

   Perturbation Methods is devoted to a more advanced treatment of the material. It is well presented and is excellent reading as a text that includes the details and the comparisons between alternative and complementary schemes of calculation. It contains a thorough discussion of the method of multiple time scales and of the method of averaging, including examples calculated through the second order of perturbation theory. In addition, it contains a discussion of a variety of presently less popular methods. Finally, it has a clear and thorough discussion of Lie series and Lie transforms that is otherwise relatively unavailable.

   Nonlinear Oscillations is a detailed treatment of this subject that is more oriented toward the research worker than the student.
8. Bender, C. M., and S. A. Orszag, *Advanced Mathematical Methods for Scientists and Engineers*, McGraw-Hill, New York, 1978. This fine text is rather advanced but contains problems that are discussed at a level that should be comprehensible to anyone who can follow the material we present.

9. Andronov, A. A., E. A. Leontovich, I. I. Gordon, and A. G. Maier, *Theory of Bifurcations of Dynamic Systems on a Plane*, Israel Program for Scientific Translations, Jerusalem, 1971. This is a comprehensive and rigorous text that contains a lot of material that is presented at a more elementary level in texts 5 and 6. In addition, it has a number of interesting examples, including a very brief discussion of the quadratic oscillator due to Bautin on page 264.
10. Jordan, D. W., and P. Smith, *Nonlinear Ordinary Differential Equations*, Oxford University Press, Oxford, 1977. This book contains a lot of general material on oscillations at the level of our text. Although, it is a slim volume, it covers a broad spectrum of topics with many interesting examples and exercises. It is suitable as a text on the subject.
11. Struble, R. A., *Nonlinear Differential Equations*, McGraw-Hill, New York, 1962. This is a most readable book that takes you step by step through a systematic discussion of many aspects of nonlinear problems. It also presents an interesting and intuitively clear way to look at the Poincaré perturbation method through a variational approach.
12. Kevorkian, J., and J. D. Cole, *Perturbation Methods in Applied Mathematics*, Springer-Verlag, New York, 1981. This text contains a wealth of material on perturbation methods, with particular emphasis on the method of multiple time scales and its variants. It has a nice discussion of Morrison's fundamental paper in which the methods of averaging and multiple time scales were shown to be identical through the second order in the perturbation expansion. There is also a very readable discussion of scaling.
13. Sanders, J. A., and F. Verhulst, *Averaging Methods in Nonlinear Dynamical Systems*, Springer Verlag, New York, 1985. This advanced text emphasizes the averaging method, although there is a discussion of multiple time methods. It also mentions that the general, or complete, equivalence between averaging and multiple time scales is not expected to be true. Rather, their equivalence has been shown for systems that need only two time scales. The book has some interesting examples that help to illuminate the pitfalls of some of the crude averaging techniques.
14. Smith, D. R., *Singular Perturbation Theory*, Cambridge University Press, New York, 1985. This book is not difficult to follow and has a clear discussion of both the averaging and multiple time techniques. It has some elementary examples that are worked in sufficient detail so as to give the reader additional insight. In addition, there is an extensive presentation of error analysis on which our treatment is based.
15. Landau, L. D., and E. M. Lifshitz, *Mechanics*, Pergamon Press, Bristol, 1960. This exceptional text on mechanics includes a brief discussion of anharmonic oscillators and associated nonlinear systems.
16. May, R. M., *Stability and Complexity in Model Ecosystems*, 2nd Ed., Princeton University Press, Princeton, 1973. This book is based primarily on a variety of articles by Professor May. It is easy to read and provides an excellent survey of ecology in the early 1970s and the mathematical techniques that were being used. It contains an extensive bibliography.
17. Lin. C. C., and L. A. Segel, *Mathematics Applied to Deterministic Problems in the Natural Sciences*, Macmillan, New York, 1974. Reprinted by Dover, New York, 1987. This book contains a comprehensive discussion of mathematical methods. It

places a strong pedagogical emphasis on the importance of carefully formulating the mathematical aspects associated with the scientific problem under study. There is an excellent discussion of singular perturbation methods and multiple scales.

18. Tabor, M., *Chaos and Integrability in Nonlinear Dynamics: An Introduction*, Wiley, New York, 1989. At a certain point it becomes important to take a giant step and learn more about the modern theory of dynamical systems. No book on this subject is easy reading, but this one contains sufficient introductory material so that the reader should, at a minimum, be able to grasp the essential principles. It also treats at some length a variety of difficult topics of current research interest and contains many references to the literature.

CHAPTER 16

ONE-DIMENSIONAL ITERATIVE MAPS AND THE ONSET OF CHAOS

There has been a significant set of developments associated with the study of nonlinear difference equations that interface with the ideas that are central to nonlinear differential equations. Although these equations have different structures, each has the capacity to exhibit chaotic behavior. In this chapter, we consider nonlinear difference equations and it would be a magnificent understatement to simply say that most of the new behavior we are observing was totally unexpected by previous investigations. Indeed, we cannot help but feel that something profound and exciting is being revealed to us. In order to get a feel for what is happening, it is essential to use some numerical or graphical scheme to establish and verify the results that are obtained. Passive observation is unsatisfactory.

We have organized the chapter by beginning with a long introduction and orientation to the concept of nonlinear difference equations. We then follow this by a detailed derivation of the Feigenbaum numbers, δ and α, that characterize one-dimensional quadratic maps. The discussion is strongly motivated by the pioneering work of Robert May, Mitchell Feigenbaum, and others who have attempted to unravel the mysteries associated with chaos.

16.1 INTRODUCTION

When we construct models that describe physical phenomena, we attempt to keep them as simple as possible while retaining the basic features of the problem under consideration. This has led, in great part, to the development of linear models in the form of differential or difference equations that are classifiable. Thus, for example, an equation of the form

$$\frac{dy}{dx} + \frac{y}{1-x} = 0; \qquad y(0) = c, \quad y(x) = c(1-x) \tag{16.1}$$

is known to have a solution that is free from singularities because of the structure of the equation. Also, if the equation were to take the form

$$\frac{dy}{dx} + \frac{y}{(x-1)^2} = 0; \qquad y(0) = c \tag{16.2a}$$

we anticipate in the solution

$$y(x) = c \exp \frac{x}{x-1} \tag{16.2b}$$

the singularity at $x = 1$. These conclusions follow because we have a classification scheme for the solutions of linear differential equations that tells us when singularities are present. On the other hand, we observe that a nonlinear differential equation of the form

$$\frac{dy}{dx} - y^2 = 0; \qquad y(0) = c, \quad y(x) = \frac{c}{1-cx} \tag{16.3}$$

has a *spontaneous singularity* at $x = 1/c$ associated with the initial condition $y(0) = c$. Our experience in classifying and solving linear differential equations provides inadequate preparation for the study of nonlinear differential equations. A parallel situation is present in the analysis of difference equations. For example, let us say that we model the growth of an isolated species with population density $x(t)$ by a linear differential equation of the form

$$\frac{dx}{dt} = bx; \qquad x(t) = x(0) \exp bt \tag{16.4}$$

The quantity b is called the birth-minus-death rate and it determines the ultimate fate of the species. We mean by this that if $b > 0$, the species grows, and if $b < 0$, the species declines. The corresponding linear difference equation is given by

$$x(n+1) - x(n) = bx(n)$$

or

$$x(n+1) = (b+1)x(n); \quad x(n) = (b+1)^n x(0) \tag{16.5}$$

where the quantity $x(n)$ represents the population density of a species at time $t = n$. Then we see from our solution that the species grows if the coefficient b is either greater than 0 or less than -2. (There is no biological significance to situations in which b is < -1.) Even though the solutions of Equations (16.4) and (16.5) have a different behavior for different ranges of the parameters, there is a *systematic* comparison method that relates the solutions. This is discussed in a variety of texts, for example, see the book by Bender and Orszag.

In this chapter, we consider the nonlinear generalization of Equation (16.5) that is analogous to the logistic differential equation

$$\frac{dx}{dt} = bx(1 - x) \tag{16.6}$$

in the form of a nonlinear difference equation

$$x(n + 1) = (b + 1)x(n) - bx(n)x(n) \tag{16.7}$$

Although this model has been studied for more than a century, its complex structure has only recently been revealed, in great part due to the effective use of computers and computer graphing techniques. We have coined a new word, chaotic, to describe the apparently random behavior of some of the solutions of this deterministic equation. Furthermore, one aspect of our experience in observing a normal dynamical system concerns its eventual settling down. In other words, it is our expectation that if we wait sufficiently long, the transients will die out and the system will be characterized as being in a periodic or steady state. By contrast, a chaotic system appears to be trapped in a permanent transient state whose structure is highly sensitive to initial conditions.

16.2 ORIENTATION

This chapter differs from all the others that we have studied in that it presents a discussion of a subject that is not as yet well defined. Until recently, we thought that there was a reasonably distinct difference between random and deterministic systems. For example, we learn early in our studies of mechanics that given the forces, the positions, and the momenta of the particles in a deterministic system, we could, in principle, compute to an arbitrary accuracy its future development. Of course, we knew that this might be an impossible task to actually carry out, but we were confident that we understood what we were doing. On the other hand, if our system has random elements, we expected that one could not give a detailed and accurate forecast of the development of the system. There were some shady or overlap areas, but the essence of the argument was always thought to be correct. However, during the past ten years or so, it has become clear, in great part due to excellent use of computers, that deterministic systems can exhibit behavior that is so complex and unpredictable that their motion has been called chaotic. Thus, in contrast to our previous approach, such systems are treated by abandoning a search for a detailed description of the motion and instead developing a qualitative picture of what is happening. Furthermore, as might be expected, discrete and continuous systems exhibit their approach to chaos quite differently. This field of research is presently extremely active and it will be some time before a systematic presentation is possible.

With these thoughts in mind, we organize our discussion so as to show that simple one-dimensional quadratic maps or difference equations exhibit complex and unexpected behavior. We develop, in reasonable detail, the analysis that is necessary to take us to the threshold or onset of chaos in discrete systems. This can be accomplished while keeping the mathematics at an elementary level. Unfortunately, we omit a discussion of the chaotic behavior of continuous systems, since this subject requires the development of some specialized mathematics. (Many of the texts in the Bibliography treat this aspect of mathematical chaos.) Finally, one should ask whether there is a direct physical significance or connection between chaos and the real world. For example, does it have a role to play in helping us uncover the key factors in the onset and spatial character of turbulence or in weather prediction? Maybe yes or maybe no, but this should not prejudice someone regarding the study of this new subject. It is like so many things in life; if you "don't have the tools, you can't play the game," and it seems as though a great many of our future investigations will require some basic understanding of the principles presently under investigation.

To accomplish our goals, we analyze the quadratic difference equation given by Equation (16.8) that relates the value of some function x at the $(n + 1)$st step to its value at the previous step:

$$x(n+1) = ax(n)[1 - x(n)]; \qquad 0 < a \leq 4, \quad 0 \leq x \leq 1 \qquad (16.8)$$

We then look at the behavior of the solutions of Equation (16.8) as the parameter a changes in value. This leads to the concept of fixed points, stability, multipliers, and period doubling. It enables us to obtain, in an approximate sense, the so-called Feigenbaum numbers, δ and α, that characterize any quadratic one-step map. It will become clear as we proceed that many of the procedures are unfamiliar in part because this is a new field of research and in part because so much of our training is in the study of differential rather than difference equations. With this in mind, we proceed slowly and repeat ourselves sufficiently often to keep in mind what we are doing.

Of the many new and unfamiliar words (i.e., concepts) introduced in this chapter, the most interesting is "chaos." It arises in discrete models as follows: we consider a deterministic system that is modeled by a difference equation. There are no random variables or parameters, so it should be our expectation that we could compute the trajectories (or behavior) of the system analytically or numerically. Let us see what we mean by this. Consider a difference equation given by Equation (16.8).

Pick an initial value for x, say $x(0)$, and then slowly change the parameter a and study the resulting behavior. As will be shown, for small values of the parameter a, the quantity $x(n)$ is attracted to the origin. This means that for any initial condition, the origin is an attractor of the system. Such behavior is familiar from our studies of attractive fixed points of differential equations.

We then increase the parameter a a bit more and notice that the original stable fixed point has become unstable and a new fixed point enters that acts as an attractor for the system. This is not very much different from the behavior of the solution to a nonlinear differential equation of the logistic type. We mean by this that the system has attracting and repelling fixed points.

Now, if we increase the parameter a bit more, we find that the system no longer has fixed points, but rather that it has what is called a period-2 cycle. It hops back and forth between two attracting points. Although this is a new phenomenon, it still makes some sense. (For example, you can think of it as a discrete analogue of a limit cycle.)

If we increase the parameter a still more, we find that the system now has a period-4 cycle. But we notice that this occurred when we changed the parameter a just a very little bit.

As we continue to increase our parameter, we notice that the system continues to have ordered cycles of higher and higher period. This all occurs when we change our parameter by smaller and smaller steps. Finally, no pattern emerges. We call this new regime "chaos."

Search for an Analogy. Consider a system that consists of a ball that must keep bouncing, confined to a box whose size is decreased by adjusting a parameter. We follow the motion of the ball, and as we make the box smaller and smaller, the trajectory of the ball becomes more and more difficult to follow since it must keep changing its velocity over smaller and smaller intervals of time. Now we are on safe ground if we believe in deterministic systems, for we say that given the initial conditions and the appropriate force laws and boundary conditions, we could predict with arbitrary accuracy the trajectory of the ball for all times. It is our expectation that systems with arbitrarily close initial conditions or arbitrarily close values of appropriate parameters follow similar trajectories.

Even though it might be difficult to realize these conditions for our ball problem, it is our experience that if the problem is linear, everything proceeds as expected. We would then say that we have a normal or deterministic system. After all, there are no random variables!

Now let us consider a related nonlinear problem in which we introduce interactions by making our box into a pinball machine. The system is deterministic and the forces are nonlinear. We find that tiny changes in the parameters (e.g., position of the pins) or in the initial conditions make an enormous difference in the results. In addition, we find that an error in the initial condition can be magnified exponentially through a parameter. This means that trajectories associated with arbitrarily close initial points diverge exponentially, and, thus, systems that have arbitrarily close initial conditions or arbitrarily close values of their parameters behave strikingly differently. We find that our system has the capacity to exhibit behavior as though random elements had entered into the system.

This is all unfamiliar to us from our study of linear systems. We say that we are in a chaotic regime and in many ways the word "chaos" signifies our helplessness in trying to maintain a deterministic characterization of a certain class of nonlinear systems. Chaos is a new aspect of the complicated and strange behavior of nonlinear systems. It takes on different forms in discrete and continuous systems and we are just struggling to understand its basic elements.

In this chapter, we limit our discussion to the analysis of quadratic maps. It will become clear that they have unusual properties amongst which is the universal behavior Feigenbaum discovered. He found that all smooth quadratic maps can be characterized by two universal constants, δ and α, called Feigenbaum numbers.

We conclude these introductory remarks with a long quote from the pioneering article by Robert May in *Nature*, Vol. 261, 459 (1976) entitled "Simple Mathematical Models with Very Complicated Dynamics" (reprinted with the permission of Professor May). I strongly urge that this article be required reading by anyone interested in studying discrete nonlinear problems. May develops a careful presentation of the logistic-model difference equation and explains its unexpected behavior, concluding with a plea to us all to learn more about nonlinear phenomena:

> The elegant body of mathematical theory pertaining to linear systems (Fourier Analysis, orthogonal functions, and so on), and its successful application to many fundamentally linear problems in the physical sciences, tends to dominate even moderately advanced university courses in mathematics and theoretical physics. The mathematical intuition so developed ill equips the student to confront the bizarre behavior exhibited by even the simplest discrete nonlinear systems, such as [the logistic equation]. Yet such nonlinear problems are the rule, not the exception, outside the physical sciences. I would therefore urge that people be introduced to, say, [the logistic equation] early in their mathematical education. This equation can be studied by iterating it on a calculator, or even by hand. Its study does not involve as much conceptual sophistication as elementary calculus. Such study would greatly enrich the student's intuition about nonlinear

systems. Not only in research, but also in the everyday world of politics and economics, we would all be better off if people realized that simple nonlinear systems do not necessarily possess simple dynamical properties.

16.3 REVIEW OF THE LOGISTIC DIFFERENTIAL EQUATION

Recall the basic concepts associated with the logistic differential equation that we introduced in Chapter 7. We write the equation in the form given by Equation (16.6):

$$\frac{dx}{dt} = bx(1 - x); \qquad x > 0 \tag{16.6}$$

The solution is

$$x(t) = x(0)\frac{\exp bt}{1 + x(0)(\exp bt - 1)} \tag{16.9}$$

The origin is unstable for $b > 0$ and is stable for $b < 0$. Conversely, the point $x = 1$ is an attractor for positive values of b and a repeller for negative values. (Refer to Figure 7.2.) Note that there is a smooth approach to the equilibrium point $x^* = 1$ for all positive values of b as the time t increases. Some aspects of this behavior are typical of nonlinear differential equations of this type, namely, those that possess a linear term and a counterbalancing nonlinear term.

We now turn our attention to the associated *logistic* difference equation, given by Equation (16.8), that models the growth of a species that takes place in discrete steps. We show that for particular ranges of the parameter a, we obtain quite different and unexpected results if we compare our solution to Equation (16.9). We say that these two equations are associated or related because they are developed to model the same systems. The difference equation assumes that the growth process takes place in discrete steps or time intervals, whereas the differential equation assumes a continuous growth pattern. Then, if we make a correspondence between the derivative and the difference, we say that Equations (16.6) and (16.8) are related. We expect the difference and the differential equations to have similar solutions as long as the step time is small enough so that the population $x(n)$ does not change substantially in this interval. During the course of our analysis, we find that this condition cannot be satisfied when the parameter a in the logistic difference equation is sufficiently large. We see, for example, for all positive values of the parameter b, the solution of the differential equation given by Equation (16.9) smoothly tends toward the stable equilibrium point that indicates a saturation population at $x = 1$. Furthermore, if we change the parameter b by a very small amount, we see no real change in the resulting

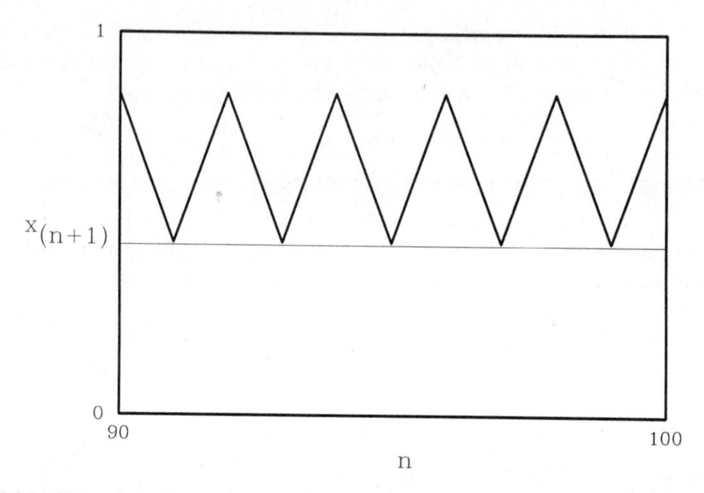

Figure 16.1 We plot $x(n + 1) = 3.4x(n)[1 - x(n)]$ with $x(0) = 0.55$. Observe that after 90 steps, the system has settled to a period-2 cycle. If we change the initial condition, we observe the same pattern. Try it.

trajectory. By way of contrast, the difference equation given by Equation (16.8) has solutions that have the capability to exhibit strikingly different behavior if we alter, ever so slightly, either the initial conditions or the value of the parameter a. This becomes clear through the analysis given in this chapter.

Example 3.1 Consider our logistic differential equation given by Equation (16.6). We can plot the solution and be confident it does not change much if

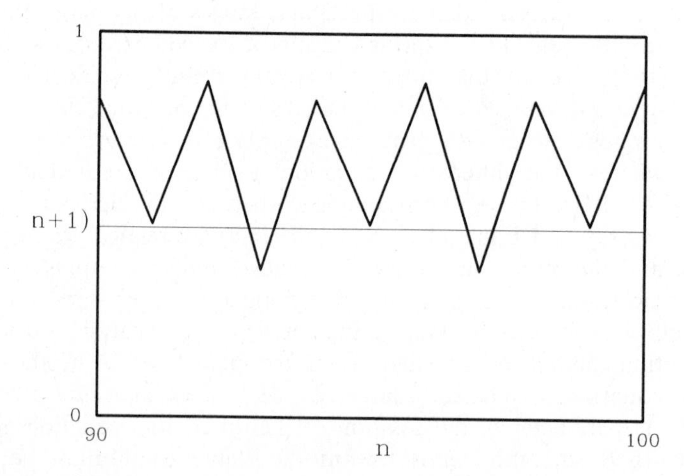

Figure 16.2 We plot $x(n + 1) = 3.5x(n)[1 - x(n)]$ with $x(0) = 0.5$. We obtain a period-4 cycle that is insensitive to a change in the initial condition.

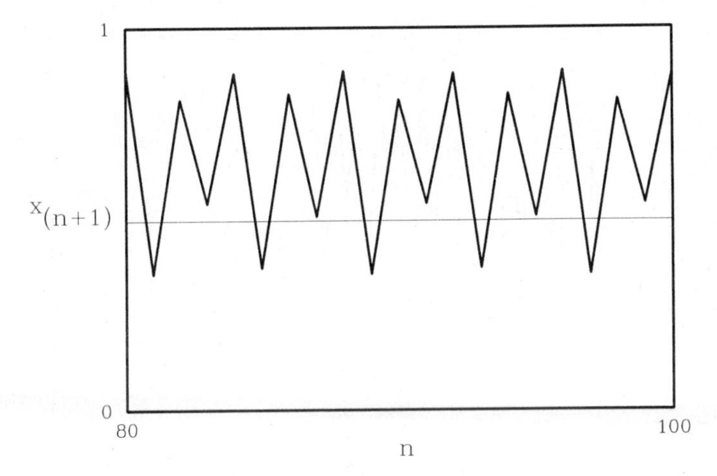

Figure 16.3 We plot $x(n + 1) = 3.55x(n)[1 - x(n)]$ with $x(0) = 0.5$. Observe that we now have a period-8 cycle. Vary the initial condition and conclude that the periodicity of the systems is unaffected.

we change either the parameter or the initial conditions by a small amount. (This can be easily verified.) Consider the logistic difference equation given by Equation (16.8). We want to explore various combinations of initial conditions and values of the parameter a. This is done in Figures 16.1 to 16.5.

We observe that in Figures 16.1–16.3, nothing very unusual has happened, meaning that the motion appears to have a regular structure to it.

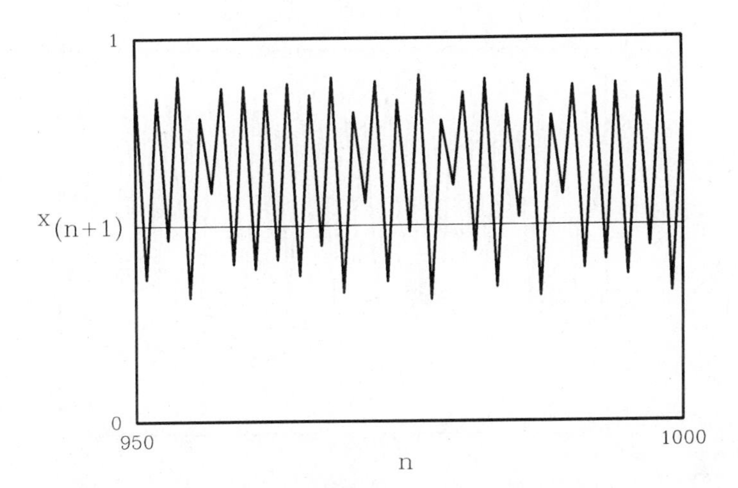

Figure 16.4*a* We plot $x(n + 1) = 3.61x(n)[1 - x(n)]$ with $x(0) = 0.5$. It is possible to conclude that no regular pattern emerges.

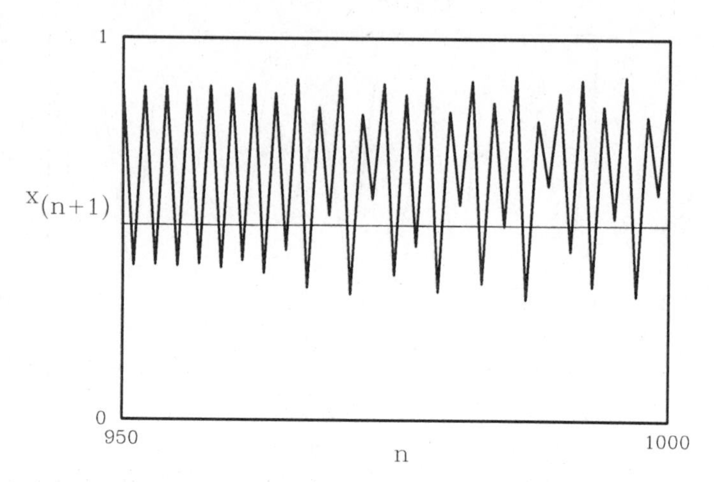

Figure 16.4*b* We plot $x(n+1) = 3.6101x(n)[1 - x(n)]$ with $x(0) = 0.5$. Observe that a small change in the parameter a leads to a figure that is different from Figure 16.4*a*. Practice plotting various intervals and convince yourself that the patterns are truly different.

On the other hand, we notice that in Figures 16.4*a*–16.4*d* and 16.5*a*–16.5*b*, small changes in either the parameter value or the initial condition lead to substantial changes in the trajectory. The system is in the chaotic region. We can continue practicing by choosing various values for our parameter and initial condition and, in this way, we gain some small measure of intuition. However, we have to develop our analytical techniques a bit in order to begin to appreciate the richness of the newly encountered phenomena. Finally, it is amusing to note that different software programs such as *Lotus 1-2-3* and

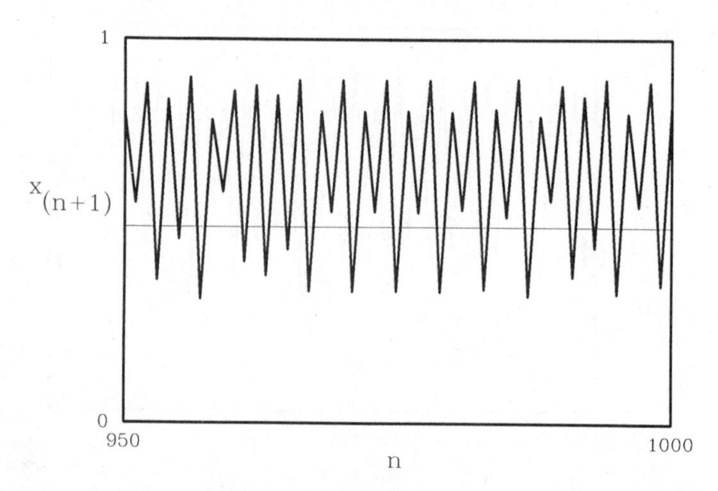

Figure 16.4*c* We plot $x(n+1) = 3.6101x(n)[1 - x(n)]$ with $x(0) = 0.501$. Here we have slightly changed the initial condition and obtained a figure that differs from Figure 16.4*b*.

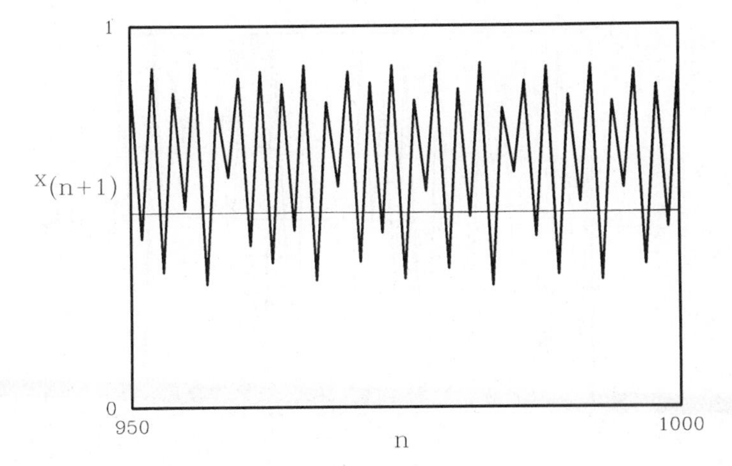

Figure 16.4*d* We plot $x(n+1) = 3.61x(n)[1 - x(n)]$ with $x(0) = 0.501$. Observe that this figure differs from Figures 16.4*a*–16.4*c*.

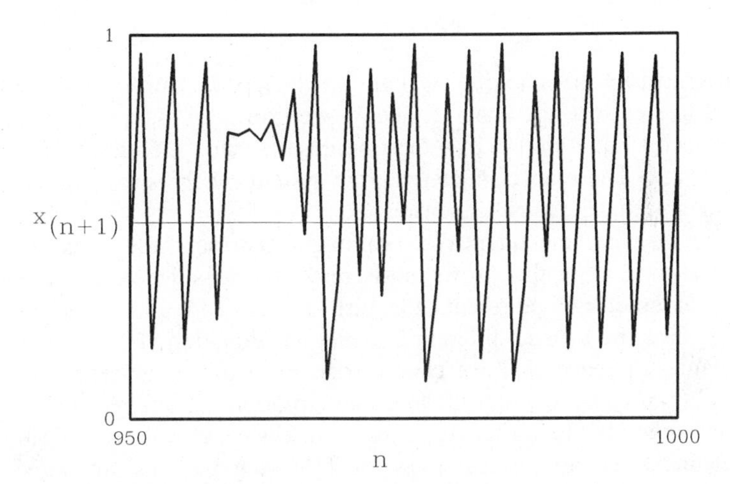

Figure 16.5*a* We plot $x(n+1) = 3.9x(n)[1 - x(n)]$ with $x(0) = 0.5$.

Math Cad yield different chaotic trajectories. This puts a new wrinkle in the comparison of results obtained doing different computer experiments. (See the concluding remarks, Section 16.9.) ■

16.4 INTRODUCTION OF A MAP

We begin by introducing the concept of a map. We say that we have a one-dimensional smooth map T that yields $x(n+1)$ if we are given $x(n)$. We write $f[x(n)] = x(n+1)$. Alternatively, we can say that if our system is

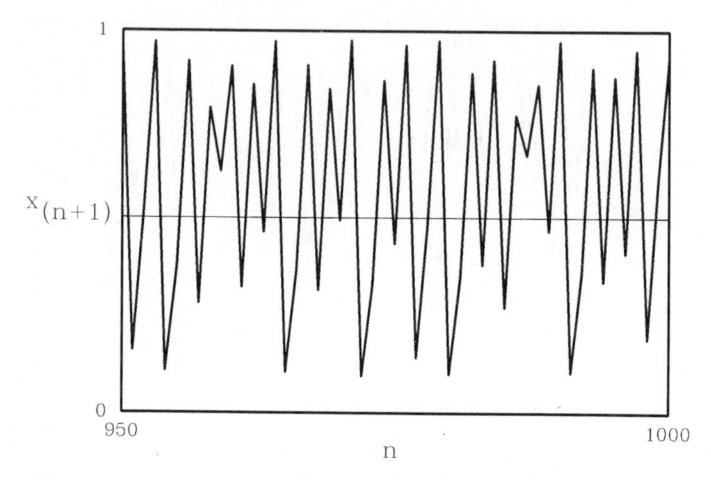

Figure 16.5*b* We plot $x(n + 1) = 3.90001x(n)[1 - x(n)]$ with $x(0) = 0.5$. Observe that this figure differs from Figure 16.5*a* even though we have changed the parameter a ever so slightly. This is one of the aspects of chaos.

characterized by the quantity $x(n)$ at its nth step, we can find its value at the $(n + 1)$th step via the map T. Thus, we call our difference equation as expressed by Equation (16.8) a map because it maps successive values of the parameter x. The map characterizes the system by the single variable $x(n)$; thus, we say that it is one-dimensional.

As mentioned, the quantity $x(n)$ might represent the density of some species at time $t = n$ that grows in discrete time steps denoted by the variable n. The parameter a represents the birth-minus-death rate. It is completely analogous to the logistic differential equation, Equation (16.6), except for the important difference that we characterize its behavior in terms of discrete values. It can be shown that the logistic map occurs in different forms and for various ranges of the parameters. But it is always effectively the same map. The elements of our discussion in the following sections are based on the analysis given in Lauwerier's article in the book edited by Holden. He, in turn, gives appropriate credit to the articles by May, Feigenbaum, and Helleman.

16.5 FIXED POINTS, ATTRACTORS, REPELLERS, AND MULTIPLIERS

Consider the logistic map given by Equation (16.8). Since the range of the variable x is $0 \leq x \leq 1$, we now show that the parameter a is restricted to the range $0 < a \leq 4$. We pick any point to start our steps. Let us begin at $x(0)$ and call its value β. Since we require that $0 \leq x(1) \leq 1$, we have

$$x(1) = a\beta(1 - \beta) \leq 1$$

Then introduce the auxiliary function

$$g(\beta) = \beta(1 - \beta)$$

We want to find the maximum of $g(\beta)$ since this gives us the maximum value of a. Taking the derivative of $g(\beta)$ with respect to β, we find

$$g'(\beta) = 1 - 2\beta; \qquad g'(\beta) = 0 \quad \text{when } \beta = \tfrac{1}{2}$$

$$g_{\max} = \tfrac{1}{4}$$

So we see that $a \leq 4$ in order that $x(n) \leq 1$ for any initial point in the range $[0, 1]$. As a parallel to our analysis of fixed points of nonlinear differential equations, we introduce the concept of a fixed point of our map. We say that x^* is a fixed point of the map $f[x(n)] = x(n+1)$ if, when $x(n) = x^*$, $x(n+1) = x^*$, or $f(x^*) = x^*$. Thus, the transformation, or the map, leaves the system at the fixed point.

Example 5.1 If $x(n+1) = x(n)$ and $x^* = ax^*(1 - x^*)$, then we have as solutions $x^* = 0$ and $x^* = 1 - 1/a$. If $a = \frac{3}{4}$, $x^* = 0$ is the only fixed point in the range. It is easy to convince oneself that if $a < 2$, the approach to the fixed point is monotonic. By contrast, if $a > 2$, say $a = 2.9$ and $x(0) = \frac{1}{2}$, we find that the approach to the fixed point $x^* = 1.9/2.9$ is oscillatory. The first few values are 0.725, 0.578, 0.707, 0.600, 0.696, 0.614, 0.687,...(see Figure 16.6). ■

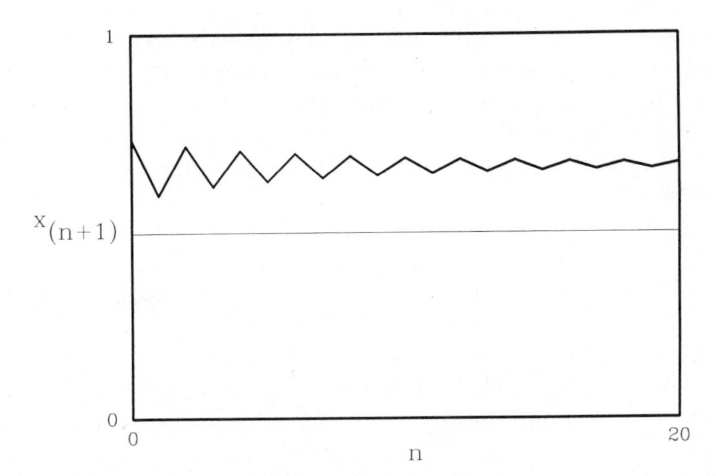

Figure 16.6 We plot $x(n+1) = 2.9x(n)[1 - x(n)]$ with $x(0) = 0.5$. Observe that the system has an oscillatory approach to its fixed point.

We now want to locate and study the stability of the fixed point as a function of the parameter a. This means that we want to know whether small deviations from the fixed point lead the system back to it or away from it.

We say that we have an attractive fixed point at $x = x^*$ if, for a value close to it, the next iterate is still closer. Consider $x(n)$ close to x^*. Then we write

$$x(n) = x^* + \varepsilon(n)$$

where $\varepsilon(n)$ is small. Then we compute $x(n+1)$ and we write it as

$$x(n+1) = x^* + \varepsilon(n+1)$$

We say that we have an attractive fixed point of our map at $x = x^*$ if

$$\frac{|\varepsilon(n+1)|}{|\varepsilon(n)|} < 1 \tag{16.10}$$

If this is the case, then succeeding steps will be closer and closer to x^*. This means that x^* is a stable or attractive fixed point. Alternatively, we can write

$$x(n+1) = x^* + \varepsilon(n+1) = f[x(n)] = f[x^* + \varepsilon(n)]$$

or, expanding in a Taylor series,

$$x^* + \varepsilon(n+1) \approx f(x^*) + \varepsilon(n)f'(x^*) = x^* + \varepsilon(n)f'(x^*)$$

where we have neglected higher-order terms. Then, if we are to satisfy the inequality given in Equation (16.10), we require the absolute value of the derivative $|f'(x^*)|$ to be less than 1.

We cannot conclude from this analysis that the fixed point is stable against all perturbations or, in other words, that it is a global attractor. We are only making a statement about the local stability of x^*. Furthermore, it might take the system a while to settle down. We mean by this that one might have to wait until the transient behavior has passed before the true nature of the motion is revealed.

Now consider a situation where $x(n)$ is close to a fixed point and the succeeding value $x(n+1)$ is further away. Then

$$|\varepsilon(n+1)|/|\varepsilon(n)| > 1$$

and we say that the fixed point is unstable. If the value of $x(n)$ did not change in position

$$|\varepsilon(n+1)| = |\varepsilon(n)|$$

we say that the point is neutrally stable. Finally, if, to the lowest order, x

returns to the fixed point, so that $\varepsilon(n+1) = 0$, we say that the fixed point is superstable. Our procedure is equivalent to taking a Taylor series of the function or map $f(x)$ in the neighborhood of $x = x^*$.

Example 5.2 Consider Equation (16.8). We have

$$f(x) = ax(1-x); \qquad f'(x) = a(1-2x); \qquad f'(0) = a$$

Since $|f'(0)| < 1$ when $a < 1$, we have a stable fixed point at the origin.

It is also possible to conclude that the origin is an unstable fixed point when $a > 1$. ■

We introduce the multiplier λ that is equal to the derivative of $f(x)$ at the fixed point x^*. Thus, $f'(0) = a = \lambda$. It follows that a stable fixed point has $|\lambda| < 1$ and an unstable fixed point has $|\lambda| > 1$.

Question: What is the stability of the fixed point $x^* = 1 - 1/a$ of Equation (16.8)? This fixed point is outside of the range of the variable x until the parameter a is greater than 1. So if $0 < a < 1$, we have a single stable fixed point at $x = 0$.

EXERCISE 5.1 Plot $x(n)$ versus n for $n = 0$ to 20, where $a = 0.5$ and 0.8, and $x(0) = 0.2$, 0.4, and 0.8.

Convince yourself that $x^* = 0$ is a fixed point that is approached monotonically for each of these values of a. ■

Now let us increase the parameter a until it crosses the value $a = 1$. The origin becomes unstable since $f'(0) = a > 1$. The second fixed point, at $x^* = 1 - 1/a$, has now come within the range of the variable x. The multiplier at that point is found by writing

$$f'\left[x = 1 - \frac{1}{a}\right] = \lambda = -a + 2; \qquad \text{and} \qquad |\lambda| < 1 \text{ if } 1 < a < 3$$

Thus, our fixed point, $x^* = 1 - 1/a$, for $1 < a < 3$, is a stable fixed point. We find that when the parameter a is greater than or equal to 3, there are no stable fixed points of our map. We discuss, in Section 16.6, the change in the behavior of the map when $a \geq 3$.

EXERCISE 5.2 Consider

$$x(n+1) = 2.5x(n)[1 - x(n)]$$

Choose $x(0) = 0.3$, 0.4, and 0.9. Plot $x(n)$ for $n = 50$ steps and observe the approach to the fixed point at $x = 3/5$.

Develop the same plots for $a = 2$, 2.8, and 2.95. Observe that the fixed point may be approached through damped oscillations. This behavior contrasts with the monotonic approach to the fixed point of the logistic differential equation given by Equation (16.6). ■

EXERCISE 5.3 It is of value to learn how to use a graphical technique to follow the approach to the fixed point. Examine Figure 16.7.

We have constructed a square box with sides equal to 1. Each of the successive values of $x(n)$ fall inside the box. In Figure 16.7, we choose an initial value $x(0)$ and plot

$$y = f(x) = x(n+1) = ax(n)[1 - x(n)]; \qquad \text{with } a = 2.5$$

We also plot the line $y = x$. Then $f(x)$ has its maximum value equal to 0.625 at $x = \frac{1}{2}$. The intersection of the straight line and the curve denotes the location of the fixed point at $x^* = 0.6$. The slope of the curve, in the neighborhood of the origin, is greater than 1. (It is convex to the straight line.) This is a graphical way of indicating that the multiplier $f'(0)$ is greater than 1. On the other hand, the slope of the curve at the intersection $x = 0.6$ has an absolute value less than 1. So, given any initial point $x(0)$ close to x^*, the system approaches this fixed point. Note that the approach is oscillatory. For example, choose $x(0) = 0.75$. Draw a vertical line from the abscissa to the curve $f(x) = 2.5x(1 - x)$. The intersection denotes the appropriate value of $f(x)$ and is equal to 2.5 (0.75) (0.25) = 0.47 = y. Rather than recording this value on the ordinate, we draw a line parallel to the x axis and locate the intersection with the line $y = x$. At this intersection, $x = 0.47$. We then draw a line parallel to the y axis and note the next intersection with the curve at $x = 0.62$. What are we doing? Once we have found the value $x(1) = 0.47$, we want to use it as input to find the value $x(2)$. This requires transferring the value we obtain for the ordinate, $y = f(x)$, to its equal value on the abscissa. The straight line $y = x$ helps us accomplish this because it has its y and x

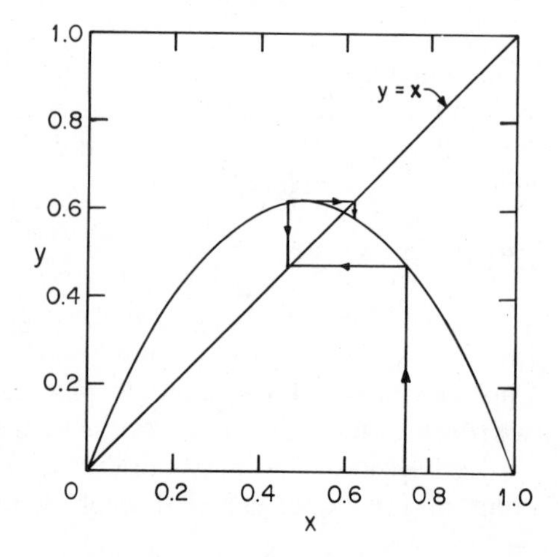

Figure 16.7 We plot a parabola whose maximum occurs at $x = 0.5$ and has the value 0.625. It intersects the line $y = x$ at $x = x^* = 0.6$.

values equal to one another. We can then develop a graphical algorithm or routine to follow the approach to the fixed point. Successive values are 0.62 and 0.59, suggesting that for our choice of the parameter a, there is an oscillatory approach to the fixed point. Repeat this analysis for $a = 2.9$ and $x(0) = 0.75$. Can you convince yourself, by a graphical argument, that the approach to x^* continues to be oscillatory? Finally, choose a value of $1 < a < 2$ and use a graphical construction to convince yourself that the approach to the fixed point is not oscillatory. ■

16.6 PERIOD-2 CYCLES AND THE PATHWAY TO CHAOS

16.6.1 Preview

We have learned in Section 16.5 that when the parameter a is greater than 3, the fixed point $x^* = 1 - 1/a$ has lost its stability since the multiplier

$$\lambda = a(1 - 2x^*) = a\left[1 - 2\left(1 - \frac{1}{a}\right)\right] = 2 - a$$

is greater in absolute value than 1. Thus, small deviations from the fixed point grow. There are no other fixed points in the range $0 \leq x \leq 1$. Instead of stable fixed points x^*, we find that for a range of the parameter a, there exists a pair of values of $x(n)$, denoted by x_1 and x_2, such that $f(x_1) = x_2$ and $f(x_2) = x_1$. After an initial settling in, the system oscillates back and forth between these points. We characterize this new behavior as a period-2 cycle.

EXERCISE 6.1 Consider the logistic map given by Equation (16.8): $x(n + 1) = ax(n)[1 - x(n)]$, where the parameter a is 3.1. Show that $x^* = 1 - 1/3.1 = 2.1/3.1$ is a fixed point. Now take an initial value $x(0)$ close to x^*, say $x(0) = 0.69$. Apply our map and plot $x(n)$ versus n for $n = 0$ to 100. See Figure 16.8 and Exercise 6.6. ■

In Exercise 6.1, we observe that the character of our map has changed in a fundamental way and we have lost, thereby, the connection with the associated logistic differential equation. The resulting period-2 cycle is analyzed by squaring our map and finding its fixed points. Alternatively, we write $f[f(x_i)] = f^2(x_i) = x_i$, $i = 1, 2$. This means that we seek a pair of points, x_1 and x_2, such that $f(x_1) = x_2$ and $f(x_2) = x_1$. We can also say that x_1 and x_2 are fixed points of an iterated or squared map. Then we can think of observing our system at each step and seeing the period-2 cycles, or, using the stroboscopic viewpoint, we can take our observations at intervals corresponding to two steps and observe the fixed-point behavior. (See Section 16.6.3.)

The period-2 cycles are found to be stable as long as the parameter a remains in an appropriate range. When it exceeds a critical value, the period-2

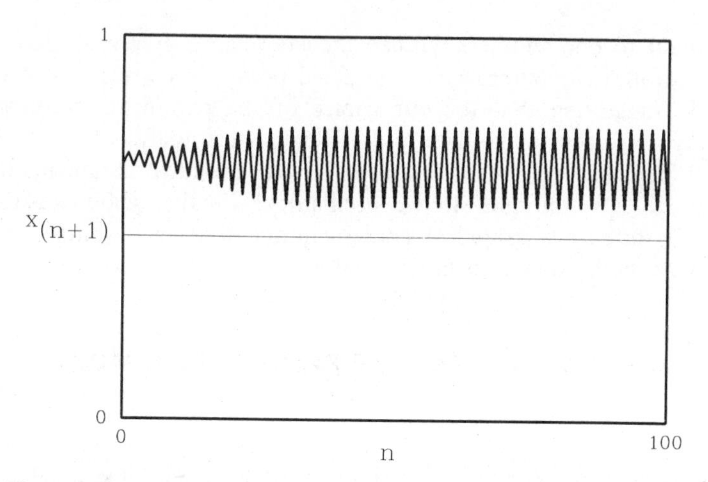

Figure 16.8 We plot $x(n+1) = 3.1x(n)[1 - x(n)]$ with $x(0) = 0.69$.

cycle becomes unstable and we find that a stable period-4 cycle emerges. This, in turn, loses its stability, and as the parameter a is increased further, cycles of higher and higher period are created. It turns out, although completely unexpectedly, that each higher-order map retains characteristics of the period-2 cycle, with an appropriate change of scale. Specifically, it was discovered by Mitchell Feigenbaum that the period-doubling scheme has a universal rate of convergence δ and a universal scaling α characteristic of any one-dimensional quadratic map. The resulting numbers δ and α are the Feigenbaum numbers. It is possible, using a powerful hand calculator or a computer, to find the fixed points of these iterated maps to a high accuracy. This is done, for example, in the papers by Feigenbaum.

Rather than proceed with an exact analysis, we choose to introduce an approximation, or truncation, of our expansion, retaining, only the linear and quadratic terms at each step of the iteration process. This enables us to continue our development with a minimum of algebraic complexity while still retaining both the essence of the calculational technique and of the conclusions. With this truncation understood, we refer to the iterations of our map without stating at each point that we are really referring to an approximate map. This should not cause any confusion, and our end result is to obtain an approximation for the Feigenbaum numbers rather than their correct values. In the analysis that follows, it is important to observe that the onset of new behavior takes place closer and closer to the fixed point as we perform successive squarings of our map. Where several squarings are required, the difference between an exact map and its truncated approximation is small. (We expect the linear and quadratic terms to be dominant and the cubic and quartic terms to enter as corrections.) Note that the truncation procedure we follow is the standard one given in expository discussions of this subject. (See the Bibliography.)

It is possible to show that any quadratic map of the form

$$s(n+1) = \sigma + \beta s(n) + \mu s(n)^2 \tag{16.11}$$

with a fixed range of the variable $s(n)$ can be transformed by a sequence of translations and changes of scale into the logistic map given by Equation (16.8). Thus, it is sufficient to derive all of our results for this particular map. (We have used the standard notation $s(n)^2$ to indicate the product $s(n)s(n)$.)

EXERCISE 6.2 Consider the map

$$x(n+1) = ax(n) - bx(n)^2 \tag{16.12}$$

Show that if we let $y = (b/a)x$, we obtain the logistic map for $y(n)$ corresponding to Equation (16.8). ■

EXERCISE 6.3 Consider the map

$$y(n+1) = 1 - \sigma y(n)^2 \tag{16.13}$$

Show that it can be transformed into the map given by Equation (16.8). What is the required range for the variable $y(n)$ and the parameter σ? This map is analyzed in the book by H. Bai-Lin. ■

EXERCISE 6.4 Consider the map

$$z(n+1) = -(1+\mu)z(n) + z(n)^2 \tag{16.14}$$

Transform this map into the map given by Equation (16.8). What ranges of the variable $z(n)$ and parameter μ correspond to the range of the parameter $3 \le a \le 4$? This form of the logistic map is analyzed by Lauwerier in the book edited by Holden. ■

16.6.2 Shift of the Origin

It is convenient to shift the origin of our map to the fixed point at $x^* = 1 - 1/a$ since it is this point whose stability we are assessing. Thus, we write

$$x(n) = 1 - \frac{1}{a} + z(n) \tag{16.15}$$

The map given by Equation (16.8) becomes

$$1 - \frac{1}{a} + z(n+1) = a\left[1 - \frac{1}{a} + z(n)\right]\left\{1 - \left[1 - \frac{1}{a} + z(n)\right]\right\}$$

This is simplified to read

$$z(n+1) = (2-a)z(n) - az(n)^2 \tag{16.16}$$

We call this our z map and it is the basis of all of our future analysis. It has fixed points $z^* = 0$ and $z^* = 1/a - 1$ corresponding to $x^* = 1 - 1/a$ and $x^* = 0$, respectively. We concentrate our attention on the fixed point $z^* = 0$, and see what happens after the parameter a crosses the value $a = 3$. We have to determine the range of the variable z associated with the range of the variable x given by Equation (16.8). We have $0 \le x \le 1$ and $3 \le a \le 4$.

Using Equation (16.15), we determine that the range of the variable z is

$$\frac{1}{a} - 1 \le z \le \frac{1}{a} \tag{16.17}$$

Notice that the range of z is a-dependent.

EXERCISE 6.5 Take values of $a = 3.2$ and 3.4 and a variety of initial values $z(0)$. Using a calculator, verify that $z(n)$ stays within the range specified by Equation (16.17). ■

16.6.3 Calculation of the Period-2 Cycle

Consider the map given by Equation (16.16) in which we write $f[z(n)] = z(n+1)$. We seek the fixed points of a period-2 cycle, denoted by p and q. We have

$$f(p) = q \quad \text{and} \quad f(q) = p \tag{16.18}$$

Substituting Equation (16.16) into Equation (16.18), we have

$$p = (2-a)q - aq^2 \tag{16.19a}$$

$$q = (2-a)p - ap^2 \tag{16.19b}$$

First, we subtract Equation (16.19a) from Equation (16.19b) and obtain

$$(q-p) = (2-a)(p-q) - a(p^2 - q^2)$$

or

$$p + q = \frac{3-a}{a} \tag{16.20}$$

We now multiply Equation (16.19a) by p and multiply Equation (16.19b) by q and subtract to obtain

$$p^2 - q^2 = apq(p-q)$$

or

$$pq = \frac{p+q}{a} = \frac{3-a}{a^2} \tag{16.21a}$$

Eliminating q between Equations (16.20) and (16.21a), we have

$$a^2p^2 - (3-a)ap + (3-a) = 0 \tag{16.21b}$$

Solving, we find

$$p = \frac{1}{2a}\left[(3-a) - \sqrt{(a-3)(a+1)}\right] \tag{16.22a}$$

$$q = \frac{1}{2a}\left[(3-a) + \sqrt{(a-3)(a+1)}\right] \tag{16.22b}$$

EXERCISE 6.6 Consider $a = 3.1$. Show that the values we obtain for p and q given by Equations (16.22a) and (16.22b) are $p = -0.1194$ and $q = 0.0871$. These are the coordinates of the z map. If you want to obtain the corresponding values for the x map given by Equation (16.8), you must recall the shift of the origin given by Equation (16.15). Thus, show that the coordinates of p and q as given by the x map are $p = 0.558$ and $q = 0.765$. Verify explicitly that $f[f(p)] = p$; $f[f(q)] = q$. See Figure 16.9. ■

16.6.4 Stability

We are now in a position to examine the stability at the points p and q. To evaluate the stability of a fixed point of a general squared map, we differentiate $f[f(s)]$ with respect to s and then put $s = p$ and q. At this point, it is convenient to introduce the auxiliary function $g(s)$ whose role is simply to clarify the steps in the differentiation process. Thus, we write

$$f[f(s)] = f[g(s)] \tag{16.23}$$

with

$$g(s) = f(s)$$

We differentiate f and obtain

$$\frac{d}{ds}[f(g(s))] = g'(s)f'[g(s)] \tag{16.24}$$

Or, since $g(s) = f(s)$, we have

$$\frac{d}{ds}[f(f(s))] = f'(s)f'[f(s)] \tag{16.25}$$

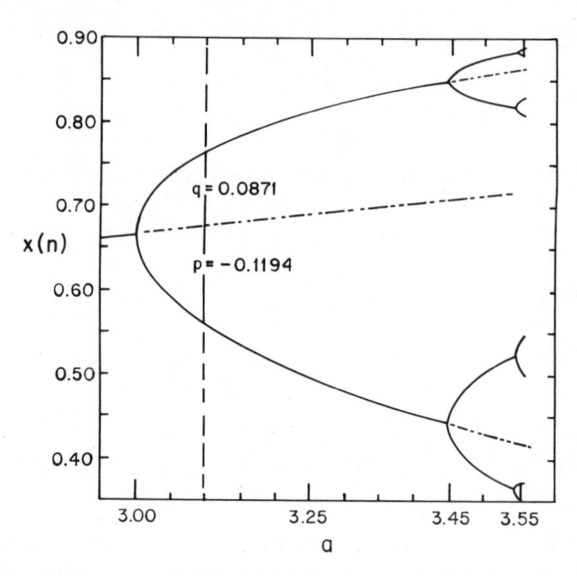

Figure 16.9 We plot $x(n)$ versus the parameter a. Observe the two stable branches that emerge when a exceeds 3. The broken central lines indicate the locus of unstable fixed points.

We want to evaluate this expression at the points $s = p$ and q in order to determine if the multiplier λ is between -1 and 1. After noting that $f(p) = q$ and $f(q) = p$, we then obtain

$$\frac{d}{ds}[f(f(s))] = f'(p)f'(q); \qquad \text{with } s = p \text{ or } q \tag{16.26}$$

Thus, the multiplier λ associated with our squared map is constructed from products of the derivative of our original map taken at the points associated with our cycle of period 2. At a later point in the analysis, we will be finding the properties of maps of higher and higher order. It is easy to show in this regard that when one considers, for example, fixed points of squared maps taken twice, or quartic-maps, that a corresponding multiplier is obtained. We mean that if we have a four cycle consisting of the distinct points g, h, i, and j such that $f(g) = h$, $f(h) = i$, $f(i) = j$, and $f(j) = g$, then we have a quartic map with multiplier λ equal to

$$\frac{d}{ds}[f(f(f(f(s))))] = f'(g)f'(h)f'(i)f'(j) \tag{16.27}$$

where we have taken $s = g$, h, i, or j.

Now, let us consider our z map given by Equation (16.16) and evaluate:

$$f'(p)f'(q) = [(2-a) - 2ap][(2-a) - 2aq] \tag{16.28}$$

Substituting the values for p and q given by Equations (16.22a) and (16.22b), respectively, we evaluate the multiplier

$$f'(p)f'(q) = \lambda = 1 - (a^2 - 2a - 3) \tag{16.29}$$

Thus, the fixed point pair (p, q) is a stable point of our squared map as long as the inequality

$$-1 < a^2 - 2a - 4 < 1 \tag{16.30}$$

or

$$3 < a < 1 + \sqrt{6} = 3.4495 \tag{16.31}$$

is maintained.

We have rejected the term with the minus sign in front of the square root since it takes us out of the range of the parameter a.

16.7 THE ITERATION PROCESS

In Equation (16.31), we observed that when our parameter a exceeds the critical value $1 + \sqrt{6}$, there are neither stable fixed points nor stable period-2 cycles. What happens is that the map connects its nth and $(n - 4)$th steps and develops a stable period-4 cycle. In order to analyze this behavior, we develop an iteration scheme that generates an associated map that is similar to the one previously encountered. This procedure involves some algebraic manipulation that begins with the iteration process.

Let us return to the situation where $3 \leq a \leq 1 + \sqrt{6}$. Our immediate goal is to develop a map associated with a period-2 cycle that looks like a map with a fixed point. To accomplish this, we initially square our original map given by Equation (16.16) and express its value after $n + 1$ steps in terms of its value after $n - 1$ steps:

$$\begin{aligned} z(n+1) = (2-a)\left[(2-a)z(n-1) - az(n-1)^2\right] \\ - a\left[(2-a)z(n-1) - az(n-1)^2\right]^2 \end{aligned} \tag{16.32}$$

Now we are going to forget our original map and concentrate on the squared map, and in doing so, we pass over the intermediate step and think of directly connecting the $(n + 1)$th and the $(n - 1)$th steps. Thus, we expand the terms in Equation (16.32) and write our squared map as

$$z(n+1) = b^2z(n) - (ab + ab^2)z(n)^2 + 2a^2bz(n)^3 - a^3z(n)^4 \tag{16.33}$$

where we introduce the coefficient $b = 2 - a$ in an attempt to simplify the expression. (Since we skip the intermediate step hereafter, we have relabeled

$n - 1$ as n.) We think of Equation (16.33) as a one-step map of fourth degree connecting the value of z at the $(n + 1)$th and nth steps. Points p and q are the fixed points of our new map.

EXERCISE 7.1 Verify that p and q are fixed points by substituting the values of p and q in terms of the parameter a as given by Equations (16.22a) and (16.22b). This involves some algebra. ■

We now concentrate our attention on the fixed point p and make a transformation of the origin of our coordinate system so that it is located at $z^* = p$. This is the first step in our iteration procedure. In what follows, keep in mind that whatever is valid for point p is valid for point q. We let

$$z(n) = p + v(n) \tag{16.34}$$

in Equation (16.33), where we expect $z(n)$ to be near the fixed point $z^* = p$, and, hence, $v(n)$ should be a small quantity. After some manipulations, including the introduction of the quantity q given by Equation (16.19b), we obtain

$$\begin{aligned} v(n+1) = {} & \left[b^2 - 2ab(p+q) + 4a^2pq\right]v(n) - \left[ab(1+b) - 6a^2q\right]v(n)^2 \\ & + O\left[v(n)^3\right] \end{aligned} \tag{16.35}$$

Note that we are neglecting terms that are of higher order than linear and quadratic. Thus, our map is no longer exact. We use Equations (16.20) and (16.21a) to simplify Equation (16.35) and obtain

$$v(n+1) = (4 + 2a - a^2)v(n) - aC_1(p)v(n)^2 \tag{16.36}$$

where we neglect terms $O[v(n)^3]$ and where the parameter $C_1(p)$ is

$$\begin{aligned} C_1(p) &\equiv b(1+b) - 6aq \\ &= (a-3)(a+1) - 3\sqrt{(a-3)(a+1)} \end{aligned} \tag{16.37a}$$

Thus, we end up with a quadratic map, given by Equation (16.36), of almost the same form as our original z map, given by Equation (16.16). We still have to adjust our coefficients.

EXERCISE 7.2 Consider Equation (16.34); instead of making a transformation to the fixed point at $z^* = p$, introduce

$$z(n) = q + w(n)$$

Perform the same sequence of operations that led to Equation (16.37a) and obtain the same map given by Equation (16.36) with $C_1(q)$ replacing $C_1(p)$, where

$$C_1(q) \equiv b(1+b) - 6ap$$
$$= (a-3)(a+1) + 3\sqrt{(a-3)(a+1)} \quad \blacksquare \quad (16.37b)$$

EXERCISE 7.3 Consider Equations (16.22a) and (16.22b). Show that when $a = 1 + \sqrt{6}$, we have

$$p = -0.27014$$
$$q = 0.13984$$

Using Equations (16.37a) and (16.37b), find

$$C_1(p) = -2.24$$
$$C_1(q) = 6.24$$

In the x map, the values corresponding to p and q are found from Equation 16.15. See Figure 16.9. ■

The maps $v(n)$ and $w(n)$ have the same structure and the coefficient of the linear term is the same. However, observe that although Equations (16.37a) and (16.37b) have the same form, they differ because p is not equal to q.

In obtaining Equation (16.36), we have performed the first crucial steps in the iteration procedure. They are

(i) squaring of our map given by Equation (16.32);
(ii) shifting the origin to a fixed point of the squared map and expanding about this point as given by Equation (16.34); and
(iii) truncating the shifted squared map, neglecting cubic and quartic terms.

From Equation (16.31), recall that we need the parameter a to be in the range

$$3 < a < 1 + \sqrt{6}$$

in order that the fixed point p be a stable fixed point of the squared map. (Equivalently, we want the multiplier λ to be in the interval between -1 and $+1$.) We now proceed to transform the v map, given by Equation (16.36), into the same form as our z map, given by Equation (16.16). (This is the very point of the mapping-iteration process.)

In order to carry out this transformation, we have to adjust the coefficient of the linear term. To accomplish this, we introduce an iterated parameter a_1 such that

$$2 - a_1 \equiv 4 + 2a_0 - a_0^2$$

or

$$a_1 = \text{the first iterate} \equiv a_0^2 - 2a_0 - 2 \tag{16.38}$$

where our original parameter a is now denoted by a_0. Then Equation (16.36) becomes

$$v(n+1) = (2 - a_1)v(n) - a_0 C_1(p)v(n)^2 \tag{16.39}$$

Our next step is to write the coefficient of the quadratic term in Equation (16.39) in the same form as we have in our original z map given by Equation (16.16). This is called a "rescaling." We introduce

$$s(n) = \frac{a_0 C_1(p)}{a_1} v(n) \tag{16.40}$$

Equation (16.39) then becomes

$$\frac{s(n+1)}{a_0 C_1(p)/a_1} = \frac{(2 - a_1)s(n)}{a_0 C_1(p)/a_1} - \frac{a_0 C_1(p)s(n)^2}{[a_0 C_1(p)/a_1]^2}$$

Eliminating the common factors, we have

$$s(n+1) = (2 - a_1)s(n) - a_1 s(n)^2 \tag{16.41}$$

Observe that Equation (16.41) is of exactly the same form as our z map, given by Equation (16.16), with a_1 in place of a. Keep in mind that we dropped our cubic and quartic terms. Also, refer to Equation (16.37a) and observe that the parameter $C_1(p)$ depends upon the parameter a. The parameter $C_1(p)$ changes slowly as p and a increase.

EXERCISE 7.4 Use the values of $C_1(p)$ and $C_1(q)$ obtained in Exercise 7.3 to show that when $a_0 = 1 + \sqrt{6}$, we have

$$\frac{a_0}{a_1} C_1(p) = -2.58$$

$$\frac{a_0}{a_1} C_1(q) = 7.17$$

■

Since the s map is equivalent to the z map, we know that it has a fixed point at $s^* = 0$ that becomes unstable when the parameter $a_1 = 3$. We then have a period-2 cycle of the s map or a fixed point p of the squared s map. This point is stable as long as the parameter a_1 satisfies the inequality

$$3 \leq a_1 < 1 + \sqrt{6} \tag{16.42}$$

Referring to Equation (16.31), note that our original parameter $a \equiv a_0$ creates a stable period-doubling regime as soon as it is equal to 3. This situation remains in force until it reaches the *numerical* value $a_0 = 1 + \sqrt{6}$. It is natural to ask: What is the value of the parameter a_0 of the z map when a_1 attains the value $1 + \sqrt{6}$? To answer this question, we recall the definition of the parameter a_1 given by Equation (16.38). When

$$a_0 = 1 + \sqrt{6} = 3.4495; \qquad a_1 = 3$$

and when $a_1 = 1 + \sqrt{6}$, we have, from Equation (16.38),

$$1 + \sqrt{6} = a_0^2 - 2a_0 - 2$$

or

$$a_0 = 1 + \sqrt{4 + \sqrt{6}} = 3.5396 \tag{16.43}$$

16.8 A DERIVATION OF THE APPROXIMATE FEIGENBAUM NUMBERS

16.8.1 The Scaled Map

In Equations (16.40) and (16.41), we developed the first step of the iteration scheme that generates successive values of the parameter a_i and scaling parameter C_1. We now want to use these results to follow the generation of higher-order maps. We repeat the process of shifting our map to a new origin associated with the fixed point of the period-2 cycle of the s map and then introducing a scale transformation of the same form to obtain a new map. We suppress the argument of C_2 since it has the same form for p and q. At the second iteration, we have

$$k(n+1) = \left(4 + 2a_1 - a_1^2\right)k(n) - a_1 C_2 k(n)^2 \tag{16.44}$$

We then introduce

$$a_2 = a_1^2 - 2a_1 - 2 \tag{16.45}$$

and

$$r(n) = \frac{a_1 C_2}{a_2} k(n) \tag{16.46}$$

The newly scaled map is

$$r(n+1) = (2 - a_2)r(n) - a_2 r(n)^2 \tag{16.47}$$

We observe that this is the same map as our original z map and we know that the fixed point $r^* = 0$ becomes unstable when the parameter $a_2 \geq 3$. The resulting period-2 cycle becomes unstable when

$$a_2 = 1 + \sqrt{6} \tag{16.48}$$

corresponding to

$$a_1 = 1 + \sqrt{4 + \sqrt{6}} = 3.5396 \tag{16.49}$$

$$a_0 = 1 + \sqrt{4 + \sqrt{4 + \sqrt{6}}} = 3.5573 \tag{16.50}$$

Notice that our true parameter $a = a_0$ changed very little during the process of the creation of the period-8 cycle. This period-8 cycle is conviently studied from the second iteration, or squaring, of our original map. Note that each of the original stable points of the period-2 cycle, p and q, have spawned period-2 cycles of their own. This doubling process continues to generate higher-order cycles.

16.8.2 The Relationship Between the Numbers A_i and the Parameters a_i

We organize the iteration by introducing

$$a_{i+1} = a_i^2 - 2a_i - 2 \tag{16.51}$$

The subscript i identifies our place in the iteration process. There is a corresponding iteration of the parameter $(a_i/a_{i+1})C_{i+1}$ that generates the successive scalings. However, the scaling changes slightly in response to the change in the parameter a_i. In addition, we must keep in mind that $C_i(p) \neq C_i(q)$. It is convenient to postpone a discussion of this point until Section 16.8.4.

In this section, we limit our attention to following the increase of the parameter a associated with the iteration scheme. As we have mentioned, we have to be careful not to confuse the variable a with the *numerical* value it takes at the onset of the next period doubling. To avoid confusion, we introduce the numbers A_i, which are the *numerical* values that the parameter a takes as it initiates successive period doublings. What happens? As the parameter a_0 crosses 3, we have our first period doubling, so $A_1 = 3$. Then a_0 grows and when it reaches $a = 1 + \sqrt{6}$, we have a period-2^2 cycle. Then $A_2 = 1 + \sqrt{6}$. Thus the onset of period doublings is given by the sequence of numbers

$$A_i; \qquad i = 1, 2, \ldots, n, \ldots$$

We expect that the numbers A_i converges to a limit A_∞. This is the value of the parameter a for which we have a cycle whose period is so high that from a practical point of view, we can treat as infinite. This result is obtained in Section 16.8.5. The technique of scaling and rescaling of variables enables us to obtain, in the new variables, an equation of the same form as we had in the original variables. This is called a renormalization procedure.

Let us assume that our system has just undergone its kth period doubling. We are at the threshold of a region in which

$$a_{k+1} = 3; \qquad a_k = 1 + \sqrt{6} \tag{16.52}$$

This means that we are about to rescale our a variable using Equation (16.51):

$$a_{k+1} = a_k^2 - 2a_k - 2; \qquad k = 0, 1, 2, \ldots \tag{16.53}$$

What is the value of A_k? It is obtained by working the recursion relation given by Equation (16.53) backwards so as to recover the value of the a parameter after its kth iteration. Namely, we have

$$A_i = A_{i+1}^2 - 2A_{i+1} - 2; \qquad i = 1, 2, \ldots, k \ldots . \tag{16.54}$$

Example 8.1 Consider a situation in which $k = 3$ in Equation (16.54), indicating that the map has a stable cycle of period 2^3. We have

$$A_1 = 3 = A_2^2 - 2A_2 - 2 \Rightarrow A_2 = 1 + \sqrt{6}$$
$$A_2 = A_3^2 - 2A_3 - 2 \Rightarrow A_3 = 1 + \sqrt{4 + \sqrt{6}}$$

So, if $k = 3$, our a parameter has the *numerical* value $= A_3 = 1 + \sqrt{4 + \sqrt{6}} = 3.539585$. ■

Solving Equations (16.53) and (16.54) recursively, we obtain successive values of the parameters a and A_i:

$$\begin{aligned}
a &= 3 = A_1 = 3 && (2^1 \text{ cycle})\\
a &= 1 + \sqrt{6} = A_2 = 3.449490 && (2^2 \text{ cycle})\\
a &= 1 + \sqrt{4 + \sqrt{6}} = A_3 = 3.539585 && (2^3 \text{ cycle})\\
a &= 1 + \sqrt{4 + \sqrt{4 + \sqrt{6}}} = A_4 = 3.557261 && (2^4 \text{ cycle})\\
a &= \cdots = A_5 = 3.560715 && (2^5 \text{ cycle})\\
a &= \cdots = A_6 = 3.561389 && (2^6 \text{ cycle})\\
a &= \cdots = A_7 = 3.561521 && (2^7 \text{ cycle})\\
a &= \cdots = A_8 = 3.561547 && (2^8 \text{ cycle})
\end{aligned}$$

This doubling process continues indefinitely, creating cycles of order 2^k, $k = 1, 2, 3, \ldots$.

It will be shown in the analysis that follows that new stable patterns associated with fixed points of higher and higher-order maps emerge when our parameter a changes by just a very tiny bit. This situation is maintained until we reach a limiting value of the parameter at $a \approx 3.5616$, after which no stable patterns emerge. (An analysis without truncation yields the value $a \approx 3.5699$. It is obtained, for example, in the article by Feigenbaum.) This means that after this value of the parameter is exceeded, we are no longer able to characterize our trajectories by saying that they are cycles of known or calculable periodicity. (In other words, there is no cycle of period 2^k in this region.) As a approaches 3.5616, we can think of the period-doubling process as correlating the behavior of the system at its nth step with its position in the more and more distant past. It is as if it has lost its ability to go forward until it checks its location at earlier and earlier locations. (This long-range correlation plays a role in the path to chaos that is not yet completely understood.) The backward-looking process eventually breaks down. If you follow the trajectories for a long time, they may appear to be periodic, but this is an illusion. If you continue watching them, the pattern disappears. In other words, it was an error to identify the trajectory as periodic with some well-defined period. This is a most subtle point and it is both too difficult and involves too much analysis for us to demonstrate this property.

16.8.3 The Feigenbaum Number δ

We begin by expressing the quantity A_n given by Equation (16.54) in terms of its lower-order iterate. In other words, we reorganize Equation (16.54) to read

$$(A_{n+1} - 1)^2 = 3 + A_n$$

or

$$A_{n+1} - 1 = \sqrt{3 + A_n} \tag{16.55a}$$

and

$$A_n - 1 = \sqrt{3 + A_{n-1}} \tag{16.55b}$$

Thus, we can establish that

$$A_{n+1} - A_n = \sqrt{3 + A_n} - \sqrt{3 + A_{n-1}} \tag{16.56a}$$

and

$$A_{n+1} + A_n - 2 = \sqrt{3 + A_n} + \sqrt{3 + A_{n-1}} \tag{16.56b}$$

Multiplying Equation (16.56a) by (16.56b), we obtain

$$(A_{n+1} - A_n)(A_{n+1} + A_n - 2) = A_n - A_{n-1}$$

or

$$A_{n+1} + A_n - 2 = \frac{A_n - A_{n-1}}{A_{n+1} - A_n} \tag{16.57}$$

We observe that the r.h.s. of Equation (16.57) generates a geometrically converging sequence. In order to find the limit of this sequence given by Equation (16.57), we introduce

$$\delta_n \equiv \frac{A_n - A_{n-1}}{A_{n+1} - A_n} \tag{16.58}$$

Then Equation (16.57) is, equivalently,

$$\delta_n = A_{n+1} + A_n - 2 \tag{16.59}$$

Our goal is to obtain the limiting value of the numbers δ_n as n tends to infinity.

Note: The Feigenbaum number δ is defined by Equation (16.58) as n tends to infinity. The relation satisfied by the approximate Feigenbaum number $\lim_{n \to \infty} \delta_n$, as given by Equation (16.59), is a result of our truncation procedure. This equation is not satisfied by the Feigenbaum number that results from the exact iteration. In solving Equation (16.59), we therefore obtain an approximate value for δ.

We seek the limit of the sequence of numbers A_n as n tends to infinity. For the limiting value, we have

$$A_{n+1} = A_n = A_\infty$$

Then, using Equation (16.54), we have

$$A_\infty = A_\infty^2 - 2A_\infty - 2$$

or

$$A_\infty^2 - 3A_\infty - 2 = 0 \tag{16.60}$$

We solve Equation (16.60) and obtain

$$A_\infty = a_\infty = \frac{3 + \sqrt{17}}{2} = 3.5616 \tag{16.61}$$

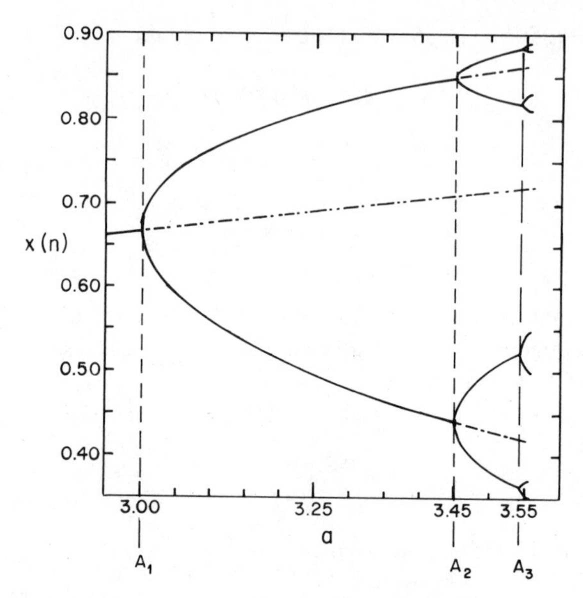

Figure 16.10 We have the same plot as given in Figure 16.9 in which we have indicated the values A_1, A_2, and A_3 that mark the onset of period doublings.

Using this value in Equation (16.59), we obtain

$$\delta_{\text{approx.}} = 2A_{\infty} - 2 = 1 + \sqrt{17} = 5.12 \tag{16.62}$$

This is an approximate value for δ. The exact value is

$$\delta_{\text{exact}} = 4.669202\ldots \tag{16.63}$$

EXERCISE 8.1 Use the recursion relation for δ_n given by Equation (16.59) together with the values of A_i given in Section 16.8.2 to obtain values

$$\delta_2 = 5.0 \qquad \text{and} \qquad \delta_3 = 5.1$$

■

See Figure 16.10.

16.8.4 The Feigenbaum Number α

Recall that we introduced a scaling of our variables in Equations (16.37a) and (16.37b) in order to have the range, at each step of the iteration, coincide with the range of the original z variable. It was important to keep in mind that the scaling associated with the transformation to the fixed point p is different

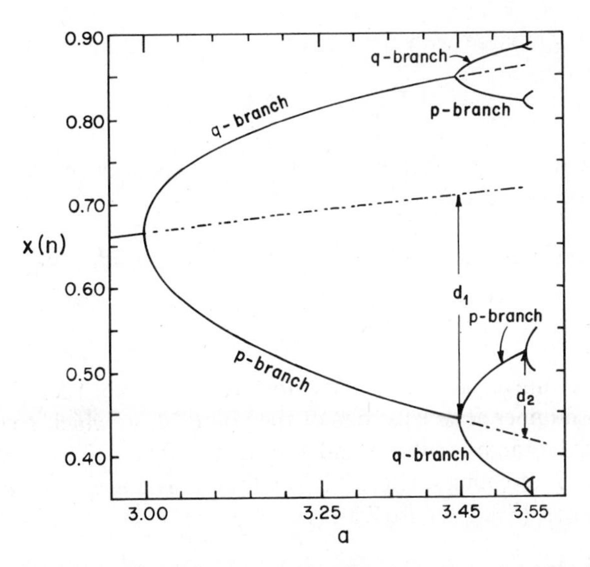

Figure 16.11 We have the same plot as in Figures 16.9 and 16.10. The p and q branches have been indicated. Convince yourself that the p branches "flip."

than the scaling to the fixed point q. The Feigenbaum number α is the limiting value of the scale factor $a_i C_{i+1}(p)/a_{i+1}$. This is the one associated with the flipping of the trajectories in Figure 16.11.

Refer to Figure 16.11. In the region $3 < a < 1 + \sqrt{6}$ the deviation of the upper branch from the unstable fixed point (given by the broken line) is equal to q with values given by Equation (16.22b). We, therefore, call this branch the q branch. Similarly, the lower branch is the p branch, with values given by Equation (16.22a). At the critical value $a = 1 + \sqrt{6}$, the branches split. In terms of the transformed variables given by Equations (16.34) and (16.40), new p and q branches are formed. The points shown in Figure 16.11 are in terms of the original z variables. They are related to the transformed variable s by a shift and a rescaling. On the p branch, the rescaling is negative, and as a result, the new p and q branches are flipped. Referring to Figure 16.11, we note that the first scaling is given by the ratio $-d_1/d_2$.

Finally, we conclude that

$$\alpha \equiv \frac{a_\infty C_\infty(p)}{a_\infty}$$

$$= (a-3)(a+1) - 3\sqrt{(a-3)(a+1)} \tag{16.64}$$

The parameter a has as its limiting value $a_\infty = (3 + \sqrt{17})/2$ given by Equation (16.61). Then Equation (16.64) yields:

$$\alpha_{\text{approx.}} = -2.24 \tag{16.65}$$

This is to be compared with the exact value:

$$\alpha_{\text{exact}} = -2.50\ldots \tag{16.66}$$

Observe that the scaling factor $a_i C_{i+1}(p)/a_{i+1}$ changes very little during the entire iteration process. (See Exercise 7.3.)

EXERCISE 8.2 We introduced $C_1(q)$, Equation (16.37b). Show that

$$C_\infty(q) = 7.36$$

This is another universal scale factor. However, it has not been designated as a Feigenbaum number. The locations of the points A_i at which new instabilities originate are the same for the p and q branches. The only difference between the two is the rescaling factor. Notice that it changes slightly during the iteration process. (See Exercise 7.3.) ■

The exact values of the Feigenbaum numbers are obtained in Feigenbaum's paper cited in the Bibliography.

16.8.5 The Convergence of the *A* Sequence

We want to show that the A sequence converges geometrically, that is, that

$$A_\infty = A_n + S\delta^{-n} \qquad \text{as} \quad n \to \infty \tag{16.67}$$

where S is a constant to be determined.

We begin by writing

$$A_\infty - A_n = (A_{n+1} - A_n) + (A_\infty - A_{n+1}) \tag{16.68}$$

It is our plan to express the quantity $A_{n+1} - A_n$ in terms of lower-order terms. Then we work Equation (16.68) upwards and repeat the process. We are confident that if we increase the index n sufficiently, we will have a value of A_n that is close enough to its limiting value A_∞ that we can terminate the procedure. With these thoughts in mind, we use Equation (16.58) to write

$$A_{n+1} - A_n = \frac{A_n - A_{n-1}}{\delta_n} = \frac{A_{n-1} - A_{n-2}}{\delta_n \delta_{n-1}} \tag{16.69}$$

Continuing, we get

$$A_{n+1} - A_n = \frac{A_2 - A_1}{\delta_n \cdots \delta_2} \tag{16.70}$$

With this result in hand, we return to Equation (16.68), and write it as

$$A_\infty - A_n = \frac{A_2 - A_1}{\delta_n \cdots \delta_2} + (A_{n+2} - A_{n+1}) + (A_\infty - A_{n+2}) \quad (16.71)$$

We repeat the iteration process that was used in Equations (16.69) and (16.70) to write Equation (16.71) as

$$A_\infty - A_n = \frac{A_2 - A_1}{\delta_n \cdots \delta_2} + \frac{A_2 - A_1}{\delta_{n+1} \cdots \delta_2} + \frac{A_2 - A_1}{\delta_{n+2} \cdots \delta_2} + \frac{A_2 - A_1}{\delta_{n+3} \cdots \delta_2} + \ldots \quad (16.72)$$

We write Equation (16.72) in the form

$$A_\infty - A_n = \frac{A_2 - A_1}{\delta_n \cdots \delta_2}\left(1 + \frac{1}{\delta_{n+1}} + \frac{1}{\delta_{n+1}\delta_{n+2}} + \ldots\right) \quad (16.73)$$

We observed in Exercise 8.2 that the δ sequence converges rapidly. Thus, for large n, we make little error by approximating each δ_i in Equation (16.73) by its limiting value δ. With this done, Equation (16.73) becomes

$$A_\infty - A_n = \frac{(A_2 - A_1)\delta}{\delta^{n-1}(\delta - 1)} = \frac{(A_2 - A_1)\delta^2}{\delta^n(\delta - 1)} \quad (16.74)$$

in the limit as n tends to infinity. We are now in a position to find the value of the constant S introduced in Equation (16.67), which is a measure of the rate of convergence of the geometric sequence. We know from the list of A values given in Section 16.8.2 that $A_2 = 1 + \sqrt{6} = 3.45$ and $A_1 = 3$; and from Equation (16.62), we have $\delta = 5.12$. With this information, we have

$$S_{\text{approx.}} = \frac{(A_2 - A_1)\delta^2}{\delta - 1} = 2.9 \quad (16.75)$$

This compares well with the exact value of this constant

$$S_{\text{exact}} = 2.6327\ldots \quad (16.76)$$

16.9 CONCLUDING REMARKS

In conclusion, it is important to note that after our parameter crosses the value A_∞, there are no stable 2^k cycles. It has been established in the literature that in this new region, there is an infinite number of fixed points with an infinite number of different periodic cycles. (See the paper by Yorke and Yorke.) One finds totally aperiodic, or nonrepeating, patterns that are called chaotic. At the same time, for many values of the parameter a greater than the critical value, the system is not sensitive to small deviations in its initial conditions. But, for most of the initial conditions, the system is highly sensitive to this variation. In this regard, note that different computer programs treat the last term in a number differently. Or they lead to different round-off errors. Thus, the sensitivity to initial conditions can lead to strikingly different trajectories. If a physical system is too sensitive to initial conditions, the initial data are of no use. We lose, thereby, an essential ingredient in the characterization of the future of the system. It is too difficult for us to establish these results or to give a proof of the onset of chaos since it requires much more development. However, with the material presented in this chapter, it is possible to continue a study of this subject through a study of the material presented in the Bibliography. What is perhaps most fascinating is that the subject of period doubling leading to chaos is at a primitive state of development and promises in the years to come to provide us with many interesting insights into nonlinear phenomena. We owe a great debt to Feigenbaum, May, and others for uncovering many remarkable properties of quadratic maps.

BIBLIOGRAPHY

1. May, R., "Simple Mathematical Models With Very Complicated Dynamics," *Nature 261*, 459 (1976). This article is strongly recommended as the first step toward learning about chaos in discrete systems. It was one of the first review articles and thus it really assumes that the reader knows nothing about nonlinear problems. It also has an excellent bibliography. May is a prolific contributor to this field and his other papers are easily found from a computer search or by consulting the extensive bibliography given in the next bibliographic entry.
2. Bai-Lin, H., *Chaos*, World Scientific, Singapore, 1984. This text is a collection of papers that treats a great many of the theoretical and experimental aspects of chaos in discrete and continuous systems. It has an extensive bibliography of papers published prior to 1985.
3. Lauwerier, H. A., "One-Dimensional Iterative Maps," in A. V. Holden, ed., *Chaos*, Princeton University Press, Princeton, 1986. This article motivated much of the discussion presented in Chapter 16. It includes a wealth of technical details and is followed in the edited volume by a second article dedicated to "Two-Dimensional Iterative Mappings."
4. Feigenbaum, M., "Universal Behavior of Nonlinear Systems," *Los Alamos Science*, Summer, 4 (1980). This article is a reasonably nontechnical presentation of discrete maps and a derivation of the Feigenbaum numbers.

5. Ruelle, D., "Strange Attractors," *Math. Intelligencer 2*, 126 (1980). This is a popular exposition of many aspects of chaos, including the notion of strange attractors.
6. Li, T. Y., and J. A. Yorke, "Period Three Implies Chaos," *Am. Math Monthly 82*, 985 (1975). This article presents the first discussion of the role of odd-period cycles in the development of chaos in discrete maps.
7. Yorke, J. A., and E. D. Yorke, "Chaotic Behavior and Fluid Dynamics," in H. L. Swinney and J. P. Gollub, eds., *Hydrodynamic Instabilities and the Transition to Turbulence*, Springer Verlag, New York, 1981. This is a most readable article on many aspects of chaos in discrete and continuous systems. It includes a discussion regarding the difficulties associated with unraveling the behavior of discrete systems after they have entered into the chaotic domain.
8. Berge, P., Y. Pomeau, and C. Vidal, *Order and Chaos*, Hermann, Paris, and Wiley, New York, 1984. This book provides a comprehensive introduction to dissipative dynamical systems. The discussion is at such a level that advanced undergraduates and graduate students should be able to understand at least the thread of the arguments. This book is especially important since the field of chaos and dynamical systems is so new that there is a lack of established literature.
9. Helleman, H. G., and R. S. MacKay, "One Mechanism for the Onset of Large-Scale Chaos in Conservative and Dissipative Systems," in C. W. Horton, Jr., L. E. Reichel, and V. G. Szebehel, eds., *Long-Time Prediction in Dynamics*, Wiley, New York, 1983. This article contains a wealth of information regarding the onset of chaos in discrete nonlinear systems. It complements other sources and outlines in considerable detail the steps that are required to put the equations into standard forms. For example, it gives the explicit calculations required for one-period doubling bifurcation for a map in the standard form. The style is such that the details of the arguments are relatively easy to follow.
10. Ekeland, I., *Mathematics and the Unexpected*, University of Chicago Press, Chicago, 1988. This is a beautifully written brief exposition that touches on many difficult (and also interesting) aspects of nonlinear phenomena such as homoclinic points, instability of trajectories, and Bernoulli shifts. The level of the presentation and the leisurely pace of the discussion helped me to better appreciate the effort that will be required of all of us as we try to understand the richness of nonlinear problems. Particular attention is given to the debt we owe to Poincaré.

APPENDIX

A DISCUSSION OF EULER'S CONSTANT

We encountered Euler's constant γ in Chapter 4 when we considered the divergent sum

$$\sum_{j=1}^{n-1} \frac{1}{j} \to \ln n + \gamma; \qquad n \to \infty \tag{A.1}$$

(See Equation 4.64c.)

It was equal to the finite part of the sum. It also appears in a totally different context when we consider derivatives of the gamma function. Thus, if we use the integral definition of the gamma function that we introduced in Chapter 2, we have

$$\Gamma(n) \equiv \int_0^\infty x^{n-1} \exp(-x)\, dx \tag{A.2}$$

This is differentiated to obtain

$$\Gamma'(n) = \int_0^\infty x^{n-1} \ln x \exp(-x)\, dx \tag{A.3}$$

This type of integral occurs in diverse investigations in pure and applied science. For example, it arises in summation problems in solid-state physics. If $n = 1$, we have

$$-\Gamma'(1) = -\int_0^\infty \ln x \exp(-x)\, dx \tag{A.4}$$

Our goal in this appendix is to show that $-\Gamma'(1)$ given by Equation (A.4) is equal to the same constant γ that occurs in Equation (A.1). At this point, we could establish the desired equivalence with just a few manipulations that are justified by the theory of complex variables. However, the text does not treat functions of a complex variable and, therefore, keeps the discussion of various topics on an elementary level. But, on occasion, you have to pay the price. And this is just such a moment. It will become clear as we proceed that we seem to be reaching our goal by a series of tricks. These are necessary precisely because we must avoid results that follow from the theory of complex variables. We begin with the introduction of a Frullani integral of the form:

$$\ln x = \int_0^\infty \frac{\exp(-\alpha) - \exp(-\alpha x)}{\alpha}\, d\alpha \tag{A.5}$$

Since Equation (A.5) is not obvious, we have to show that it is correct. (Frullani integrals are discussed in the text by Jeffreys and Jeffreys.)

We divide the integral into two parts: $[0, \delta]$ and $[\delta, \infty)$. The break point δ is chosen to be arbitrarily small. We consider the infinitesimal interval and expand the integrand to obtain

$$\int_0^\delta \frac{\exp(-\alpha) - \exp(-\alpha x)}{\alpha}\, d\alpha \to \int_0^\delta (x-1)\, d\alpha \to 0 \qquad \text{as } \delta \to 0 \tag{A.6}$$

We thus can limit our attention to the exterior region $[\delta, \infty)$. We divide the integrand into two parts:

$$\int_\delta^\infty \frac{\exp(-\alpha) - \exp(-\alpha x)}{\alpha}\, d\alpha = \int_\delta^\infty \frac{\exp(-y)}{y}\, dy - \int_{x\delta}^\infty \frac{\exp(-y)}{y}\, dy \tag{A.7}$$

In the first integral, we have simply renamed the variable α as y. In the second integral, we changed variables, $y = \alpha x$. We combine the integrals and have

$$\int_\delta^{x\delta} \frac{\exp(-y)}{y}\, dy \to 1 \cdot \ln x; \qquad \delta \to 0 \tag{A.8}$$

So Equation (A.5) is established. Now we replace the term $\ln x$ inside the integral in Equation (A.3) with its form as given by Equation (A.5). Thus, we have

$$\Gamma'(n) = \int_0^\infty \int_0^\infty x^{n-1} \exp(-x) \frac{\exp(-\alpha) - \exp(-\alpha x)}{\alpha}\, d\alpha\, dx \tag{A.9}$$

We interchange the orders of integration and integrate over the variable x to obtain

$$\frac{\Gamma'(n)}{\Gamma(n)} = \int_0^\infty \left(\frac{\exp(-\alpha)}{\alpha} - \frac{1/\alpha}{(\alpha+1)^n} \right) d\alpha \tag{A.10}$$

We need the value of $\Gamma'(1)/\Gamma(1)$, which is given by Equation (A.10) with $n = 1$:

$$\frac{\Gamma'(1)}{\Gamma(1)} = \int_0^\infty \left(\frac{\exp(-\alpha)}{\alpha} - \frac{1/\alpha}{\alpha + 1} \right) d\alpha \tag{A.11}$$

We are nearly done. We subtract $\Gamma'(1)/\Gamma(1)$ from both sides of Equation (A.10) to obtain

$$\frac{\Gamma'(n)}{\Gamma(n)} - \frac{\Gamma'(1)}{\Gamma(1)} = \int_0^\infty \frac{1}{\alpha} \left[\frac{1}{\alpha + 1} - \frac{1}{(\alpha + 1)^n} \right] d\alpha \tag{A.12}$$

Make the change of variables: $u = 1/(\alpha + 1)$. Then Equation (A.12) becomes

$$\frac{\Gamma'(n)}{\Gamma(n)} - \frac{\Gamma'(1)}{\Gamma(1)} = \int_0^1 \frac{1 - u^{n-1}}{1 - u} du \tag{A.13}$$

Almost finally, we recognize the integrand in Equation (A.13) as the geometric series

$$\sum_{j=0}^{n-2} u^j \tag{A.14}$$

Substituting Equation (A.14) into Equation (A.13) and integrating, we have

$$\begin{aligned} \frac{\Gamma'(n)}{\Gamma(n)} - \frac{\Gamma'(1)}{\Gamma(1)} &= 1 + \frac{1}{2} + \frac{1}{3} + \cdots + \frac{1}{n - 1} \\ &= \sum_{j=1}^{n-1} \frac{1}{j} \end{aligned} \tag{A.15}$$

We have one last step. We need the asymptotic form for $\Gamma'(n)$ and $\Gamma(n)$. In Chapter 5, Equation (5.73), we found that

$$\Gamma(n + 1) = n\Gamma(n) \sim \sqrt{2\pi n}\, n^n \exp(-n) \tag{A.16}$$

Taking the logarithmic derivative of Equation (A.16), we find that the leading term is

$$\frac{\Gamma'(n)}{\Gamma(n)} \sim \ln n \tag{A.17}$$

We note from Equation (A.3) that

$$\frac{\Gamma'(1)}{\Gamma(1)} = \int_0^\infty \ln x \exp(-x)\, dx \tag{A.18}$$

Now consider Equations (A.15)–(A.18). We rearrange terms to obtain

$$\sum_{j=1}^{n-1} \frac{1}{j} = \ln n - \int_0^\infty \ln x \exp(-x)\, dx = \ln n + \gamma \tag{A.19}$$

Thus, we have finally arrived at our goal to establish that Euler's constant γ, which is defined by Equation (A.1) is equal to the same constant obtained by Equation (A.4), or

$$\gamma = -\int_0^\infty \ln x \exp(-x)\, dx$$

In my opinion this circuitous route to establishing our result should truly give anyone pause to consider learning the basics of complex variable theory.

EXERCISE A.1 Consider

$$K(z) = \int_0^\infty t^{z-1} \ln t \exp(-t)\, dt$$

Show that $K(2) = 1 - \gamma$. ■

BIBLIOGRAPHY

1. Jeffreys, H., and B. S. Jeffreys, *Methods of Mathematical Physics*, 3rd Ed., Cambridge University Press, London (1962). Frullani integrals are discussed on page 406. There is also an extensive discussion of gamma functions using the methods of complex variable theory.
2. Lebedev, N. N., *Special Functions and Their Uses*, Prentice-Hall, Englewood Cliffs, New Jersey, 1965. Revised and reprinted by Dover, New York, 1972. The first chapter has a derivation of Euler's constant that is closely related to the one that we have given.
3. Bowman, F., *Introduction to Bessel Functions*, Dover, New York, 1958. Here we find Euler's constant introduced in connection with Bessel functions. The derivation is almost the same as the one that we have given and it, together with a few related exercises, is given on pages 68–70.

INDEX